STRUCTURE AND BONDING is issued at irregular intervals, according to the material received. With the acceptance for publication of a manuscript, copyright of all countries is vested exclusively in the publisher. Only papers not previously published elsewhere should be submitted. Likewise, the author guarantees against subsequent publication elsewhere. The text should be as clear and concise as possible, the manuscript written on one side of the paper only. Illustrations should be limited to those actually necessary.

SPRINGER-VERLAG		SPRINGER-VERLAG NEW YORK INC.
D-6900 Heidelberg 1	D-1000 Berlin 33	
P. O. Box 1780	Heidelberger Platz 3	175, Fifth Avenue
Telephone (06221) 4 91 01	Telephone (0311) 822001	New York, N. Y. 10010
Telex 04-61 723	Telex 01-83319	Telephone 673-2660

STRUCTURE AND BONDING

Volume 12

Editors: J. D. Dunitz, Zürich
P. Hemmerich, Konstanz · J. A. Ibers, Evanston
C. K. Jørgensen, Genève · J. B. Neilands, Berkeley
Sir Ronald S. Nyholm †, London · D. Reinen, Marburg · R. J. P. Williams, Oxford

With 37 Figures

Springer-Verlag
Berlin Heidelberg GmbH 1972

ISBN 978-3-540-05901-1 ISBN 978-3-540-37580-7 (eBook)
DOI 10.1007/978-3-540-37580-7

Originally published by Springer-Verlag Berlin · Heidelberg · New York in 1972

Contents

Electron Correlation in the Seventies. P. v. Herigonte 1

Ligand Field Splittings in Copper (II) Compounds. D. W. Smith 49

Phenomenological Approach to Cation-Solvent Interactions. U. Mayer, and V. Gutmann 113

Acid Salts of Carboxylic Acids, Crystals with some "Very Short" Hydrogen Bonds. J. Clare Speakman..................... 141

Phase-fixed 3-Γ Symbols and Coupling Coefficients for the Point Groups. S. E. Harnung, and C. E. Schäffer 201

Real Irreducible Tensorial Sets and their Application to Ligand-Field Theory. S. E. Harnung, and C. E. Schäffer 257

Electron Correlation in the Seventies

Pasquale v. Herigonte

European Group of Theoretical Molecular Physics (EGTMP),
Roma, Via Cornelio Celso 7, Italia

Table of Contents

1. Introduction 2

2. Evidence for the Approximate Constancy of Correlation Energy 4
   a) The Inversion Barrier of Ammonia 5
   b) The SCF/CI Dilemma and Molecular Symmetry 6
   c) The Rotation Barrier of Ethane and Related Topics 8
   d) Variation of Correlation Energy with Interatomic Distances 11

3. Correlation and Prediction of Molecular Properties 12
   a) Molecular Parameters Related to the Total Energy 12
   b) Classification of Observables with Regard to Correlation 12
   c) One-Electron Properties: Electric-Dipole Moments 13
   d) Comparison of Different IPM Approximations 15

4. Correlation in Configuration and Perturbation Treatments 16
   a) Independent-Particle Models for Closed-Shell and Open-Shell Systems 16
   b) Electron-Nuclei Hyperfine Coupling Constants 17
   c) First-Order and Second-Order Properties 18
   d) Example of Nuclear Spin-Spin Coupling Constants 19

5. Energy and Space Dependence of the Correlation Content of Wave Functions 24
   a) Error Potential and Double Perturbation Theory 24
   b) Direct and Indirect Dependence of Molecular Properties on Correlation 26
   c) Correlation Functions and Correlation Coefficients 27

6. Inclusion of Correlation into an Approximate Wave Function 31
   a) Computational Considerations on the SCF Scheme 32
   b) Computational Considerations on the CI Scheme 33
   c) Introduction of Correlation by Multiconfigurational SCF Methods 35

7. Physical Significance of Independent-Particle Models and Correlation 38
   a) The Role of Independent-Particle Models in the Interpretation of Molecular Properties and the Charge-Cloud Picture 38
   b) The Independent-Particle Model and the Particle Concept 40

8. Conclusions 42

9. References 45

1. Introduction

Several reviews mainly devoted to the methodological and computational aspects of the electron correlation problem have appeared in recent years. An attempt to complete the picture by systematically analyzing the validity and significance of that problem was made during discussions among several quantum chemists. Their remarks and questions, together with the work and the references they presented to substantiate them, are reported here.

Looking at the history of correlation from the fifties to the seventies, one may be led to ask whether correlation has been a scientific fashion or a real problem. Twenty years ago, almost everybody seemed to accept the idea that the simple molecular orbital method (MO) must be completed by configuration interaction (CI), in order to obtain reliable prediction for the physical properties of atoms and molecules. Ten years ago, electron correlation was considered as the central problem of Quantum Chemistry (*1*). Nowadays, about 90% of the quantum-mechanical calculations on molecules are performed by the self-consistent-field method (SCF) using more or less extended sets of basis functions, without any consideration of the possible effects of correlation.

Of course, this is chiefly due to the availability of efficient SCF programs which can be used by anyone as soon as computer time is available. Work of that sort had the merit of showing that certain disagreements between experiment and "precomputational" theory which were attributed to correlation may in fact have little to do with it. For instance, the "reduction" of Coulomb integrals in the Parr-Pariser-Pople theory of π-electron systems was considered in the past as produced by σ- and π-electron correlation. The determination of optimized basis orbitals in atomic valence states (*2*) and in simple polyatomic molecules (*3*) suggests that the $2p\pi$ orbitals have smaller "effective nuclear charges" (orbital exponents) in a planar system than in a spherical free atom. (For details, see (*23*)).

On the other hand, certain phenomena, either known for a long time or recently discovered, can hardly be explained within the framework of the independent-particle model (IPM) on which the SCF method is based. The study of electron correlation therefore remains a valuable research subject, even though, for a variety of reasons, there has been a change of emphasis during recent years towards other topics.

The predominance of SCF methods in present-day calculations is partly due to the difficulty of calculating correlation effects. It is far less complicated to treat many-electron systems without correlation. Not knowing whether or not correlation will be important in a given problem, one is easily tempted to rationalize results by the statement that "correlation energy is probably unimportant in this problem". Such a statement, made on the basis of nothing better than a SCF calculation of a certain molecule, is clearly excessive. On the other hand, it is unfortunate that the importance of correlation — if any — can be decided only *after* a calculation going far beyond the Hartree-Fock scheme. Thus, an urgent task should be to try to extract from the available computational evidence some general rules which could indicate whether a "better than SCF" calculation would be necessary for a given problem or not.

One puzzling fact is that SCF treatments, with no allowance for correlation, have been unexpectedly successful in many studies of molecular processes. The most plausible explanation of this is that the correlation energy ΔE_c does not change much with certain changes in geometry or in chemical rearrangements. Consider the schematic energy diagram of Fig. 1, where a distinction is made between actual SCF calculations (which are approximate) and ideal ones.

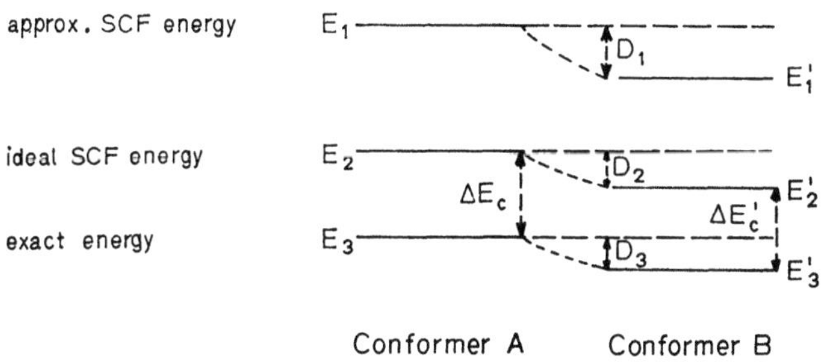

Fig. 1. Effect of correlation on the energy of conformers

By definition, the barrier height is D_3, but in practice only D_1 is available. If the correlation energy is nearly constant, *i.e.* $\Delta E_c \approx \Delta E_c'$, D_2 is a good estimate of D_3; but we have no guarantee that $D_1 \approx D_2$. In this equality, the choice of the atomic basis functions plays an important role: even if "saturation" with respect to one kind of basis extension has been obtained, for instance if the "*s*" or "*p*" limit is reached, this does not imply a true *limit* where $E_1 \rightarrow E_2$; such an assumption is often made, but, as will be mentioned later, the difference E_1—E_2 is probably many times larger than the estimated barrier height D_2, so that its foundation may be very weak.

As an example of difficulties arising from the choice of the basis, we recall the inadequacy of the Gaussian functions used in most SCF calculations to represent cusps, particularly for the inner-shell orbitals. Since the latter give very large contributions to total energies, Gaussian-type orbitals (GTO's) should be poorer than Slater-type orbitals (STO's) as far as energy properties are concerned. However, in many cases the predictions made from GTO calculations are not affected by this defect, simply because the "cusp errors" are not much altered when the nuclei are moved into new positions.

In order to state the "correlation problem" in its relation to chemical problems of the present and the near future, one can put it into the form of the three following questions:

1. Is it really established that correlation energy is insensitive to certain changes in geometry or structure?

2. Are there experimental properties which depend directly on electron correlation?

3. Is it really necessary to go through the calculations associated with the independent-particle model (which is the basis for the introduction of the correlation concept) as an essential step towards highly accurate wave functions?

The same questions can be formulated in various alternative ways. For example:

1. Is the correlation energy substantially the same in different systems with the same number of "similar" electron pairs? If this is the case, it need not be calculated in discussing the energy difference for a "reaction" in which one system changes into the other.

2. If statement 1 is true with reference to the *energy*, are there other properties which cannot be properly understood without taking account of correlation effects on *wave functions* and *electron distribution functions*?

3. Is it necessary or desirable always to start from a one-determinant SCF function, to be improved by CI or other expansion methods? Would another starting point be better (*e.g.* a valence-bond function, which is essentially of many-determinant form)?

Such alternative statements are merely developments of the three preceding basic ideas. We shall now try to review the present status of the correlation problem following these three lines and making reference to specific work.

2. Evidence for the Approximate Constancy of Correlation Energy

We start with two examples where it is undecided whether correlation plays a role at all, namely the barrier to nitrogen inversion in ammonia and aza-compounds and the barrier to internal rotation about a single bond in organic molecules. In the last years the study of such problems by quantum-mechanical methods has become so popular that it gave rise to a new branch of a theoretical chemistry, namely "theoretical conformation analysis" (*4*).

a) The Inversion Barrier of Ammonia

The inversion barrier of ammonia is the energy difference between two extreme conformations, pyramidal and planar, during the out-of-plane motion of the nitrogen atom. Experimentally, this quantity is equal to 5.77 kcal/mol, which is a very small part of the total energy of the molecule (about 35.500 kcal/mol). Recently, two *ab initio* calculations in fairly good agreement with experiment have been presented. Unfortunately, their conclusions are rather contradictory.

The first one by *Rauk et al.* (*5*), is a pure SCF calculation with an extended basis set of GTO's including polarization orbitals on nitrogen and hydrogen atoms and optimization of the geometry. In the opinion of the authors, the success of their calculations (a barrier of 5.08 kcal/mol) definitely proves that an independent particle model of the SCF type can account for inversion barriers, provided that polarization orbitals are included and the basis is sufficiently flexible otherwise.

The second by *Pipano et al.* (*6*) is a SCF calculation, in a non-minimal but less complete basis set of STO's (*e.g.* without any *d*-orbital on nitrogen), followed by configuration interaction including all singly and doubly excited configurations. Whereas their SCF calculation gives too small a barrier (1.4 kcal/mol), the introduction of configuration interaction increases this value to 5.2 kcal/mol. Consequently, the main part of the barrier is tentatively attributed by the authors to electron correlation. As a matter of fact, configuration interaction improves the ground state wave functions since the total energy is lowered by 100 kcal/mol with respect to the initial SCF calculation and 75 kcal/mol with respect to the approximate Hartree-Fock limit, and the wave function of *Pipano et al.* is much better than that of *Rauk et al.*, as far as a simple energy criterion goes.

Clearly, the two interpretations are contradictory as concerns the origin of inversion barriers. An option can be made only by answering the general question: How valuable are SCF calculations in the prediction of inversion barriers, and are they really meaningful in the absence of correlation corrections? This is a crucial point, even though the calculations are not yet conclusive. (For example, the correlation energy coming from the CI treatment is only 44% or 37% of the total correlation eneıgy, according as the primitive SCF result or the Hartree-Fock limit is chosen as a reference value.)

b) The SCF/CI Dilemma and Molecular Symmetry

The radically different interpretation given in the above-mentioned calculations for the inversion barrier of ammonia illustrates the difficulties of the situation depicted by the diagram of Fig. 1: if the expansion basis is too limited, it is difficult to distinguish between correlation energy, in the true sense, and the result of a CI calculation in which the energy lowering can largely be ascribed to the improvement of the charge density with respect to its SCF form. As a simple example, let us consider the hydrogen molecule. The one-determinant function built from the bonding MO ($1s_a + 1s_b$) is greatly improved by CI with the function formed by promoting both electrons to the anti-bonding MO ($1s_a - 1s_b$), but the energy change may be attributed to the improvement of the charge density (at equilibrium distance, the overlap population decreases by about 10%) and not to departure from the independent particle model. Starting with a limited basis, we use as a wave function a determinant which is an exact eigenfunction of a "projected" Hamiltonian operator H_0 describing an independent-particle model within the space spanned by the basis. If the true Hamiltonian is written $H = H_0 + \Delta$ with a correction term Δ including both one- and two-electron contributions, we may say (in the language of perturbation theory) that both the charge density and the pair distribution function could in principle be corrected to all orders. Therefore, it is not astonishing that CI starting from calculations made with minimal basis sets may lead to numerical results which could also be obtained by improving the basis set of the SCF treatment.

An important point put forward by the study of ammonia is the choice of a well-balanced basis set. For the calculation of effects such as barriers it is important to use a basis as "saturated" as possible — or, if this cannot be achieved, at least equally well adapted to the different conformations to be considered. This requirement may be a cause of trouble, especially if the different geometrical configurations belong to different symmetry point groups. By symmetry reasons, the mixing of some atomic orbitals in a given molecular orbital may be forbidden for certain conformations and not for others.

For example, in methylene (CH_2) σ, π_y and π_z-orbitals are separated by symmetry in the linear form, whereas they are not in the bent form. In the case of ammonia, the more symmetrical planar form D_{3h} seems to be favoured with respect to the pyramidal one C_{3v} by SCF calculations with basis sets limited to s and p orbitals, and the inversion barrier may be found to be negative (7).

In conformation problems, polarization orbitals (e.g. $2p$ orbitals for hydrogen atoms, $3d$ orbitals for carbon etc...) are considered to be essential, because they increase the flexibility of SCF variational wave functions in the vicinity of the atoms involved in deformation motions.

But inclusion of basis orbitals with appropriate symmetries can also improve the charge distribution by a specific CI mechanism. In fact, consider a molecule belonging to a certain point group and suppose that some irreducible representations do not appear in the set of occupied molecular orbitals. Now suppose that it is possible to construct virtual molecular orbitals belonging to such representations; then, diexcitations towards these orbitals improve the total charge density in a non-local way. These excitations have to be included in the CI treatment in order to obtain the whole correlation energy contained in the primitive basis set. Typical examples are the well-known *s-limit* of atoms as He (*8*), or the *σ-limit* of planar systems as H_3^+ (*9*). In the case of ammonia in the pyramidal form C_{3v}, the ground-state configuration is $(1a_1)^2$ $(2a_1)^2$ $(1e)^4$ $(3a_1)^2$, and no molecular orbital of the symmetry species a_2 is found in the set of virtual orbitals as far as the basis set is limited to *s* and *p* orbitals on nitrogen and *s* orbitals on the hydrogens; one virtual orbital with symmetry a_2 appears if a set of polarization orbitals p_x p_y p_z is added on each hydrogen.

To clarify the question of electron correlation in ammonia, it is convenient to study the effect of the atomic basis set used to construct the set of occupied and virtual MO's by a perturbation method. Recently, a second-order calculation has been performed for the pyramidal and planar forms of NH_3, starting with a medium-size set of Gaussian functions reduced by contraction to atomic orbitals of double-zeta type for *s* and *p* orbitals of nitrogen as well as for *s* orbitals of hydrogens and gradually improved by polarization functions of *d* type on nitrogen and of *p* type on hydrogens (*10*). At the SCF step, the inclusion of polarization functions increases the total energy more strongly in the pyramidal form than in the planar one, as is found in similar calculations (see e.g. (5)); the effect of *p* functions on the hydrogens is larger than that of *d* fonctions on nitrogen, but the combined effect of $(p + d)$ fonctions is hardly larger than that of *p* functions only. The final values for the SCF total energy of planar and pyramidal forms of NH_3 are —56.1828 a.u. and —56.1961 a.u. respectively, i.e. about 0.03 a.u. above the Hartree-Fock limit.

The correlation energy which can be gained by a second-order perturbation treatment carried out from the preceding SCF canonical MO's of the ground state, increases almost linearly when the basis functions are enlarged first to the set of *p* orbitals on the hydrogens, next to the set of *d* orbitals on nitrogen, and finally to both of them. The final second-order correction to the total energy is —0.2363 a.u. for the planar form and —0.2424 a.u. for the pyramidal one, i.e. 72% of the difference between the estimated values of the Hartree-Fock limit (—56.225 a.u.) and the non-relativistic limit (−56.563 a.u.). As concerns the inversion barrier, the energy difference between the two forms of NH_3 is increased by the second-order corrections, so that the too high value found at the SCF step is not amended. The polarization functions of the hydrogens or of nitrogen have a large effect on these values if they are separately considered, but they neutralize each other in the largest ($p + d$) basis set, leading to a correction of 0.0061 a.u. to the barrier as compared to 0.0063 a.u. in the case of the original double-zeta set. Diexcitations involving the MO of a_2 symmetry have no appreciable effect on the inversion barrier, since their contributions to the second-order energy of the planar and pyramidal forms of NH_3 is —0.0047 a.u. and —0.0043 a.u. respectively. It should also be added that analogous results have been obtained by the group-function method (*11*).

c) The Rotation Barrier of Ethane and Related Topics

Many theoretical works have been devoted to the problem of the rotation barrier of ethane (see e.g. (*4*)). The first successful attack by *ab initio* methods was made by *Pitzer* and *Lipscomb* (*12*); using a minimal basis set of STO's, they found a value of 3.26 kcal/mol for the difference between the SCF total energies of the eclipsed and staggered form (exp. 2.93 kcal/mol). As in the case of ammonia, the barrier is obtained as a difference of two large numbers (about 49,500 kcal/mol) whose error with respect to the exact total energy is assumed to be fairly constant in both conformers. In agreement with this conjecture, it was shown later that the amount of correlation energy to be recovered by CI does not modify the predictions of the SCF calculation to a large extent: according to a second-order perturbation treatment, the effect of diexcitations is to lower the SCF barrier by less than 5% (*13*).

It is possible to analyse the preceding results in terms of chemical concepts, provided that the ground and excited configurations of the perturbation calculation are built from localized orbitals equivalent to the standard molecular orbital. The various types of excitations are indicated in Table 1.

Table 1. *Contribution of various excitation types to the rotation barrier of ethane (in kcal/mol)*

		next-neighbour excitations	long-range excitations
intra-bond excit.	N° 1: 0.163		
inter-bond excit. (charge transfers)		N° 2 } : —0.241	N° 4 } : 0.203
dispersion excit.		N° 3 }	N° 5 }

For both conformations, terms of type 1 give half of the total second-order correction at least, but their variation from one form to the other is completely balanced by the variation of the terms of type 2 and 3. Dispersion terms of type 5 (type 4 is negligible) decrease the barrier slightly; however, their effect is very small (0.2 kcal/mol) because of the large distance between the CH_3 groups and of the high symmetry of the molecule. For instance, in a rotation of 60°, the dispersion type interaction between two CH bonds varies by an amount equal to 0.3 kcal/mol.

It is interesting to compare the results of the second-order perturbation calculations and those of the McWeeny pair function method in the case of ethane. The latter formalism includes correlation terms of type 1 only and leads to the conclusion that the eclipsed form should be more stable than the staggered one (*14*). This unrealistic result may be ascribed to the fact that the pair functions are subjected to the so-called "strong orthogonality condition"; otherwise, the pair function method should be equivalent to a CI including terms of types 1, 2 and 3 and hence should give a good value (i.e. as good a result as the SCF method).

The reason of the failure of the pair functions is clarified by the following simple model (*15*): the physical situation in a rotation barrier problem is pretty well simulated by two hydrogen molecules, one being rotated with respect to the other (Fig. 2).

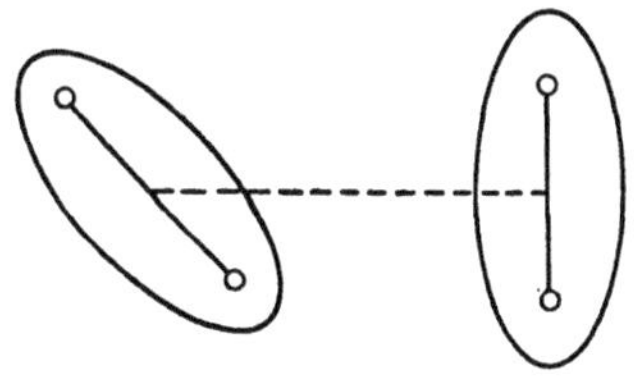

Fig. 2. The H_2—H_2 model for rotation barriers

On this model three types of calculation can be performed:

1. A complete CI calculation based on a set of four 1s orbitals;

2. A pair-function calculation, where each electron pair is associated to a quasi-localized bond function built from Löwdin orthogonalized 1s orbitals. This is equivalent to the treatment of ethane mentioned above (*14*);

3. A calculation of type (2) without any strong orthogonality condition, i.e. starting with wavefunctions of isolated hydrogen molecules without any preliminary orthogonalization.

In the last two cases, the various localized pairs may be written in the form of either MO or Heitler-London functions. The interaction energy of the two molecules was investigated as a function of geometry, calculation (1) being taken as a reference. The main conclusions are that (3) reproduces the energy variations of the ($H_2 + H_2$) system over a wide range of angles and distances, the Heitler-London "covalent + ionic" function for each electron pair always giving about 95% of the interaction energy, but that (2) is unsatisfactory because the behaviour of the rotating group is altered by the orthogonalization process. Actually, performing a calculation of type (3) with two MO's localized in each molecule is mathematically equivalent to a one-determinant function for the whole system, which again suggests that a SCF function (without any orthogonality restriction) may be rather good. The Heitler-London functions introduce intra-bond correlation, but this does not change the results appreciably. On the other hand, the poor behaviour of (2) can be corrected by adding configurations formally associated with "charge transfer".

One can also say that use of non-orthogonal orbitals automatically introduces a large part of the CI needed in a calculation starting with orthogonal bond orbitals. In this respect, orthogonalized MO's are a bad starting point, whereas non-orthogonal orbitals seem to give a good description of the localized bonds and their interactions. These remarks apply to calculations previously described for the ethane molecule and indicate the importance of formal "charge-transfer" configurations whose effects are included in the "inter-bond excitations" of Table 1. On the other hand, correlation effects introduced in a complete CI calculation seem to be unimportant (perhaps fortuitously) for the interpretation of the rotation barrier.

To summarize, it can be stated that only a small change in correlation energy during internal rotation occurs, because the intra-pair correlation terms are almost identical, while the inter-pair contributions between nearest neighbours remain essentially constant. Only small long-range contributions are different from one form to another, but they balance each other to some extent for symmetry reasons.

In spite of the above considerations concerning the approximate constancy of correlation energy in ethane, it is dangerous to draw general conclusions concerning to origin of rotation barriers. Because of the high symmetry of the molecule, many effects may be constant or cancel each other when one passes from the staggered to the eclipsed form without changing the remaining geometry. Allowance for such changes was made in some recent SCF calculations on ethane and hydrogen peroxide (*16*): in this way, a very good agreement with experiment is obtained, suggesting that one should associate, at least partially, the rotation barrier to appreciable variations of bond distances and bond angles between conformers.

d) Variation of Correlation Energy with Interatomic Distances

We now consider the dependence of correlation energies on interatomic distances, especially distances between directly bonded atoms. This dependence may not be significant for the barrier problem but can play an important role in other cases (e.g. in the calculation of energy surfaces for reactions).

The variation of correlation energy with distance is illustrated by the example of H_2 : ΔE_c is practically constant (about 0.042 a.u. (*17*)) between R=0 and R=1.6 a.u., but then increases quite fast towards its asymptotic value of 0.285 a.u. In most typical molecules ΔE_c changes very little in the vicinity of the equilibrium distance, and so the SCF method leads to good equilibrium geometries and force constants. In some cases, however, as in molecules like Li_2 where the equilibrium distance is very large, one is no longer in the region where correlation energy remains constant and its variation influences the geometry and force constants.

In examining figures for the variation of correlation energy with distance, it should be remembered that the definition of correlation energy is somewhat arbitrary and when the large-distance limit is approached ΔE_c may become a purely formal quantity with little physical meaning; for instance, in the case of H_2 at $R = \infty$, ΔE_c is equal to the difference between the exact non-relativistic value and the mean value of the Hartree-Fock energy obtained for $(H^+ + H^-)$ and $(H + H)$ in a variational treatment using the *same* orbital for the hydrogen atom and its ion (*1*). On the other hand, if the usual *ad hoc* definition is accepted, the "correlation energy" is simply all that cannot be accounted for in the restricted Hartree-Fock approximation (with the exception of relativistic effects).

The variation of correlation energy with geometry might be of the utmost importance in the theoretical analysis of chemical reactions, even if only the reactants and products rather than the whole energy surface are considered. However, it is widely assumed that changes in correlation energy can be neglected for interaction between systems in closed-shell states where the number of electron pairs is not changed. The study of chemical reactivity by SCF methods (*18, 19*) invokes this assumption. Unfortunately, there are counterexamples; for instance, it was found that 2/3 of the dimerisation energy of BH_3 is due to a change in correlation energy (*20*). Actually the intra-bond contributions do not change, but in B_2H_6 there are 11 nearest-neighbour bond pairs in contrast to 3 in BH_3, which leads to an increase of the overall inter-pair correlation energy.

This suggests the following modification in the criterion for approximate constancy of correlation energy, namely that both the number of electron pairs and the number of nearest-neighbour interbond terms should not change. This criterion is satisfied in the ethane rotation. At present there are no completely convincing theoretical arguments for such a criterion, but the calculations to date seem to support its validity; therefore, it may be worth adopting as a plausible working rule in trying to assess the need for introducing correlation corrections in any given situation.

3. Correlation and Prediction of Molecular Properties

The problem we want to examine now regards the experimental properties which are known or expected to be affected by correlation. It does not seem likely that inclusion of correlation in molecular calculations will lead to a revision of the general qualitative interpretation of most molecular properties and processes (except, perhaps, in the "large-molecule limit", i. e. when the electronic levels form a very dense spectrum). On the other hand, inclusion of correlation may bring about important quantitative corrections even in the calculation of properties of small and medium-size molecules.

a) Molecular Parameters Related to the Total Energy

It is possible to classify the corrections due to electron correlation (in other words the "correlation effects") according to various criteria. A first distinction is between effects which are directly related to the amount of correlation taken into account in the wave function and effects which arise from the interdependence of correlation energy and molecular geometry. The latter were already mentioned in a different context at the end of the previous section; they include changes in equilibrium distances and force constants. Usually, these molecular constants are considered to be insensitive to correlation, but such a statement is only true within limits.

Table 2 gives an idea of the importance of correlation for equilibrium distances and force constants in diatomic molecules. It shows that the correlation effects are much larger than the experimental errors but are still minor corrections on the SCF results.

b) Classification of Observables with regard to Correlation

With respect to the general correlation problem, molecular properties can be classified in another way. First, it may be convenient to separate one-electron properties, which depend on one-electron operators, and two-electron properties, which depend on operators involving the relative

Table 2. *SCF Calculations of observables in diatomic molecules*

Molecule	Method[a]	R_e[b]	ω_e[c]	Ref.
H_2	SCF	1.39	4561	(*21*)
	OVC	1.40	4398	
	exp.	1.40	4400	
Li_2	SCF	5.26	326	(*21*)
	OVC	5.09	345	
	exp.	5.05	351	
F_2	SCF	2.50	1257	(*21*)
	OVC	2.70	750	
	exp.	2.68	892	
BH	SCF	2.312	$k=3.50$	(*22*)
	IEPA	2.317	$k=3.23$	
	exp.	2.325	$k=3.04$	

[a]) OVC stands for "optimum-valence configuration", IEPA for "independent-electron-pair approximation"
[b]) Equilibrium distances (in a. u.)
[c]) Frequencies of oscillation in the harmonic limit (in cm^{-1}) or force constants (in mdyn/Å)

positions of electrons. Second, among one-electron properties, it is useful to distinguish properties directly associated with the expectation values of certain operators (like electric dipole moments, which are obtained by calculating the expectation value of the position operator for a given electronic state, usually the ground state), and those involving a second-order response to a one-electron perturbation operator, e.g. an external electric or magnetic field (like nuclear spin-spin coupling constants). The calculation of the latter class of one-electron properties is complicated by the fact that it involves the wave functions and energies of virtual excited states; it will be considered in Section 4 in connection with the application of the perturbation theory to the study of correlation. Finally, the properties peculiar to a given state, and the properties characterizing a transition between two states, give rise to different kinds of problems. Although a great number of efforts have been made in order to clear up the role of correlation for the properties of electronic transitions (see e.g. (*23*)), we do not discuss them in more detail because a separation between true correlation effects and other factors, for instance a change of geometry in the excited state, is not always easy to prove.

c) One-Electron Properties: Electric Dipole Moments

Generally speaking, one-electron expectation values obtained by the SCF method are expected to be in good agreement with experiment, because the first-order correlation corrections to them vanish in virtue of the Brillouin theorem (*24*), and doubly excited configurations arising in the second-order terms enter the density matrix only with the square of their expansion coefficient. However, this statement cannot be accepted without reservation for the following reasons:

1. in its primitive form, the Brillouin theorem applies to closed-shell systems only. Its extension to open-shell systems is not straightforward, because the extension of the SCF method itself can be done in different ways, leading to various restricted, unrestricted or extended Hartree-Fock theories (see e.g. (*25*)).

2. for one-electron properties given by a perturbation formula, there is a relatively large freedom in constructing the appropriate excited states, since they are not necessarily given by a variational calculation,

3. if the magnitude of a physical property in a certain molecule is small with respect to its usual value in standard compounds, the SCF method may give spurious results, even for the sign (if any) of the observable, because the calculation becomes too sensitive to the approximations made, e.g. those related to the nature and the size of the basis set etc., and the second-order correlation correction may be larger than the SCF zero-order value.

The electric dipole moment of carbon monoxide is a good example of an exceedingly small quantity. Experimentally, it was definitely proved that the dipole moment of CO is $\mu = 0.11\,\mathrm{D}$ ($C^- — O^+$) (*26*). No reliable conclusions concerning the sign of μ can be drawn from the LCAO—SCF calculations made on CO, even the ones which are close to the Hartree-Fock limit for energy (*27*). On the contrary, configuration interaction gives a correct dipole moment with a positive end on oxygen (*28—30*). In fact, the only conclusion to be drawn from SCF calculations is that the dipole moment of carbon monoxide is nearly zero; and this first approximation is indeed a good approximation, since CO is isoelectronic with N_2.

The point that correlation is important if the absolute value of the property under study is small suggests to reexamine the role of correlation, even in the case of dipole moments which are not close to zero. As a matter of fact, even the dipole moments of highly polar compounds are small compared with those which would result from the displacement of a whole electron from one nucleus to the other in a standard chemical bond (i.e. a quantity equal to about 7D). In general, the moments of the electronic charge distributions in molecules are not small; but a very large part of them, which would continue to exist even if the atoms were simply juxtaposed is balanced by the moment of the nuclear charges. However, as is suggested by X-ray and neutron diffraction studies (*31*, *32*), a molecular dipole moment is highly sensitive to the distorsion of the electron cloud due to chemical binding and hence correlation may play an important role.

In the case of typical molecules, the coefficient of the first "weakly occupied" natural orbital is about 0.1, so that the corresponding occupation number, with which it contributes to the electron density is $(0.1)^2 = 0.01$. Therefore, one may expect changes of about 1% in the one-particle density arising from correlation. Such changes correspond to variations in the dipole moments of the order of 0.1 D. Actually, studies on LiH indicate that the main part of the variation comes from singly excited configurations coupled to the ground state configuration through doubly excited configurations: the SCF dipole moment (6.088 D, exp. 5.83) is not modified if single or double excitations are separately considered, because the mixing of the first ones with the SCF ground state is forbidden by virtue of the Brillouin theorem and the contributions of the second ones to the expectation value of any one-electron operator vanish by reason of orbital orthogonality; but the moment is appreciably reduced (5.85 D) if both the single excitations and a number of double excitations are included (*33*, *30*).

Considerations similar to those made about electric dipole moments apply to other one-electron properties, for instance the nuclear spin-spin coupling constants between non-bonded hydrogen atoms in molecules like methane. These quantities are approximately equal to zero in the simple molecular orbital theory, as it is easily proved by using equivalent orbitals corresponding to the CH bonds instead of the usual delocalized MO's (*34*). Actually, the nuclear spins of protons cannot interact *via* the electrons, since a localized MO cannot be large on two hydrogens at the same time, and correlation should be primarily responsible for all coupling constants, except perhaps for those observed for directly bonded atoms (see Sec. 4).

d) Comparison of Different IPM Approximations

In evaluating correlation corrections to one-particle densities and one-electron expectation values, one must realize that single excitations only vanish to first order, but not to higher orders. This implies that the first natural orbitals differ somewhat from the *Hartree-Fock* orbitals, and since the singly excited configurations enter linearly in their expansion coefficient, they can have more influence on the density than doubly excited ones, even if their coefficients are smaller.

The admixture of singly excited configurations (with respect to the SCF determinant) *via* their interaction with doubly excited ones is exactly equivalent to a replacement of *Hartree-Fock* by *Brueckner* (best-overlap) orbitals. Expectation values of simple one-electron operators like x^2, r^2, $(3z^2 - r^2)/2$ etc., have been calculated (*35*) for H_2 in its equilibrium geometry using SCF and *Brueckner* determinants as wave functions and compared to the more exact values given by a CI function in its natural expansion form. It is found that the SCF results are in error to about 1—2% with respect to the CI results. For most operators (but not for all of them), a wave function of *Brueckner* type leads to more correct values.

4. Correlation in Configuration and Perturbation Treatments

To analyze correlation effects, it is convenient to refer to a standard independent-particle model. Remembering that the occupation numbers n_i of the exact natural orbitals are either close to 2 and sometimes 1 (strongly occupied orbitals and unpaired-electron orbitals) or close to 0 (weakly occupied orbitals), the following process seems to be the simplest way: construct an IPM from a set of MO's whose occupation numbers are exactly $n_i = 2$ or 1, i.e. construct an approximate wave function from doubly and possibly singly occupied MO's and supplement it by a set of virtual orbitals to be used for studying the effect of correlation by CI.

a) Independent-Particle Models for Closed-Shell and Open-Shell Systems

If the SCF theory is applied in order to obtain the "best" IPM for the system under study, the concept of doubly and singly occupied MO's, in which the same space orbital can be associated with an α spinfunction and a β spinfunction, introduces a constraint in the variational treatment. The result is that the energy lowering produced by a subsequent CI calculation is not a genuine correlation effect. However, in closed-shell systems, there is still a connection between configuration interaction and correlation, because the mixing of the ground state configuration with singly excited configurations is formally equivalent to a standard SCF treatment (*86*); in order to take correlation into account, it is necessary to include doubly excited configurations at least. The situation is more involved in the case of open-shell systems (free radicals, triplet states, etc.) because a SCF treatment preserving the concept of doubly and singly occupied orbitals does not prevent a mixing of the ground-state determinant with the monoexcited determinants containing three unpaired electrons (four in the case of triplets) and involving spin reversal of the original unpaired electron (*40*). This kind of CI is considered to be responsible for the so-called "spin-polarization effects" in free radicals. However, spin polarization cannot be unambiguosly referred to as correlation, since it is due to the special form assumed for the wave function of the ground state and could be incorporated from the beginning in an appropriate extended independent-particle model in atoms. In fact, spinpolarization is a correlation effect only if the restricted *Hartree-Fock* scheme previously defined (RFH) is taken as the refeıence independent particle model.

As a matter of fact, the choice of the best IPM for open-shell systems raises a number of problems: there are many possible extended forms of the *Hartree-Fock* theory in addition to the standard RHF method. All these schemes include both spin polarization and correlation to some extent, but in an ill-defined way. The best way to include spin polarization is probably to take RHF as a starting point and to admit single excitations afterwards.

A spin-correct description of a wave function perturbed by a one-electron operator leads to a many-determinant function containing certain single excitations in addition to the RHF determinant. A definite reference state for the discussion of dynamic correlation effects would be one which includes all singly excited configurations coupled directly to the RHF function by a perturbing Hamiltonian, which may be the exchange operator of an unpaired electron as well as an external field or a nuclear moment.

b) Electron-Nuclei Hyperfine Coupling Constants

A CI treatment limited to the monoexcited configurations responsible for spin polarization is not always sufficient to account for the magnitude and sign of the hyperfine coupling constants in atoms and free radicals, just because it does not include any excitation admitting for electron correlation. For the lithium atom ^{7}Li, the restricted ground state and monoexcited configurations all together give 91% (71% and 22% respectively) of the experimental value; on the other hand, the hyperfine splitting of nitrogen ^{14}N (exactly equal to zero in the s^2p^3 RHF approximation) is still 19% of the experimental value, and both double excitations (positive contribution) and triple excitations (negative contribution) have to be included (*36*). Similar results are obtained for oxygen (*37*) and phosphorus (*38*).

As concerns molecules, recent *ab initio* calculations suggest that spin polarization is the prevailing mechanism for the so-called "indirect" part of electron-proton coupling constants in organic free radicals. In the methyl radical, where the RHF "direct" contribution is zero by symmetry reasons, the coupling constant a_H was found to be equal to —29.1 G if all the monoexcited configurations coming from a minimal basis set of STO's are taken into account and —33.1 G if diexcited configurations are included (*39*) (a_{exp} = —23.0 G). In other cases (α and β protons of vinyl radical (*41*), β protons of ethyl radical (*42*)) it turns out that the "direct" RHF contribution and the "indirect" spin-polarization contribution are of the same order of magnitude, but in some cases (α protons) of opposite signs. As concerns electron-proton splittings, the preceding conclusions are not limited to hydrocarbon radicals; they are still valid for other systems, e.g. in the isoelectronic series OH, NH_2, CH_3 (*43*), but the hyperfine splitting of other nuclei (^{17}O, ^{14}N, ^{13}C) is apparently more complicated, as in the atomic case itself.

c) First-Order and Second-Order Properties

In general, the physical properties of an electron system are defined by referring to a specific perturbation problem and can be classified according to the order of the perturbation effect. For instance, the electric dipole moment is associated with the first-order response to an applied electric field (i.e. the perturbation), the electric polarizability with the second-order response, hyperpolarizabilities with higher-order terms. In addition to dipole moments, there is a number of properties which can be calculated as a first-order perturbation energy and identified with the expectation value

$$\bar{A}_k = \int \Psi_k^{(0)} \, \mathfrak{K}^{(1)} \, \Psi_k^{(0)} \, d\tau$$

of a given one-electron operator for the unperturbed state k. Often, the complete operator $\mathfrak{K}^{(1)}$ includes a nuclear part not acting on the electronic wave function $\Psi_k^{(0)}$ and an electronic part and is calculated at a certain origin P (a given nucleus for nuclear quadrupole moments, the center of mass for molecular quadrupole moments etc...); thus its general form is

$$\mathfrak{K}_P^{(1)} = \sum_N Z_N e \, \mathfrak{O}_P(N) - \sum_\nu e \, \mathfrak{O}_P(\nu)$$

where $Z_N e$ and e are the electric charge of nucleus N and electron ν (in absolute value) and the second term is the electronic operator whose expectation value is needed. As for dipole moments, the quantity $\bar{A}_k$ is computed as a difference of two large numbers and its accuracy may be low if the wave function $\Psi_k^{(0)}$ is not good enough. For any independent-particle system, the electronic term is simply equal to the sum of the contributions of the occupied MO's and easily evaluated from LCAO—SCF results, using for example GTO basis functions (*44*). The magnitude of correlation effects on various one-electron properties has not been examined in detail, except for very small molecules, like LiH (*45*), but the correlation of electrons should work in the same way as for dipole moments. The problem of one-electron second-order properties given by an expression of the form

$$\bar{A}_k = \int \Psi_k^{(0)} \mathfrak{H}^{(1)} \Psi_k^{(1)} \, d\tau$$

where $\Psi_k^{(0)}$ and $\Psi_k^{(1)}$ are the zeroth-order and first-order wave functions of a state k perturbed by an external field $\mathfrak{H}^{(1)}$, is more complicated: the effect of the perturbation $\mathfrak{H}^{(1)}$ and the effect of the perturbation V due to the correlation of electrons on the eigenfunctions of the IPM Hamiltonian $\mathfrak{H}_0^{(0)}$ have to be simultaneously considered. In the frame of the Hartree-Fock theory various forms of perturbation treatments have been developed (uncoupled and coupled SCF-methods (*46–48*), finite perturbation SCF-method (*49*)), but none of them takes correlation into account, since without any external perturbation they return to the conventional SCF method.

If the functions $\Psi_l^{(0)}$ and $\Psi_k^{(1)}$ (or alternatively the whole set of unperturbed functions $\Psi_l^{(0)}$) are expressed in terms of configurations by the CI method, it is seen that a second order quantity should be more sensitive to correlation than a one-electron expectation value, because the matrix element of $\mathfrak{H}^{(1)}$ contains the coefficients of the CI expansion for $\Psi_k^{(0)}$ and $\Psi_k^{(1)}$ linearly. It may happen that the quantity to be calculated (e.g. a magnetic susceptibility or a magnetic shielding constant) consists of two parts, a second-order one with respect to an external perturbation $\mathfrak{H}^{(1)}$ and a first-order one with respect to another operator $\mathfrak{H}^{(2)}$:

$$\bar{A}_k = \int \Psi_k^{(0)} \mathfrak{H}^{(1)} \Psi_k^{(1)} \, d\tau + \int \Psi_k^{(0)} \mathfrak{H}^{(2)} \Psi_k^{(0)} \, d\tau$$

The second term (e.g. the diamagnetic susceptibility) is simply the expectation value of $\mathfrak{H}^{(2)}$ and may be predicted by the Hartree-Fock theory safely (*50*). On the other hand, electron correlation may alter the SCF value of the first term (e.g. the paramagnetic susceptibility) to a certain extent, so that the whole correlation effect depends on the relative values of the two parts.

d) Example of Nuclear Spin-Spin Coupling Constants

The preceding remarks concerning the effect of correlation on second-order properties are best illustrated by the calculation of indirect nuclear spin-spin coupling constants $J_{NN'}$ from molecular wave functions. The main contributions to such observables (at least for H—H coupling) comes from the contact terms associated with each nucleus and is proportional to the second-order quantity

$$A_{NN'} = \sum_t \frac{\int \Psi_0^{(0)} \Theta(N) \Psi_t^{(0)} d\tau \int \Psi_t^{(0)} \Theta(N') \Psi_0^{(0)} d\tau}{E_0^{(0)} - E_t^{(0)}}$$

where $\Psi_0^{(0)}$ and $\Psi_t^{(0)}$ are the ground-state and triplet excited wave functions of the unperturbed molecule and the summation is taken over the whole set of triplet states.

It is well known that the magnitude of the coupling constants $J_{NN'}$ is determined by the exchange integrals in valence-bond theory (*51*) as well as in molecular orbital theory (*52*). For convenience, we shall start with MO descriptions using equivalent quasi-localized orbitals (*53*). There is a striking difference as concerns the mechanism of correlation effects in the case of the total energy and the case of NMR (or ESR) coupling constants. For energy, the intra-bond terms involving occupied and virtual "bond orbitals" φ_i and φ_{i^*} are dominant, and long-range effects in polyatomic molecules are chiefly due to molecular integrals of the form $(ii^*|jj^*)$, which can be interpreted as dipole-dipole interaction terms; the role of exchange integrals $(ij|i^*j^*)$ or $(ij^*|ji^*)$ is negligible. On the contrary, the dipole-dipole terms play no role at all for coupling constants, because they vanish after integration over spin coordinates, and exchange integrals only are significant in NMR.

Let us now consider *ab initio* calculations of the spin-spin coupling constants between the directly-bonded nuclei of HD and HF molecules (*54*), and the non-bonded geminal protons of CH_4 (*55*). In both cases, the effect of electron correlation has been added by perturbation on the usual LCAO—SCF treatment.

The computation of J_{HD} and J_{HF} was performed using a large basis set of GTO's and it was checked that a further extension of the basis does not alter the final results appreciably. The uncorrelated result is obtained with unperturbed SCF wave functions including the ground-state determinant and all the monoexcited triplet functions contained in the chosen GTO basis; correlation is introduced by mixing the ground-state configuration with the diexcited singlet functions which can be constructed from the possible excitations of electron pairs in an approximate natural-orbital representation.

As shown in Table 3, the inclusion of correlation in the ground-state wave function produces a marked improvement on the second-order values, increasing J_{HD} by about 25% and restoring the agreement with experiment. However, the magnitude and sign of the correlation correction depends greatly on the choice of the zeroth-order Hamiltonian $\mathfrak{H}_0^{(0)}$: if, for instance, the MO's used for HD are the exact eigenfunctions of the molecular ion H_2^+, the final value at the fourth-order of perturbation: $J_{HD} = 42.57$ Hz (including continuum contributions) is obtained by adding the following contributions: +73.02 Hz (second-order), —35.31 Hz (third-order), +4.86 Hz (fourth-order) (*56*).

Table 3. *Values of H—D and H—F spin-spin coupling constants (in Hz) after successive correlation corrections*

	H—D	H—F
No correlation in $\Psi_0^{(0)}$	30.93	221
Intra-pair correlation[a]		
1 σ	41.73	223
2 σ	—	319
3 σ	—	243
Inter-pair correlation[a]		
1 σ—2 σ	—	272
1 σ—3 σ	—	180
2 σ—3 σ	—	319
Experiment[b]	42.7	521

[a] Giving about 75% of the total correlation energy for HD and 65% of the σ-correlation energy for HF.
[b] Including small dipole-dipole and orbital contributions

Table 4. *Successive contributions (in Hz) to the J_{gem} coupling constant of methane*

	CH_4
No correlation in $\Psi_0^{(0)}$	+ 1.505
Intra-bond correlation	+ 1.545
Inter-bond correlation[a]	—20.460
Experiment	—12.4

[a] The leading term of the inter-bond contributions (terms involving the exchange integrals $(ij^*;ji^*)$ and $(ij;i^*j^*)$ associated to the bonding and antibonding orbitals between the carbon and the two *gem* protons) is equal to —27.075 Hz

With regard to the spin-spin coupling constants between non-bonded nuclei, it may be remarked that the matrix element

$$\int \varphi_k \,\Theta(\mathrm{N})\, \varphi_l d\tau \int \varphi_m \,\Theta(\mathrm{N}')\, \varphi_n \,d\tau$$

gives an important contribution to the coupling N—N′ only if the orbitals φ_k and φ_l and the orbitals φ_m and φ_n have appreciable values on nucleus N and nucleus N′ respectively. For localized (occupied or virtual) orbitals, this implies that φ_k and φ_l encompass N, φ_m and φ_n encompass N′. Consequently, intra-bond contributions to long-range coupling constants are negligible, because localized orbitals cannot be important on two non-bonded nuclei simultaneously. As far as the electronic structure of a polyatomic molecule can be described by equivalent quasi-localized MO's, the values obtained by SCF uncorrelated calculations of the second-order in Θ are necessarily small. If now the lack of correlation of the SCF independent-particle model is also treated as a perturbation V, it can be shown that in consequence of the elimination of dipole-dipole terms the propagation of the interaction is step-wise, so that the larger the number of atoms between N and N′ is, the higher the order of the perturbation treatment should be.

The results reported in Table 4 for the protons of methane support the present analysis as concerns the relative magnitude of the second-order and third-order contributions to the J_{gem} coupling constant; they are obtained by a perturbation calculation starting with an optimized minimal basis set of STO's and including all the possible monoexcited triplet and diexcited singlet configurations (*55*). Moreoever, the negative sign found for the third-order correlation contribution can be understood, because it is simply related to the sign of the "leading term" (i.e. the term containing the bonding and antibonding MO's of the bonds CH, CH′). The deviation of the final value from experiment may be attributed to an inadequate choice of basis functions. In particular, the values of the leading terms are determined by the differential overlap of the corresponding MO's, which strongly depends upon the exponents assigned to the atomic orbitals, and there is no reason why the best exponents for the total energy should necessarily be suitable for the calculation of NMR coupling constants.

To generalize the preceding conclusions to larger molecules, it is convenient to formulate the problem of long-range coupling constants in terms of perturbation diagrams. Let us consider the chain of nuclei

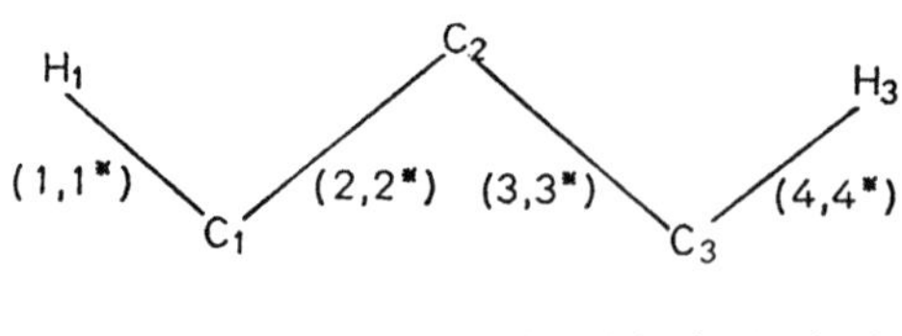

to which the occupied and virtual bond orbitals 1,1*, 2,2*, etc. correspond. In the perturbation expansion of the nuclear spin-spin coupling constants, the perturbation operator is $V + \mathfrak{R}^{(3)}$ where V corresponds to correlation and $\mathfrak{R}^{(3)}$ the contact term. In addition to the matrix elements of V which appear in the expression for the total energy, one has the matrix elements of the components of the contact term $\Theta(\mathrm{N})$ and $\Theta(\mathrm{N}')$, N and N′ referring to the two nuclear spins coupled by the electron system.

Using the notations

$$\int \varphi_i^*(\mu)\, \varphi_k^*(\nu) \frac{1}{r_{\mu\nu}} \varphi_j(\mu)\, \varphi_l(\nu)\, d\tau_\mu\, d\tau_\nu = (ij,kl) =$$

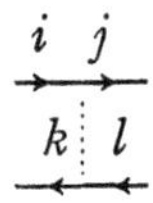

$$\int \varphi_i^*(\nu)\, \Theta(\mathrm{N})\, \varphi_k(\nu)\, d\tau_\nu = (i|\, \Theta(\mathrm{N})\, |k) =$$

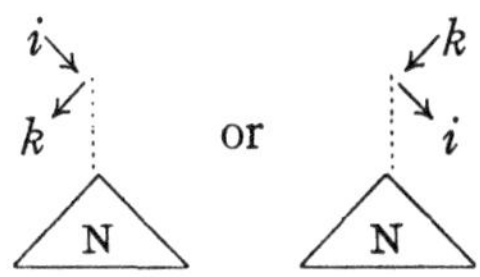

(where lines directed to the right indicate spinorbitals $|i\rangle$, $|j\rangle$ occupied in the ground-state, and lines directed to the left indicate "virtual" spinorbitals $|k\rangle$, $|l\rangle$, the expansion terms of J_{gem} that contain $\Theta(\mathrm{H}_1)$ and $\Theta(\mathrm{H}_3)$ at the first order only are represented by diagrams of the type

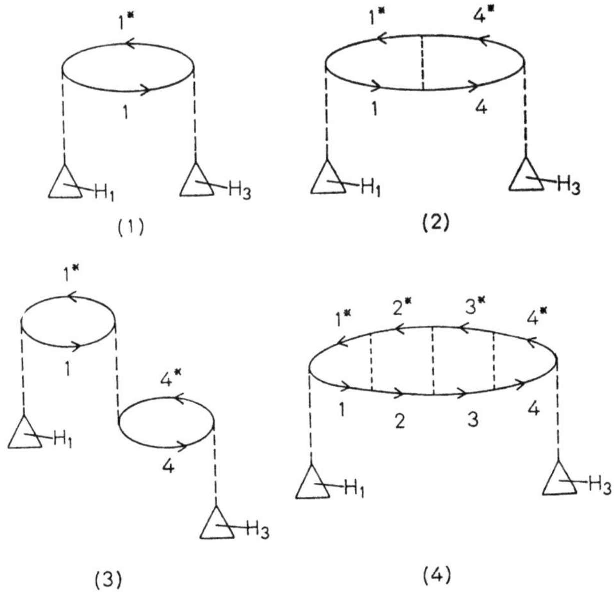

These diagrams have different weights in the perturbation expansion giving J_{HH}: types (1) and (2) are ineffective, because (1) includes only the direct action of $\Theta(\mathrm{H}_3)$ on the bond $\mathrm{C}_1\mathrm{H}_1$, and (2) involves a two-electron interaction between distant bonds C_1H_1 and C_3H_3. Diagrams of type (3) are found to be zero if integrated over spin coordinates. As a result, only terms corresponding to the propagation of perturbation effects through the chain, namely type (4), play an important role. For energy, on the other hand, the important diagrams belong to the type (5) and to a smaller extent to the type (6):

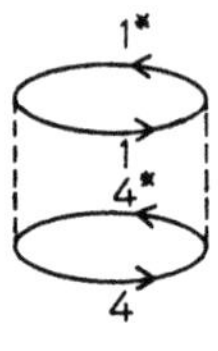

(5)

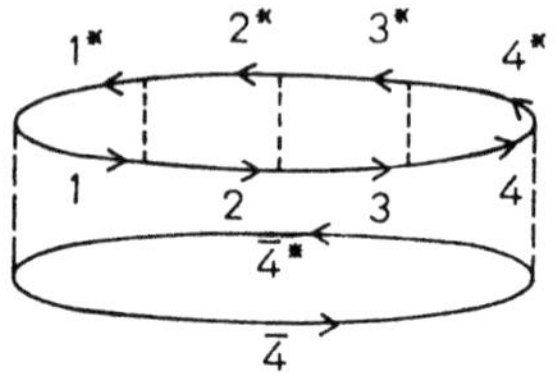

(6)

(4 and $\bar{4}$ denoting that different spin functions are associated to the space orbital 4). It should be noted that diagrams of type (3) for coupling constants and diagrams of type (5) for energy correspond to the same form of interaction, but the former vanish for the spin reasons mentioned above.

5. Energy and Space Dependence of the Correlation Content of Wave Functions

The concept of correlation used in quantum chemistry has two different meanings: it refers either to the energy lowering of an electron system in comparison to the energy of an independent-particle model, or to the spatial deviation of the probability of finding two electrons simultaneously at two given positions (diagonal elements of the second-order density matrix) with respect to the product of the probabilities of finding one electron at each of these points (diagonal elements of the first-order density matrices). In this section, we shall examine the double aspect of the correlation concept for electronic wave functions and related properties.

a) Error Potential and Double Perturbation Theory

As is illustrated by the example of nuclear spin-spin coupling constants, it is possible to estimate the consequences of correlation effects on the values of physical properties computed from IPM wave functions by mixing the perturbations due to electron correlation and external fields in a double perturbation calculation (*57—59*). In such an approach to assess the accuracy with which such properties are calculated it must be born in mind that in addition to the external field there are already two perturbations to be considered, one to compensate the truncation effects due to the limited size of the atomic basis, another to introduce the electron correlation *sensu stricto*, but both can be absorbed into a single operator Δ given by

$$\mathfrak{H} = \mathfrak{H}_0 + \Delta$$

where $\mathfrak{H}_0$ is a *model* Hamiltonian with known exact eigenfunctions (i.e. Slater determinants). The rôle of the operator Δ, called fluctuation potential by *Sinanoğlu* (*60*), was probably first recognized by *Laforgue* (*61*), who used the term "error potential", a terminology which may be preferred since Δ appears in any treatment beginning with an approximate wave function.

Perturbation theory can be formulated in terms of a complete set of functions $\Psi_0, \ldots, \Psi_k$, which may be eigenfunctions of $\mathfrak{H}_0$ (although this is not necessary and may in fact yield a poor convergence). If an external perturbation is applied, for example an electric or magnetic field, a second perturbation operator must be added to $\mathfrak{H}_0$ and the Hamiltonian becomes

$$\mathfrak{H} = \mathfrak{H}_0 + \lambda_1 \Delta_1 + \lambda_2 \Delta_2$$

where λ_1 and λ_2 are ordering parameters for the series expansion, to be put equal to one in the final formulas. Here Δ_1 is used for the actual perturbation and Δ_2 for the formal perturbation including the error potential.

The perturbation expansion of the total energy can be written in the form

$$E = E_0 + H_{00} + \sum_{k \neq 0} \frac{H_{k0} H_{0k}}{H_{00} - H_{kk}}$$

and if Ψ_0 is an eigenfunction of $\mathfrak{H}_0$ the series limited to terms bilinear in λ_1 and λ_2 reduces (for closed-shell systems) to

$$E = E_2 + \lambda_1 \int \Psi_0 \Delta_1 \Psi_0 \, d\tau + \lambda_2 \int \Psi_0 \Delta_2 \Psi_0 \, d\tau +$$

$$+ \lambda_1 \lambda_2 \sum_{k \neq 0} \frac{\int \Psi_0 \Delta_1 \Psi_k \, d\tau \int \Psi_k \Delta_2 \Psi_0 \, d\tau + \int \Psi_0 \Delta_2 \Psi_k \, d\tau \int \Psi_k \Delta_1 \Psi_0 \, d\tau}{E_0 - E_k}$$

with

$$E_k = H_{kk} = \int \Psi_k \mathfrak{H}_0 \Psi_k \, d\tau \quad (E_0 = H_{00})$$

Even in the calculation of first-order properties, i.e. properties depending on terms linear in λ_1, the inaccuracy of the wave function Ψ_0 will have repercussions through the matrix element of Δ_2. This point becomes clear by rewriting the $\lambda_1 \lambda_2$ term in the form

$$\lambda_1 \lambda_2 \int \Psi_0 \Delta_1 \Psi_0^{(\lambda_2)} d\tau + \lambda_1 \lambda_2 \int \Psi_0 \Delta_2 \Psi_0^{(\lambda_1)} d\tau$$

where $\Psi_0^{(\lambda)2}$ is the perturbation of Ψ_0 due to the correlation term $\lambda_2 \Delta_2$ alone, and $\Psi_0^{(\lambda_1)}$ the corresponding term for $\lambda_1 \Delta_1$.

The so-called "correlation effects" on the calculation of molecular properties refer to the effect of the terms in λ_1 only. Therefore, the energy has to be expanded as follows:

$$E = E_0 + \lambda_1 (E^{(10)} + \lambda_2 E^{(11)} + \ldots)$$

where the power series in λ_2 indicates the correlation (or more exactly the correlation plus truncation) corrections to the first-order property determined by the coefficient of λ_1. If one of the perturbations Δ_1 and Δ_2 is easier to handle than the other (e.g. Δ_1 involves one-electron operators only whereas Δ_2 includes two-electron operators), it may be convenient to use "interchange theorems" (*57*), which, in going to second order, avoids computing the correlation effects on the wavefunction itself, i.e. to calculate the correction $\Psi^{(\lambda_2)}$.

Thus the double perturbation theory enables to discuss the effects of electron correlation on molecular properties systematically, but there are not many numerical calculations proceeding along these lines instead of calculating the effect of Δ_2 by variation methods and subsequently dealing with Δ_1 by ordinary perturbation theory. The former procedure has some advantages, for, in principle, it should include Δ_2 to *infinite* order (in other words, Δ_1 and Δ_2 might enter on different levels of the perturbation treatment).

b) Direct and Indirect Dependence of Molecular Properties on Correlation

The statistical sense of the correlation concept leads to classify the molecular properties according to their "direct" dependence on correlation through the second-order density matrix (pair distribution functions) or their "indirect" dependence through the first-order density matrix (charge distribution functions). Properties directly determined by the pair distribution function may be expected to show a greater sensibility to improvements intended to introduce correlation in an approximate wave function. Such is the case of two-electron properties, which, by nature, are directly affected by correlation. An obvious example is the Coulomb interaction of electrons: its dependence on correlation just gives rise to the correlation energy. Other properties of this kind usually involve the electronic spins. Thus, the dipole-dipole coupling of electron spins can be written in terms of a "spin-spin coupling function" to be computed from the pair distribution function by separating terms corresponding to different spin situations (*62*). The effects of correlation on this coupling in π electron systems have been found to be important in connection with the calculation of the zero-field splitting parameters of triplet hydrocarbons and heterocycles (*63–64*).

Much more familiar is the case of one-electron properties. To the first order, they depend on the charge density or the spin density. However, to higher orders, for example in obtaining nuclear spin-spin coupling *via* the electron distribution, transition charge and spin densities also appear. The transition densities refer to *pairs* of states (of the unperturbed system) and determine such properties as transition probabilities. The sensitivity of one-electron properties with respect to improvements of the wave function arises "indirectly" in the sense that the matrix elements of the perturbation are determined by changes of charge or spin densities, though the energy expression contains these quantities with weights depending on pair functions (i.e. the matrix elements of Δ_2).

The intervention of transition densities leads back to the subject of the electron correlation in the whole set of wave functions of the unperturbed system. There have been few calculations of accurate molecular functions for excited states, and the exact importance of correlation in the calculation of ionization and excitation is still debated. For the hydrogen molecule the correlation energy of the ground state is about 0.04 a.u., while in the lowest ${}^1\Sigma_u^+$ state is about 0.01 a.u. and in the lowest ${}^3\Sigma_u^+$ state is almost negligible (0.004 a.u.). Similar changes possibly occur for other valence-shell excitations, but there are few conclusive results to date, except some calculations on ionization energies and excitation energies of light atoms (*65*) and molecules (*66, 67*); they suggest that the correlation energies of states excited in their inner shells may be unusually large because of the inadequate description of the inner open-shell by a RHF function. As matter of fact, the usual difficulties connected with the formal definition of correlation in open shells (see Section 4a) are of importance for this problem.

c) Correlation Function and Correlation Coefficients

The charge distribution function ϱ (1) and the pair distribution function π (1, 2) of a N-electron system described by a total wave function Ψ are defined as follows

$$\varrho\,(1) = \mathrm{N} \int \Psi\,(1, 2, \ldots\ldots, \mathrm{N})\, \Psi^*\,(1, 2, \ldots\ldots, \mathrm{N})\, dv_2\, dv_3 \ldots\ldots dv_\mathrm{N}\, d\sigma$$

$$\pi\,(1, 2) = \mathrm{N}\,(\mathrm{N}-1) \int \Psi\,(1, 2, \ldots\ldots, \mathrm{N})\, \Psi^*\,(1, 2, \ldots\ldots, \mathrm{N})\, dv_3 \ldots\ldots dv_\mathrm{N}\, d\sigma$$

where (1) or (1, 2) stand for the position coordinates of electron 1 or electron pair 1—2 and $d\sigma$ denotes integration over spin coordinates (*68, 69*). From the general theory of reduced density matrices of nth order whose diagonal element determines the probability of finding any n of the N electrons, in any order, at n given positions, it follows that the preceding expressions should be divided by N or N (N—1) in the case of a system of undistinguishable electrons, if one is interested in the probability densities for a specific electron or a specific pair:

$$\varrho_i = \frac{1}{N} \varrho(1) \qquad \pi_{ij}(1,2) = \frac{1}{N(N-1)} \pi(1,2)$$

To introduce correlation functions, let us consider a completely uncorrelated model, in which the probability of finding two particles at two given positions is simply

$$\pi(1,2) = \varrho(1)\,\varrho(2)$$

according to the multiplication rule of probabilities for independent events. This result, of course, is not valid for electrons subjected to Coulomb repulsive forces, but it is always possible to write

$$\pi(1,2) = \varrho(1)\,\varrho(2)\,[1 + f(1,2)]$$

where the two-electron function $f(1,2)$ can be interpreted as a "*correlation function*" giving the departure of the electron system from a completely independent-particle model (*70*).

For descriptive purposes, it is convenient to use the distribution function

$$\frac{\pi(1,2)}{\varrho(2)} = \varrho(1)\,[1 + f(1,2)] = \varrho_c(1)$$

which is the ***conditional probability*** of finding one particle at the position (1) when the other is at the position (2). Since two electrons cannot simultaneously be at the position (1), one has

$$f(1,2) < 0 \quad \text{if} \quad 1 \to 2$$

The effect of correlation can be visualized as a "hole" of depth $\varrho(1)\,f(1,2)$ in the conditional probability $\varrho_c(1)$.

Usually, the electron system under consideration has a well-defined spin state S_z and the reduced density matrices can be decomposed into several components corresponding to the various spin situation:

$$\varrho(1) = \varrho^{\alpha}(1) + \varrho^{\beta}(1)$$
$$(\text{with } \varrho^{\alpha} = \varrho^{\beta} \text{ for } S_z = 0)$$

$$\pi(1,2) = \pi^{\alpha\alpha}(1,2) + \pi^{\beta\beta}(1,2) + \pi^{\alpha\beta}(1,2) + \pi^{\beta\alpha}(1,2)$$
$$(\text{with } \pi^{\alpha\alpha} = \pi^{\beta\beta} \text{ and } \pi^{\alpha\beta} = \pi^{\beta\alpha} \text{ for } S_z = 0)$$

Thus correlation functions themselves may be defined for each spin situation by putting

$$\pi^{\alpha\alpha}(1,2) = \varrho^{\alpha}(1)\,\varrho^{\alpha}(2)\,[1 + f^{\alpha\alpha}(1,2)]$$
$$\pi^{\alpha\beta}(1,2) = \varrho^{\alpha}(1)\,\varrho^{\beta}(2)\,[1 + f^{\alpha\beta}(1,2)]$$
$$(\text{with } f(1,2) = \frac{1}{2} f^{\alpha\alpha} + \frac{1}{2} f^{\alpha\beta} \text{ for } S_z = 0)$$

For an antisymmetrized independent-particle model represented by a single-determinant wave function, only the correlation function $f^{\alpha\beta}(1,2)$ is identically equal to zero; and one has

$$f^{\alpha\alpha}(1,2) \to -1 \text{ if } 1 \to 2.$$

The functions $f^{\alpha\alpha}$ and $f^{\alpha\beta}$ describe the correlation between electrons of the same spin and between electrons of opposite spin respectively, and determine the so-called "Fermi-hole" and "Coulomb hole".

The size of correlation holes has been calculated in the case of an electron gas and their general aspect may easily be imagined in the case of atoms (*70–72*). On the other hand, almost nothing is known about molecules, because the details of the two-electron distribution function become very difficult to follow as soon as there are many attractive nuclear centers (see e.g. (*73*)). Preliminary results are available for the formaldehyde molecule (*74*): in Table 5, the values of the correlation functions at the positions of nuclei C and O are given for a minimal LCAO—SCF wave function and the corresponding multiconfigurational SCF function including all the diexcited closed-shell configurations (see section 6b). These results suggest that the Coulomb correlation in an electron system described by a minimal set of STO's does not change very much the correlation functions, except for the non-zero value of the function $f^{\alpha\beta}$.

Correlation functions give a complete description of the spatial behaviour of electrons in atoms and molecules, but they are complicated distribution functions in a six-dimensional space, and it is worthwhile trying to replace them by single numbers, even at the cost of part of the information contained in them. This is the aim of the so-called *"correlation coefficients"* which have been borrowed from the standard probability theory in order to solve the problem (*75*).

Table 5. *Correlation functions in formaldehyde*

		(1) at C (2) at C	(1) at C (2) at O	(1) at O (2) at O	(1) at O (2) at C
SCF	f	—0.5	0.0	—0.5	0.0
MC—SCF	$f^{\alpha\alpha}$	—1.0000	—0.0656	—1.0000	—0.0658
	$f^{\alpha\beta}$	—0.0659	—0.0656	—0.0660	—0.0658
	f	—0.5329	—0.0656	—0.5330	—0.0658

For two particles i and j whose individual distribution functions $\varrho_i(x,y,z)$ and $\varrho_j(x,y,z)$ and joint-distribution function $\pi_{ij}(x_1,y_1,z_1; x_2,y_2.z_2)$ are known, the correlation coefficient τ_{ij} is equal to the covariance of π_{ij} divided by the product of the variance of ϱ_i and ϱ_j, as defined by probability theory. Assuming that the particles are undistinguishable ($\varrho_i = \varrho_j$) and putting the origin of the coordinate system at the center of gravity of the distribution ($\int \vec{r}\, \varrho_i(x,y,z)\, dv = 0$), one has

$$\tau_{ij} = \frac{\int \vec{r}_1 . \vec{r}_2\, \pi_{ij}\, (x_1 . y_1, z_1; x_2, y_2, z_2)\, dv_1\, dv_2}{\int \vec{r}_1^{\,2}\, \varrho_i\, (x, y, z)\, dv}$$

The corresponding quantum-mechanical expression for an electron system is

$$\tau = \frac{1}{N-1} \frac{\int \vec{r}(1).\vec{r}(2)\, \pi(1,2)\, dv_1\, dv_2}{\int \vec{r}^{\,2}(1)\, \varrho(1)\, dv_1}$$

where $\varrho(1)$ is the usual charge distribution function and $\pi(1,2)$ the pair distribution function containing the correlation function $f(1,2)$ implicitly; hence (*76*):

$$\tau = \frac{1}{N-1} \frac{\int \vec{r}(1).\vec{r}(2)\, \varrho(1)\, \varrho(2)\, f(1,2)\, dv_1\, dv_2}{\int \vec{r}^{\,2}(1)\, \varrho(1)\, dv}$$

The quantities τ defined in this way verify the usual relation of correlation coefficients, namely

$$-1 \leq \tau \leq +1$$

and are clearly equal to zero if the position coordinates of the two electrons are uncorrelated. It should be noted that independent variables are uncorrelated, but the converse is not true, and the values actually found for the simplest two-electron system, helium atom and hydrogen molecule, are indeed small numbers:

$$\tau\,(\mathrm{He}\ {}^1\mathrm{S}) = -0.0542 \quad \tau\,(\mathrm{H_2}\ {}^1\Sigma_g^+) = -0.1002$$

A single correlation coefficient can give an exact picture of correlation only with respect to certain locations of electron pairs in space. This is seen for an atom by expressing the product $\vec{r}\,(1).\,\vec{r}\,(2)$ in spherical coordinated, which involves the angle $\vartheta\,(1,2)$ between the position vectors of electrons 1 and 2. Consequently, τ can be considered as a convenient measure for the "angular" correlation in atoms. However, by replacing the vectors $\vec{r}$ by appropriate function $g(\vec{r})$ in the general definition of τ, it is possible to define correlation coefficients corresponding to other conditions. For instance, the coefficient τ_r obtained from the modulus $|\vec{r}|$ of the position vector $\vec{r}$ ($\tau_r(\mathrm{He}\ {}^1\mathrm{S}) \simeq -0.11$) gives a criterion for the "radial" or "in-out" correlation in atoms.

Similarly, in linear molecules, the "axial" (or "left-right") correlation and the "equatorial" correlation are related to the coefficients τ_a and τ_e obtained from the projection of the position vector $\vec{r}$ on the internuclear axis and the plan perpendicular to it, respectively. For the hydrogen molcule at the equilibrium distance, one has

$$\tau_a\,(\mathrm{H_2}\ {}^1\Sigma_g^+) = -0.0723 \quad \tau_e\,(\mathrm{H_2}\ {}^1\Sigma_g^+) = -0.1561$$

The concept of correlation coefficients has not yet been applied to polyatomic molecules, except from a very qualitative point of view (*75*).

6. Inclusion of Correlation into an Approximate Wave Function

We now turn to the last of the three general questions raised in Section 1. If we are going to perform calculations beyond the Hartree-Fock theory, we have to decide whether or not we are to proceed *via* the SCF approximation. Before discussing the methods themselves, we shall examine the computational side of the problem.

a) Computational Considerations on the SCF Scheme

An *ab initio* SCF calculation using a fixed set of m basis functions includes the following steps:

1. Evaluation of integrals over the basis of atomic orbitals. The computing time of this process for the two-electron integrals $(pq;rs)$ increases as m^4 and is common to any *ab initio* calculation.

2. Linear transformations of the integrals previously listed, in order to prepare for the determination of the effective Hamiltonian and take advantage of the symmetry properties of the molecule. The computing time needed for this increases at least as m^5; even if the molecule has no symmetry at all, the presence of exchange terms in the Hartree-Fock theory makes a part of the transformation still necessary, namely the construction of the integral list.

$$(pq;rs) - \frac{1}{4}(ps;rq) - \frac{1}{4}(pr;qs)$$

3. Iterative construction of the self-consistent field. Each iteration can be assumed to require the computing time of an eigenvalue problem, which varies as m^3, but the construction of the new Hartree-Fock operator goes as m^4 and the number of iterations necessary to obtain convergence may considerably grow with the size of the molecule, especially if one uses non-minimal basis sets.

Sometimes it happens that the process (3) diverges and the work in (1) and (2) is then lost. Consequently, it might be advantageous to by-pass the SCF step, if it is only an intermediate step. This is the philosophy, for example, of the perturbation method called "PCILO" (*77*). However, in the usual cases (at least for closed-shell states), there is no material objection to starting with a standard SCF calculation including suitable devices to ensure convergence. Another practical reason, of course, is that the usual definition of correlation energy is based on the SCF model; therefore, a SCF calculation is needed in order to fix a reference point.

Apart from practical and computational sides, the SCF model has both intuitive appeal and formal mathematical advantages: the Hartree-Fock solution is a *well-defined* one, with known mathematical properties (Brillouin theorem, Hellmann-Feynman theorem etc....). As an example one may mention the invariance of the coupled perturbed Hartree-Fock solution under a translation of the gauge of an external perturbing magnetic field (*78*). Such properties give the SCF approach a certain special position.

On the other hand, if one tries to introduce correlation by perturbation methods, the SCF single determinant may not be the best starting point from the computational point of view as well as for the interpretation of the numerical results. This is illustrated by the calculation of the rotation barrier of ethane: in the work analyzed in section 2c, correlation effects have been introduced by a second-order perturbation treatment starting with SCF MO's and with *ad hoc* localized (i.e. non-SCF) MO's. Inclusion of monoexcitations alone (giving a first-order contribution to the total energy in the non-SCF case) is apt to reproduce the rotation barrier as well as a standard SCF calculation. The diexcitations (introducing correlation) make no significant difference in the two treatments. However, the interpretation is much clearer with localized MO's than with SCF ones: in the SCF case, long-range exchange terms are important in balancing the bond and nearest-neighbour contributions, but in the case of localized MO's, the bond and nearest-neighbour terms balance each other, while the long-range exchange terms are negligible. Localized orbitals are particularly convenient, because they give rise to correlation energy contributions with very different orders of magnitude and different physical meanings (e.g. *intra* and *inter*bond terms). The most significant configurations can be selected according to their physical meaning and the convergence of the perturbation series (as compared with the convergence obtained with SCF orbitals) may be better.

The case of atoms provides further examples. Using the Z-dependent perturbation theory, it is possible to compare the results obtained either from Hartree-Fock orbitals or from simple hydrogenic functions (*79*). Except for the lowest orders, the results for hydrogenic (non-SCF) functions are better. There has been some discussion about this result, because in the case of He (*80*), it makes some difference whether one takes the Hartree or the Hartree-Fock Hamiltonian as a starting point (both having the same zeroth-order wave function). Nevertheless, it can be said that the extra effort in performing first a Hartree-Fock calculation may not be worth while, at least if one intends to make a perturbation calculation subsequently.

b) Computational Considerations on the CI Scheme

At first sight, the choice of a set of starting MO's to be used in a CI treatment is not an important problem. If the total wave function is expressed in terms of Slater determinants Φ_K (1,2, ... , N) constructed from a basis of one-electron function φ_i (ν), the expansion

$$\Psi(1,2,\ldots,N) = \sum_{K} d_K \Phi_K(1,2,\ldots,N)$$

gives the best possible energy value for a function of that sort, provided that the linear coefficients d_K are variationally determined by including all the possible determinants of that basis. For the total energy and the total electron density, it does not matter how the one-electron functions enter into combination with each other to form a particular kind of MO's (e.g. canonical or equivalent quasi-localized MO's or localized non-SCF MO's). Moreover, if the basis of one-electron functions is sufficiently extended, the CI process yields practically correct wave functions in which the electron correlation is fully incorporated.

Unfortunately, such a "brute-force" technique is not really applicable to molecules, because the number of Slater determinants to be included in the CI calculation is much too large, unless one is content with studying little molecules by means of minimal basis sets giving a small part of the correlation energy only. Furthermore, the convergency of a CI expansion is usually very poor, and it is not easy to truncate the series in a satisfactory way. For instance, a system consisting of 5 electrons with only 10 basis functions generates over 15000 Slater determinants, which can be reduced to 3300 linearly independent spin-eigenfunctions for the doublet ground-state. Even if one applies the SCF method to the ground-state configuration in order to obtain the leading term Φ_0 of the CI expansion and one uses the virtual orbitals of the SCF calculation to construct the other terms Φ_K, the latter have small but significant coefficients d_K for the most part.

As a matter of fact, the main problem in the application of the CI method is to find an appropriate way of selecting configurations, and several schemes using the weight and/or the energy of the various possible basis functions Φ_K have been suggested (*81*, *82*). Ordinarily, extensive CI calculations are performed by taking a SCF starting point, because the mixing of the ground-state configuration with the monoexcited ones is then negligible by virtue of the Brillouin theorem, and by limiting the CI treatment to the diexcited ones, because they are of prime importance as concerns the lowering of the ground-state energy. In this respect, it would be useful to transform the canonical SCF MO's into equivalent quasi-localized MO's in order to make a further selection among the diexcited configurations, intra-bond and nearest-neighbour inter-bond excitations being expected to give the most significant contributions. This is confirmed by calculations based on the independent electron-pair model (*20–22*, *83–85*).

At all events, the total number of diexcitations continues to be exceedingly large for medium-size molecule: for a system of $2n$ electrons described by m basis functions, it is equal to $\frac{n(n+1)}{2} \times \frac{(m-n)(m-n+1)}{2}$, say 27720 for benzene in a minimal basis of atomic orbitals. There is no question of constructing a CI matrix of that size and solving the corresponding eigenvalue problem, but its lowest energy is approximately

$$E \simeq E_0 + \frac{|\int \Psi_0 \mathfrak{H} \Psi_K \, d\tau|^2}{E_0 - E_K}$$

where $E_0 \ldots E_K$ are the energies of the various configurations Φ_K, as given by the total Hamiltonian $\mathfrak{H}$ (*86*). This expression is just the second-order energy which is obtained by treating the electron correlation as a perturbation $\mathscr{V}$ whose matrix elements are defined according to the Epstein-Nesbet formula (*77*)

$$\int \Phi_K \mathscr{V} \Phi_L \, d\tau = (1 - \delta_{KL}) \int \Phi_K \mathfrak{H} \Phi_L \, d\tau$$

It can be computed exactly, even for molecules like benzene, (*87*), if the necessary input data are stored on direct-access units. The main part of this calculation is to transform the list of electronic integrals $(pq;rs)$ over the basis functions into a list of integrals $(ij;kl)$ over the molecular orbitals; the computing time needed for that increases as m^5.

c) Introduction of Correlation by Multiconfigurational SCF Methods

If it is necessary to drastically reduce the number of terms to be handled in a CI process, it seems to be profitable to optimize the orbitals $\varphi_i(\nu)$ from which the Φ_K's are constructed, in such a way that the leading terms in the expansion produce the maximum efficiency as concerns correlation. This optimization is the aim of the multi-configuration (MC)—SCF (in which a limited set of functions Φ_K is constructed and the expansion coefficients of the determinants and those of orbitals are optimized) and of alternative methods based on the use of natural orbitals.

To solve MC—SCF equations, several methods have been suggested, but few of them have been applied numerically to molecular calculations involving more than two configurations: one uses the Brillouin theorem as an optimization criterion (*88*), the other tries to reach the minimum point on the energy surface directly (*89*).

In the first method, variation of the orbitals φ_i, from which the Φ_K's are constructed, is made through the transformation

$$\varphi_i \rightarrow e^X \varphi_i$$

where X is an anti-hermitean operator: $X^+ = -X$. This yields an expression for the energy derivatives

$$\frac{1}{2}\left(\frac{\delta E}{\delta X_{ji}}\right) = \int (i^* j - j^* i) \, \Psi \mathfrak{H} \Psi \, d\tau$$

The X_{ij}'s are the matrix elements of X with respect to the φ's i^*, j being creation and annihilation operators for spinorbitals φ_i and φ_j. The condition for optimum orbitals is then the generalized Brillouin theorem:

$$\int (i^* j - j^* i)\, \Psi \mathfrak{H} \Psi \, d\tau = 0$$

If this condition is not fulfilled, the expression for the derivatives is used to progress towards the solution. The relationship of this treatment to other methods, involving pseudo-eigenvalue problems, can also be established; but the present method showed a better convergency in the case of simple polyatomic molecules as formaldehyde (*88*), ethylene and ethane (*90*).

In the second method, one expresses the matrix elements of the MC—SCF methods in terms of first-order and second-order transition density matrices connecting the N-electron basis functions Φ_K of the expansion of Ψ. The matrix elements of the transition density matrices with respect to the n one-electron functions i, j, k, l forming the molecular orbitals used in the CI expansion, are the coefficients of the one-electron and two-electron molecular integrals in the matrix element of the total Hamiltonian:

$$H_{KL} = \sum_{i,j} P_{ij}^{KL} (i|h|j) + \sum_{i,j,k,l} Q_{kl,ij}^{KL} (ij;kl)$$

The first-order and second-order density matrices associated with the ground state of the N-electron system itself take the form

$$\boldsymbol{P} = \sum_{K,L} d_K\, d_L\, \boldsymbol{P}^{KL} \qquad \boldsymbol{Q} = \sum_{K,L} d_K\, d_L\, \boldsymbol{Q}^{KL}$$

where the d_K's are the expansion coefficients defined in Section 6b, and the total electronic energy can be written

$$E_{elec} = \text{trace}\ \boldsymbol{C}^+ \boldsymbol{h}\, \boldsymbol{C}\, \boldsymbol{P} + \frac{1}{2}\ \text{trace}\ \boldsymbol{C}^+ \boldsymbol{Z}$$

where $\boldsymbol{C}$ is a matrix collecting the LCAO coefficients of the m atomic basis functions p, q, r, s in the n molecular orbitals of the CI expansion, and $\boldsymbol{Z}$ is an interaction matrix given by

$$Z_{pi} = \sum_{j,k,l} \sum_{q,r,s,} c_{qs}\, (pr;\, qs)\, c_{rk}\, c_{sl}\, Q_{kl,ij}$$

(note that $\boldsymbol{C}$ and $\boldsymbol{Z}$ are rectangular matrices of dimensions $m \times n$, $\boldsymbol{P}$ and $\boldsymbol{h}$ are square matrices of dimensions $m \times m$ and $n \times n$ respectively).

To minimize the total energy, successive adjustements δC of C are made using either the steepest-descent technique or the conjugate gradient technique: δC is written in the form

$$\delta \boldsymbol{C} = - \lambda \boldsymbol{X}$$

where $\boldsymbol{X}$ is a matrix of residues and λ a step parameter to be evaluated, taking the orthogonality constraints of $\boldsymbol{C}$ into account. These methods have been tested numerically in the case of lithium hydride; the iterative process was found to be smoothly convergent (*91*).

The following conclusions may be drawn from the few MC—SCF calculations that are available to date. First, convergence of the CI series expansion is greatly accelerated by careful optimization of the orbitals: for example, in the case of LiH, an energy limit reached after about 50 terms using SCF virtual orbitals can be achieved after 7 or 8 terms. Second, it appears that various "correlation limits" (depending in the choice of basis functions and configurations) can be distinguished and rapidly reached: e.g. limits corresponding to left-right correlation, angular correlation, inner-shell correlation, etc. Finally, the leading term Φ_0 of the MC—SCF expansion is not very far from the one-determinant representation of the ground state, but the weakly occupied MO's of the other terms can be strongly different from the virtual orbitals of the standard SCF theory. This suggests that the SCF independent-particle model is a natural step in the treatment of many-electron systems and that to go beyond that limit requires much more elaborated methods.

Other forms of multiconfigurational methods are based on the natural orbital (NO) concept introduced by *Löwdin* (*92*). It is very difficult to solve the integro-differential equations giving the NO's, except for two-electron systems (*93*), but it is possible to determine them iteratively by a succession of configuration interactions: given a set of approximate NO's coming from appropriate basis functions, one constructs an approximate CI wave function built on these orbitals; the diagonalization of the corresponding first-order density matrix then yields a new set of NO's from which one repeats the calculation until convergency of the NO's. The procedure has been applied successfully to helium hydride and lithium hydride (*94*).

Instead of calculating the NO's for the whole N-electron system, one can also try to determine the NO's of a part of the system by configuration interaction (for instance an electron pair for the three electrons of He_2^+ or H_3) and to use these pseudo-NO's in order to increase the convergency of the general CI treatment (*95*). The methods starting with the calculation of natural orbitals for different electron pairs in a molecule are really attractive, but such NO's are the NO's of the natural geminals rather than those of the total wave function (for an account of the various methods of this type, see (*85*)).

Last not least, the classical way of improving convergency by including the interelectronic coordinates explicitly in the variational trial functions has recently been revived by *Boys* (*96*): in its new form, this technique called "method of transcorrelated functions" is able to reproduce the total energy of lithium hydride with chemical accuracy (*97*).

7. Physical Significance of Independent-Particle Models and Correlation

So far, the main concern of this report has been with mathematical and computational considerations, but this should not be allowed to obscure the physics of the story. The question is still that of deciding whether correlation should be considered a physical effect or a purely numerical problem. We shall try to give a brief analysis of that question.

a) The Role of Independent-Particle Models in the Interpretation of Molecular Properties and the Charge-Cloud Picture

If correlation is purely computational in origin, it looses its meaning as soon as one finds a way to by-pass the IPM step in the computation. On the other hand, if the independent-particle model has a special physical significance, correlation occupies a rightful place in the interpretational scheme of molecular reality. Now, such a model has two main qualities: first, it corresponds to the Schrödinger "electron-cloud" picture of the atom, resolving the cloud into orbital contributions, and hence provides a quasi-chemical first approximation to the correct physical description of a molecule; second, it provides the rigorous form of the MO "theory" of chemistry which has rendered so much service and possesses a great heuristic value.

It may be objected that the orbitals themselves have no "existence" (they are not "buckets" into which electrons may be placed, but merely solutions of a particular canonical form of the SCF equations), and indeed unitary mixing of the occupied orbitals among themselves makes no difference whatever to the charge cloud — which is merely "resolved" into new components — or even to the many-electron wave function itself. On the other hand, the canonical orbitals (i.e. the MO's) allow us to visualize many processes in which they may acquire a unique significance. For example, Koopmans' theorem shows that ionization may be visualized by removing an electron from an orbital — provided this is defined as a canonical MO (not, for example, as an equivalent quasi-localized MO, to which the theorem does not apply). The MO's allow a corresponding rationalization of electron excitations, one which has provided a spectacularly successful basis for the whole of atomic and molecular spectroscopy.

In spite of the advantages of the (canonical) MO's in discussing excitation processes, and of orbitals in general for discussing the electron distribution, other formulations (e.g. non-Schrödinger) might also be explored. It is not really true that a first understanding of the properties of molecules can be reached *only* in terms of orbitals, even granting that we cannot avoid visualizing the concepts of the theory of matter in terms of macroscopic analogies. For instance, the theory of resonance is based on a completely different description of molecules, the building blocks being systems of bonds which correspond to special forms of many-electron wave functions. In fact, it is not even necessary to associate a particular visualization to the building blocks of a molecular eigenstate, except, say, fictitious schemes playing the same role as Feynman diagrams. In fact, it should be possible to give a theoretical interpretation of the facts of chemistry simply by treating certain reference states (say, the bond systems of resonance theory) through Dirac's formulation of quantum mechanics without specifying any functional dependence. It remains to be seen whether such an approach is really better for understanding facts.

Nevertheless, judging from the theory of atoms, it seems reasonable to adopt the Schrödinger picture for a description of electrons in molecules. As is well known, this picture consists in interpreting the square of the wave function describing an electron as the density of a cloud into which the electron is somehow smeared out. This interpretation is visualizable, of course, as long as we are dealing with a wave function depending on only three space coordinates, i.e. an orbital. This is why a molecular orbital description of molecules may appear to be particularly effective in providing a standard picture on which to base the description of physical and chemical processes in which a given molecule is involved.

In this sense any kind of orbital may be used. Only if one wants the orbital picture to include as much as possible of the correct quantum-mechanical description of a given molecule, one must refer to SCF orbitals. This is the basis for taking the Hartree-Fock model as a physical reference point, and for introducing correlation effects as those effects which cannot possibly be interpreted in terms of the charge cloud visualization.

b) The Independent-Particle Model and the Particle Concept

It is interesting to draw a distinction between the two aspects of correlation which we have considered so far in terms of the second-quantization method for systems of N identical Fermi particles. Those methods (which are but a more effective and general way of formulating CI) rest upon the occupation-number representation: given the set of all possible single-particle states (spinorbitals), one builds a complete set of N-particle states $|n_1, n_2, \ldots >$ by constructing Slater determinants or appropriate combinations thereof, each corresponding to a given assignment $n_1, n_2, n_3, \ldots$ of the N particles to the various orbitals. $|1>$, $|2>$, $|3>$, ... The N-particle states are thus specified by giving just the binary number having as the j-th digit the occupation number of the j-th spinorbital $|j>$; and second-quantization techniques start from reformulation of the Schrödinger equation in terms of those states taken as a basis, which is nothing but the configuration interaction of quantum chemists. However, this general formulation, which takes for granted a preliminary choice of the single-particle states, indicates more explicitly on basic point: the very fact of speaking of many-particle systems implies IPM reference basis states (obtained from the single-particle states), and therefore this is an obvious starting point for a theory of such systems. If the quantum-mechanical behaviour of an independent particle is understood, what matters is to find the novelties arising from the simultaneous presence of many identical particles.

The point of view of second quantization (see e.g. (*98*)) thus treats the "correlation problem" as a major problem of the physics of matter. However, correlation is seen here in a way somewhat different from that of quantum chemistry. First, no definition of the ideal reference model is required; second, periodic and limiting properties of sums of matrix elements are considered in order to draw general (qualitative) conclusions on the properties of bulk matter. If any choice of an independent-particle model is made, it is made just to introduce simplifying features in the sense of the mathematical properties just mentioned. In this way, the correlation problem loses again part of its connection with the "non-independence" of particles in many-particle systems.

The above considerations may be put in mathematical form by referring the following formulas (*100*). Let a_j^* be the creation operator for an electron in the single-particle state $|j>$; and a_j the corresponding annihilation operator. Then the Hamiltonian operator of N electrons can be written:

$$\mathfrak{H} = \mathrm{R}_0 + \sum_{k,k'} \mathrm{H}(k,k')\, a_k^* a_{k'} + \frac{1}{2} \sum_{k,k'} \sum_{l,l'} \mathrm{W}(k,k';l,l')\, a_k^* a_l^* a_{l'} a_{k'}$$

Here R_0 is a zero-point energy (say, nuclear repulsion); and $H(k,k')$ and $W(k,k';l,l')$ are appropriate matrix elements. In particular, $H(k,k')$ is the elements $<k|\mathfrak{K}_1|k'>$ of a given IPM one-electron Hamiltonian $\mathfrak{K}_1$. Recalling that

$$\overset{*}{a}_j\, a_j = n_j$$

n_j being the occupation number of $|j>$, we can also write (after some algebra):

$$\mathfrak{K} = R_0 - T_0 + \sum_{k,k'} \bar{H}(k,k')\, \overset{*}{a}_k\, a_{k'} +$$

$$+ \frac{1}{2} \sum_{k,k'} \sum_{l,l'} \bar{W}(k,k';l,l')\, \overset{*}{a}_k\, \overset{*}{a}_l\, a_{l'}\, a_{k'}$$

where

$$T_0 = \frac{1}{2} \sum_{\substack{k,l \\ k \neq l}} V(kk;ll)\, \overset{*}{n}_k\, n_l$$

$$\bar{H}(k,k') = H(k,k') + \sum_{l \neq k,k'} V(kk';ll)\, n_l$$

$$\bar{W}(kk';ll') = V(kk';ll')\,(1-\delta_{kk'})\,(1-\delta_{ll'}) + [W(kk';ll') - V(kk';ll')]$$

with V an arbitrary matrix.

The latter equation shows that there is a great freedom of choice of the one-electron operator explicitly used: any given choice can always be replaced by another by shifting part of the two-particle terms to the one-particle terms. This property was used some time ago to analyze π-electron theories (*99*), and has been recently used to analyze the energy of large molecules, especially with regard to non-bonded interactions (*100*).

Such studies are related to the correlation problem not only because of the probable connection of long-range interactions with correlation, but because they involve the choice of a suitable independent-particle model. This need not be an SCF one; indeed, for the purpose of treating on a common footing classes of related molecules, it is better to take a one-electron Hamiltonian whose form depends through simple rules on the nature and the arrangement of the atoms under consideration. Such a choice might propose, say, a scheme enjoying "additive bond properties" as a "physically significant" model; then correlation proper would be included in correction terms largely depending on readjustments of MO's upon optimization, and concurrent deviations from additivity.

Even with such a choice, the correlation contribution could still be isolated because the SCF optimization is equivalent to first-order perturbation correction of the one-electron part of $\mathfrak{K}$; the higher-order terms are the correlation correction in the sense used so far (see (*101—103*)).

The question remains open: what physical effects do the latter represent that are qualitatively different from those represented by the first-order terms?

As regards extensive use of the so-called second-quantization techniques, we recall again that they rely upon systematic use of periodicity or other general relationships between matrix elements in the preceding equations. These properties play a very important role for very large numbers of particles ($N \to \infty$); and, at any rate, matrix elements involving the IPM states still occur, and unless they are actually evaluated (using, for example, the orbitals in Schrödinger form), they become parameters whose numerical values merely define a fictitious model system (cf. the BCS model in superconductivity). Models of this kind are known (see (*23*)).

8. Conclusions

The discussion on which the present report is based started with the three questions reported in the introductory section. It appears that no final conclusion can yet be given, although many clarifying remarks and partial answers are provided by the preceding analysis.

First of all, it is clear that two different problems have gone under the name of correlation problem: the choice of an independent-particle model suitable for a general understanding of atomic and molecular phenomena, and the discussion of the importance of deviations from the *optimum* (SCF) charge-cloud picture in rigorous estimates of energies and other molecular properties.

Most of the work reported here is connected with the latter aspect of the problem. The main point is not so much to discuss the existence of correlation corrections, but the practical consequences of their existence. Now it seems that in many cases correlation does exist, but is irrelevant because it does not change when physical phenomena take place. This point and objections to it are illustrated by detailed considerations on inversion and rotation barriers as in ammonia and in ethane. For the inversion of NH_3, one can only say that the problem of the choice of the AO basis is so intermingled with that of correlation proper that no definite conclusion can be drawn. For the rotation of CH_3 groups the approximate constancy of correlation energy during rotation is almost certain. Unfortunately this does not rule out the possibility that correlation plays a role in more complicated cases. The situation with other changes in geometry or chemical reactions is very similar: it is usually assumed that correlation energy is approximately constant when systems in closed-shell states interact, but there are counterexamples, as in the case of the dimerization of BH_3.

Correlation plays a significant role even in equilibrium distances and force constants. Nevertheless, a physical picture based on results without correlation is not affected by the corrections in question, which are very small.

The same holds in general for one-electron properties which are affected by correlation indirectly. In some cases correlation is important, e.g. with electric dipole moments that are close to zero. The best example is carbon monoxide: the reason why there have been uncertainties even regarding the sign of its dipole moment is due to the fact that the only conclusion to be drawn from SCF calculations is that the dipole moment in question is nearly zero. Therefore, calculations including correlation should play a decisive role in this connection.

When one studies the effect of correlation on one-electron properties, the question of the independent-particle scheme with respect to which correlation is defined becomes very important. For instance, the Hartree-Fock orbitals might be replaced by Brueckner orbitals, and this might produce some changes in the expectation values of some operators.

We have already mentioned that the choice of an appropriate independent-particle model is a problem in itself, especially for open-shell systems. One of the effects which are correlation only in the frame of a particular definition is spin polarization. This remark is connected with the question of the role of correlation in spin densities and hyperfine splitting constants in free radicals.

Typically, the problem of correctly computing correlation effects comes into play with properties like nuclear spin-spin coupling constants, which are "second-order one-electron" properties. A perturbation study of the role of correlation in this case shows that the terms contributing to energy are not the same as those contributing to other properties. The important moral of the story is that conclusions about the importance of correlation should be drawn only by considering at the same time properties of different types, because effects on one property may be important while those on other properties are not, and *viceversa*.

The above remark suggests that the importance of correlation should be estimated directly on wave function. This can be done either by treating correlation as an effect arising from an appropriate perturbation of the independent-particle scheme (the so-called error potential) or using suitable correlation coefficients and correlation functions.

The counterpart of the problem of determining the correlation content of a wave function is, of course, the problem of including correlation into an approximate wave function. Putting aside strictly technical details, there remains the question whether the SCF treatment is an essential step in a refined calculation. If one by-passes the SCF step, the genuine definition of correlation becomes inapplicable, unless one performs separately an independent-particle calculation. Both procedures have arguments in their favour, and this means that the real question is the physical significance of correlation, or, better, of the independent-particle model.

The correlation problem thus becomes the following: is it possible (and useful) to find a physically significant independent-particle scheme? If so, is this satisfactory from the interpretation point of view? If the answers to these questions are respectively yes and no, correlation can be both defined and assigned an important role. But as the independent-particle model which satisfied the first question may not be that of Hartree-Fock, the correlation thus defined may have nothing to do with the correlation usually referred to.

A detailed discussion of the question leads to the conclusion that there are indeed an infinite number of ways in which an independent-particle model can be defined. Many such models are physically significant, because they describe particles which behave in a certain model way. Regarding the choice, very little has been done so far, the main emphasis having been on computations for the last decade or so. In this sense, the correlation problem is not yet out of date.

Strictly speaking, the question whether or not the use of an independent-particle model is important is a matter of opinion, because it is certainly not necessary for calculating the expectation values of many molecular observables. However, if one looks at the question from the point of view of usefulness in schematization and in general in understanding of physical facts, it seems that there is little doubt about a positive answer. Natural philosophers believe that, in general, simple reference models cannot be dispensed with in the process of understanding the physical world, and a model based on the concept of particle seems to be particularly suitable for that purpose in dealing with atoms and molecules.

In conclusion, the analysis of the correlation problem presented here has shown that enough is known about the quantitative importance of correlation in the traditional sense to make it clear that correlation corrections are important only in exceptional cases. On the other hand, practically everything remains to be done in connection with physical meaning. In particular, the work done so far on correlation has had the great result of showing that too little importance has been attributed to the problem of the independent-particle model and to the related question of the physical interpretation of molecular properties.

Acknowledgements: The author expresses his deep gratitude to all the participants in the Ischia 1970 Symposium on Electron Correlation whose contributions to the discussions provided the material for this report.

The support of NATO Grant N° 223 is gratefully acknowledged.

References

1. *Löwdin, P. O.:* Advan. Chem. Phys. *2*, 207 (1959).
2. *Silverstone, H. J., Joy, H. W.:* J. Chem. Phys. *47*, 1384 (1967).
3. *Switkes, E., Stevens, R. M., Lipscomb, W. N.:* J. Chem. Phys. *51*, 5229 (1969).
4. *Lehn, J. M.:* In: Computational Problems in Quantum Chemistry, edited by *G. Diercksen* and *A. Veillard*, Max-Planck-Institut für Physik und Astrophysik, München (1969).
5. *Rauk, A., Allen, L. C., Clementi, E.:* J. Chem. Phys. *52*, 4133 (1970).
6. *Pipano, A., Gilman, R. R., Bender, C. F., Shavitt, I.:* Chem. Phys. Letters *4*, 583 (1970).
7. *Kaldor, U., Shavitt, I.:* J. Chem. Phys. *45*, 888 (1966).
8. *Shull, H., Löwdin, P. O.:* J. Chem. Phys. *30*, 617 (1959).
9. *Kutzelnigg, W., Ahlrichs, R., Labib-Iskander, I., Bingel, W.:* Chem. Phys. Letters *1*, 447 (1967).
10. *Munsch, B., Vinh, J.:* (unpublished results).
11. *Robb, M. A., Csizmadia, I. G.:* Intern. J. Quant. Chem. *4*, 365 (1970).
12. *Pitzer, R. M., Lipscomb, W. N.:* J. Chem. Phys. *39*, 1995 (1963).
13. *Lévy, B., Moireau, M. C.:* J. Chem. Phys. *54*, 3316 (1971).
14. *Klessinger, M.:* Symp. Faraday Soc. *2*, 73 (1969).
15. *Magnasco, V., Musso, G, F., Mc.Weeny, R.:* J. Chem. Phys. *47*, 4617 (1967).
16. *Veillard, A.:* Chem. Phys. Letters *3*, 128 (1969); *4*, 51 (1970).
17. *Kołos, W., Roothaan, C. C. J.:* Rev. Mod. Phys. *32*, 211 (1960).
18. *Clementi, E.:* J. Chem. Phys. *46*, 3851 (1967).
19. *Snyder, L. C., Basch, H.:* J. Am. Chem. Soc. *91*, 2189 (1969).
20. *Gelus, M., Ahlrichs, R., Staemmler, V., Kutzelnigg, W.:* Chem. Phys. Letters *7*, 503 (1971).
21. *Das, G., Wahl, A. C.:* J. Chem. Phys. *47*, 2934 (1967).
22. *Gelus, M., Ahlrichs, R., Staemmler, V., Kutzelnigg, W.:* Theoret. Chim. Acta. *21*, 63 (1971).
23. *Kutzelnigg, W., Del Re, G., Berthier, G.:* Forsch. Chem. Fors. *22*, 1 (1971).
24. *Møller, C., Plesset, M. S.:* Phys. Rev. *46*, 618 (1934).
25. *Nesbet, R. K.:* Advan. Chem. Phys. *9*, 321 (1965).
26. *Oziev, I., Yi, P. N., Khosla, A., Ramsey, N. F.:* J. Chem. Phys. *46*, 1530 (1967).
27. *Huo, W. M.:* J. Chem. Phys. *43*, 624 (1965).
28. *Grimaldi, F., Lecourt, R., Moser, C.:* Intern. J. Quant. Chem. *1S*, 153 (1967).
29. *Siu, A. K. Q., Davidson, E. R.:* Intern. J. Quant. Chem. *4*, 223 (1970).
30. *Green, S.:* J. Chem. Phys. *52*, 3100 (1970); *54*, 827 (1971).
31. *Steward, R. F.:* J. Chem. Phys. *53*, 205 (1970).
32. *Coppens, P.:* Science *158*, 1577 (1967).
33. *Bender, C. F., Davidson, E. R.:* Phys. Rev. *183*, 23 (1969).
34. *Hiroike, E.:* J. Phys. Soc. Japan *22*, 379 (1967).
35. *Staemmler, V., Jungen, M.:* (unpublished results).
36. *Nesbet, R. K.:* Coll. Int. C.N.R.S. *164*, 87 (1967).
37. *Kelly, H. P.:* Phys. Rev. *173*, 192 (1968).
38. *Dutta, N. C., Matsubara, C., Pu, R. T., Das, T. P.:* Phys. Rev. Letters *21*, 1139 (1968).
39. *Chang, E. S., Davidson, E. R., Vincow, G.:* J. Chem. Phys. *52*, 1740 (1970).
40. *Lefebvre. R.:* In: Modern Quantum Chemistry, Part I, p. 215; edited by O. Sinanoğlu. New York: Academic Press (1965).
41. *Millié, Ph., Lévy, B., Berthier, G.:* Intern. J. Quant. Chem. *6*, 155 (1972).

42. *Ellinger, Y., Rassat, A., Subra, R., Berthier, G., Millie, Ph.:* Chem. Phys. Letters *11*, 362 (1971).
43. *Meyer, W.:* J. Chem. Phys. *51*, 5149 (1969).
44. *Kortzeborn, R. N.:* I.B.M. Research Report RJ 577 (1969).
45. *Arrighini, G. P., Tomasi, J., Guidotti, C.:* Theoret. Chim. Acta. *18*, 329 (1970).
46. *Dalgarno, A.:* Advan. Phys. *11*, 281 (1962).
47. *Stevens, R. M., Pitzer, R. M., Lipscomb, W. N.:* J. Chem. Phys. *38*, 550 (1963).
48. *Langhoff, P. W., Karplus, M., Hurst, R. P.:* J. Chem. Phys. *44*, 505 (1966).
49. *Pople, J. A., McIver, J. W., Jr., Ostlund, V. S.:* J. Chem. Phys. *49*, 2960 (1968).
50. *Arrighini, G. P., Tomasi, J., Petrangolo, C.:* Theoret. Chim. Acta. *18*, 341 (1970).
51. *Karplus, M., Anderson, D. H.:* J. Chem. Phys. *30*, 6 (1959).
52. *McConnell, H. M.:* J. Chem. Phys. *24*, 460 (1956).
53. *Lennard-Jones, J. E.:* Proc. Roy. Soc. (London) *A 202*, 1, 14 (1949).
54. *Meyer, W.:* Z. Physik *229*, 452 (1969).
55. *Barbier, C., Gagnaire, D., Levy, B., Berthier, G.:* J. Magn. Res. *5*, 11 (1971).
56. *Das, T. P., Dutta, C. M., Dutta, N. C.:* J. Physique *31*, C 4—99 (1970).
57. *Dalgarno, A., Stewart, A. L.:* Proc. Roy. Soc. (London) A *238*, 269 (1956); A *247* (1958); *257*, *534* (1960).
58. *Hall, G. G.:* Advan. Quant. Chem. *1*, 241 (1964).
59. *Hirschfelder, J. O., Byers-Brown, W., Epstein, S. T.:* Advan. Quant. Chem. *1*, 255 (1964).
60. *Sinanoğlu, O.:* Proc. Natl. Acad. Sci. *47*, 1217 (1961); Advan. Chem. Phys. *6*, 315 (1964).
61. *Laforgue, A.:* Compt. Rend. Acad. Sci. *238*, 1933 (1954); Thèse Revue Opt. Theo. Exp., Paris (1955).
62. *McWeeny, R., Kutzelnigg, W.:* Intern. J. Quant. Chem. *2*, 187 (1968).
63. *Levy, B., Kochanski, E.:* Theoret. Chim. Acta. *12*, 150 (1968).
64. *Daudey, J. P., Pullman, A.:* Chem. Phys. Letters *5*, 347 (1970).
65. *Chase, R. L., Kelly, H. P., Kohler, H. S.:* Phys. Rev. *3*, 1550 (1971).
66. *Meyer, W.:* (unpublished results).
67. *Pauzat, F., Ridard, J., Levy, B.:* Mol. Phys. (1972).
68. *Löwdin, P. O.:* Phys. Rev. *97*, 1474 (1955).
69. *McWeeny, R.:* Proc. Roy. Soc. (London) A *253*, 242 (1959).
70. — Rev. Mod. Phys. *32*, 335 (1960); Intern. J. Quant. Chem. *1* S, 351 (1967).
71. *Slater, J. C.:* Phys. Rev. *81*, 385 (1951).
72. *Maslen, V. W.:* Proc. Phys. Soc. (London) A *69*, 734 (1956).
73. *Jørgensen, C. K.:* Modern Aspects of Ligand Field Theory. Amsterdam: North-Holland 1971. Solid State Physics, *13*, 375 (1962); Orbitals in Atoms and Molecules, Academic Press, London, 1962; Oxidation Numbers and Oxidation States, Springer Verlag, Berlin, 1969.
74. *Suard, M., Jouhet, G.:* Atti. Soc. Nat. (Modena) *100*, 59 (1969).
75. *Kutzelnigg, W., Del Re, G., Berthier, G.:* Phys. Rev. *172*, 49 (1968).
76. *Bingel, W. A.:* Chem. Phys. Letters *5*, 367 (1970).
77. *Claverie, P., Diner, S., Malrieu, J. P.:* Intern. J. Quant. Chem. *1*, 751 (1967); Theoret. Chim. Acta *8*, 390 (1967); *13*, 1, 18 (1969).
78. *Epstein, S. T.:* J. Chem. Phys. *42*, 2897 (1965).
79. *Dickson, R. J., Sokoloff, J.:* Phys. Rev. *141*, 32 (1966).
80. *Byron, F. W., Joachim, C. J.:* Phys. Rev. *146*, 1 (1966).
81. *Pipano, A., Shavitt, I.:* Intern. J. Quant. Chem. *2*, 741 (1968).
82. *Whitten, J. L., Hackmeyer, M.:* J. Chem. Phys. *51*, 5584 (1969).

83. *Ahlrichs, R., Kutzelnigg, W.:* J. Chem. Phys. *48*, 1819 (1968). Theoret. Chim. Acta *10*, 377 (1968); Chem. Phys. Letters *1*, 651 (1968).
84. *Jungen, M., Ahlrichs, R.:* Theoret. Chim. Acta *17*, 339 (1970) .
85. *Kutzelnigg, W.:* Proceedings of the IBM Symposium on Molecular Physics, Ludwigsburg (1970).
86. *Brillouin, L.:* Actualités Scientifiques N⁰ 17, Hermann, Paris (1933).
87. *Berthier, G., Meyer, A. Y., Praud, L.:* Jerusalem Symp. Quant. Chem. Biochem. *3*, 174 (1971).
88. *Lévy, B.:* Chem. Phys. Letters *4*, 17 (1969); Intern. J. Quant. Chem. *4*, 297 (1970).
89. *McWeeny, R.:* Faraday Soc. Symp. *2*, 7 (1968).
90. *Lévy, B.:* (unpublished results).
91. *Mukherjee, N. G., McWeeny, R.:* Intern. J. Quant. Chem. *4*, 97 (1970).
92. *Löwdin, P. O.:* Phys. Rev. *97*, 1474 (1955).
93. *Kutzelnigg, W.:* Theoret. Chim. Acta. *1*, 343 (1963).
94. *Bender, C. F., Davidson, E. R.:* J. Phys. Chem. *70*, 2675 (1966).
95. *Edmiston, C., Krauss, M.:* J. Chem. Phys. *45*, 1833 (1966); *49*, 192 (1968).
96. *Boys, S. F.:* Proc. Roy. Soc. (London) A *309*, 195 (1969).
97. — *Handy, N. C.:* Proc. Roy. Soc. (London) A *311*, 309 (1969).
98. *Longuet-Higgins, H. C.:* In: Quantum Theory of Atoms, Molecules and the Solid State, p. 105; edited by *P. O. Löwdin*. New York: Academic Press 1966.
99. *Del Re, G., Parr, R. G.:* Rev. Mod. Phys. *35*, 604 (1963).
100. — CECAM Symposium, Paris (1970).
101. *Pines, D.:* Elementary Excitations in Solids. New York: Benjamin 1963.
102. *Čižek, J.:* Advan. Chem. Phys. *14*, 35 (1968).
103. *Tolmachev, V. V.:* Advan. Chem. Phys. *14*, 421 (1968).

Received March 1, 1972

Ligand Field Splittings in Copper(II) Compounds

D. W. Smith

Department of Chemistry, the University, Sheffield, Great Britain

Table of Contents

I. Introduction 50

II. Survey of Experimental and Theoretical Studies 51

A. Early Work 51

B. Copper(II)-Nitrogen Complexes 52

1. Ammonia Complexes 52
2. Complexes with Saturated Chelating Amines 54
3. Complexes with Unsaturated Chelating Amines 55
4. Complexes with Monodentate Unsaturated Nitrogen Donors 56

C. Copper(II)-Oxygen Complexes 58

1. Aquo complexes 58
2. Oxides 60
3. Carboxylate Complexes 61
4. Compounds with Diketones and Related Chelating Ligands 63
5. Compounds with Chelating Ligands Containing both N and O Donor Atoms 66

D. Fluorocuprates(II) 66

E. Chlorocuprates(II) 69

1. Tetragonal Chlorocuprates(II) 69
2. The Tetrachlorocuprate Ion (D_{2d}) 72
3. The Pentachlorocuprate Ion (D_{3h}) 74
4. Molecular $CuCl_2$ 75

F. Bromocuprates(II) 76

III. Theoretical Models for Computing Ligand Field Splitting 77

A. Crystal Field Model 77

B. The Angular Overlap Model 84

C. Semi-Empirical MO Calculations 91

D. *Ab initio* SCF-MO Calculations 93

IV. Application of the Angular Overlap Model with Crystal Field Corrections for Copper(II) Complexes 95

A. Critique of Model 95

B. Nitrogen Ligands 99

C. Oxygen Ligands 101

D. Fluoride 101

E. Chloride 102

F. Bromide 102

V. References 103

I. Introduction

The measurement and interpretation of the $d-d$ spectra of copper(II) complexes have provided a highly fruitful field for coordination chemists in recent years. There are two reasons for this: first, copper(II) has a 'one hole' d^9 electronic configuration which should render the $d-d$ spectra amenable to theoretical analysis. Secondly, the structural chemistry of copper(II) encompasses a remarkable variety of stereochemistries and coordination numbers. This can to some extent be rationalised in terms of the Jahn-Teller effect, and a vast amount of crystallographic data has been accumulated. Indeed, it is probably true to say that more structural parameters have been measured for the coordination compounds of copper(II) than for those of any other transition metal ion. Such data are, of course, a prerequisite for extensive theoretical studies. Moreover, the $d-d$ spectra are sensitive to the detailed geometry about the metal atom, and interest in the ligand field splitting naturally accompanies structural studies of copper(II) compounds.

The $d-d$ spectra of copper(II) compounds have been discussed in several recent review articles. *Hush* and *Hobbs* (1968) give a detailed account of single crystal spectra measured up to the end of 1967. *Hathaway* and *Tomlinson* (1970) have reviewed copper-ammonia complexes, with much emphasis on their electronic properties. *Hathaway* and *Billing* (1970) discuss the electronic properties and stereochemistry of copper(II) compounds, and summarise a great deal of experimental data, particularly single crystal spectra. *Ferguson* (1970), in a masterly account of the spectra of the first transition series ions, discusses some aspects of the $d-d$ spectra of copper(II) compounds. Other recent sources of information are the review by *Hatfield* and *Whyman* (1969) on copper complexes, and the book by *Lever* (1968) on inorganic electronic spectroscopy. These articles concentrate on the assignment of *d–d* spectra, and on the relationship between spectra and stereochemistry. The present article is intended to complement these, by presenting a critical account of the theoretical models which have been applied to ligand field splittings in copper(II) compounds, and in particular the empirical methods used recently by the present author.

In Sect. II we give a brief survey of experimental and theoretical studies of copper(II) spectra. In Sect. III the various theoretical models are critically assessed, and in Sect. IV we present tabulations of the parameters required for calculations of ligand field splittings of the type advocated in our recent publications. These should enable the reader to calculate the *d–d* spectrum of any copper(II) complex of known (or suspected) geometry.

II. Survey of Experimental and Theoretical Studies

A. Early Work

The colours of copper(II) compounds have fascinated chemists for a very long time. *Mellor* (1923) cited over seventy papers of the pre-Schroedinger era dealing with the colour and spectra of cupric chloride solutions alone. Many of these investigations sought to correlate the colour with the chemical constitution of the chromophore, but a proper understanding of these phenomena had to await the development of the crystal field theory. The earliest applications of crystal field theory to copper(II) systems, for example the work of *Johrdahl* (1934), concentrated on the magnetic properties of cupric salts. In these studies, splittings of the order of 20 kK were postulated in order to explain the magnetic properties, but little appeal was made to optical data, although the *d–d* spectra of many copper(II) complexes had been measured. *Bjerrum* (1932) performed a detailed analysis of the visible spectra of copper(II) ammine complexes in aqueous solution, and this work was to provide the foundation for the first attempts to interpret the *d–d* spectra of copper(II) complexes in terms of crystal field theory, after its 'rediscovery' by *Ilse* and *Hartmann* (1951) and *Orgel* (1952). *Bjerrum et al.* (1954) used a point dipole model to discuss the relationship between molecular geometry and the electronic spectra of complexes such as $[Cu(NH_3)_x(H_2O)_y]^{2+}$ and were able to explain the effects of increasing ammonia concentration on the electronic spectrum of Cu(II) in aqueous solution. They were able to show that only an elongated octahedral chromophore was consistent with the observed spectra, and they were able to explain the 'pentammine effect', the red shift of the main absorption band on the addition of a fifth ammonia molecule. *Ballhausen* (1954) published the underlying theory in more detail. Although some aspects of this work are now outdated, it stands as a pioneering study which has exerted a great influence on all subsequent studies of this nature.

B. Copper(II)-Nitrogen Complexes

1. Ammonia Complexes

Following the pioneering work of *Bjerrum et al.* (1954), little was reported on the d–d spectra of copper(II)-ammonia complexes until fairly recently. *Elliot* and *Hathaway* (1966) measured the powder reflectance spectra of the compounds then formulated as hexamminecopper(II) salts, $Cu(NH_3)_6X_2$, and interpreted them on the basis of a tetragonally elongated octahedral cation. The bands were assigned on the assumption of the relative energies $d_{x^2-y^2} > d_{z^2} > d_{xy} > d_{xz,xz}$. With the somewhat dubious assumption that the octahedral e_g and t_{2g} orbitals split in such a way as to preserve a baricentre, they showed that a value of 10 Dq comparable to that found for $Ni(NH_3)_6^{2+}$ was required. However, later work (*Tomlinson* and *Hathaway* (1968a)) on the pentammines showed that the 'hexammines' are probably best formulated as pentammines, with an uncoordinated ammonia molecule, notwithstanding crystallographic evidence for the hexammine (*Distler* and *Vaughan* (1967)). It is therefore rather unfortunate that the apparently nonexistent $Cu(NH_3)_6^{2+}$ ion has been the subject of two molecular orbital calculations. *Roos* (1966) applied a rather sophisticated model based on the Pariser-Parr-Pople method with the ZDO (zero differential overlap) approximation. His calculated d–d transition energies were in excellent agreement with experimental values, assuming a tetragonally distorted octahedral chromophore. More significantly, perhaps, he showed that the tetragonal elongation was energetically favourable. *Shupack* (1967) reported a semi-empirical MO calculation based on the SCCC (self consistent charge and configuration) method, and claimed good agreement with experiment for the experimental d–d transition energies, assuming the configuration proposed by *Distler* and *Vaughan* (1967).

Tetramminecopper(II) salts are much better characterised, and began to attract interest at about the same time as the 'hexammines'. *Smith* (1966) measured the polarized crystal spectrum of $NH_4Cu(NH_3)_5(ClO_4)_3$, which was reported (*Bukovska* and *Porai-Koshits* (1960)) to contain $Cu(NH_3)_4^{2+}$ square coplanar cations, with the fifth ammonia molecule uncoordinated at a distance of 3.75 Å from the metal. The d-orbital energies were found to be $d_{x^2-y^2} > d_{z^2} > d_{xy} > d_{xz,yz}$. However, there is now some doubt concerning the crystal structure of this compound (*Tomlinson* and *Hathaway* (1968a)). *Tomlinson et al.* (1969) measured the polarized crystal spectra of several tetramminecopper(II) compounds, and were able to assign the *d–d* bands on the basis of the same splitting pattern as that proposed by *Smith* (1966). They also studied the ESR spectra of the tetrammines, which were consistent with the ordering $d_{xy} > d_{xz,yz}$, assuming an isotropic orbital reduction factor, as is reasonable for a complex where π-bonding is unlikely to be important (*Smith* (1966); *McGarvey* (1966); *Smith* (1970a)). *Smith* (1969a) was able to explain the ligand field splitting pattern by use of the angular overlap model, with corrections for purely crystal field effects. *Shupack* (1967) performed a semi-empirical MO-SCCC calculation on $Cu(NH_3)_4^{2+}$ and his calculated *d–d* transition energies were in good agreement with the experimental data of *Smith* (1966). His wave functions indicated that the nitrogen 2*s*-orbitals were of greater importance than the 2*p*-orbitals in the metal-ligand binding, in contrast to the assumptions of *Smith* (1969a) who considered only the nitrogen 2*p*-orbitals. However, this may be an artefact of Shupack's method of choosing the diagonal matrix elements and the basis set for the ligand atoms. *Hathaway* and *Stephens* (1970) redetermined the crystal structures of the compounds $Na_4[Cu(NH_3)_4][Cu(S_2O_3)_2]_2 \cdot L$, where $L = H_2O$ and NH_3. The presence of the L molecules had previously remained unnoticed. The two compounds had significantly different properties, which suggested that in the H_2O compound the water molecule was essentially uncoordinated, whereas in the NH_3 compound the ammonia molecule was deemed to be 'semi-coordinated', and leads to a considerable destabilization of the d_{z^2} orbital. *Baglio* and *Vaughan* (1970) have further discussed the spectra of the $Cu(NH_3)_4^{2+}$ ion in $Cu(NH_3)_4(CuBr_2)_2$ and in $Cu(NH_3)_4(CuCl_2)_2 \cdot H_2O$. The *d–d* bands of the latter compound are significantly lower in energy that those in the former compound, consistent with 'semi-coordination' of the water molecule; however, it is not certain that the water molecule is absent in the former compound (*Hathaway* and *Stephens* (1970)).

There now seems to be general agreement that the relative energies of the *d*-orbitals in the tetrammines is $d_{x^2-y^2} > d_{z^2} > d_{xy} > d_{xz,yz}$. However, it must be pointed out that the chromophore can rarely, if ever, be regarded as square coplanar CuN_4; there are usually one or two axial ligands which are certainly quite strongly bound to the metal (*Smith* (1970b)). Their influence on the *d–d* spectrum is apparent from the results of *Hathaway* and *Stephens* (1970).

We have already referred to the work of *Tomlinson* and *Hathaway* (1968a) on the pentammines of copper(II). Their *d–d* spectra can be interpreted on the basis of C_{4v} symmetry, and appear at lower energies than in the tetrammines, presumably a consequence of a longer equatorial Cu—N bond length and (possibly) a small displacement of the metal out of the equatorial plane.

Tomlinson and *Hathaway* (1968b) have reported the spectra of several diamminecopper(II) complexes, and were able to explain satisfactorily the *d*—*d* spectra of those compounds whose structures were known; the structures of other compounds were predicted from their spectra. Finally *Hathaway* and *Tomlinson* (1970) discuss the electronic properties of copper-ammonia complexes in some detail, and give extensive tabulations of spectroscopic data.

2. Complexes with Saturated Chelating Amines

The most familiar ligand in this category is ethylenediamine, en. *Bjerrum et al.* (1954) discussed the spectra of ethylenediamine complexes of copper(II) in aqueous solution, and interpreted them in much the same way as for the ammonia complexes. *Yamada* and *Tsuchida* (1956a) measured the polarized crystal spectra of several compounds $Cu(en)_2X_2$, but their spectra were rather poorly resolved and no interpretation in terms of ligand field theory was offered. A systematic study was not made until the work of *Procter et al.* (1968) who measured the powder reflectance spectra of several compounds of the type $Cu(en)_2X_2$. The spectra were resolved into three gaussian components. On the basis of the variation of the band maxima in the series of compounds studied, and also the ESR *g*-values, the relative energies of the *d*-orbitals were given as $d_{x^2-y^2} > d_{z^2} > d_{xy} > d_{xz,yz}$, assuming approximately D_{4h} symmetry. *Tomlinson* and *Hathaway* (1968c) reported powder reflectance data for some monoamine adducts of $Cu(en)_2^{2+}$; their spectra were very similar to those obtained for the pentammines, and were assigned on the basis of C_{4v} symmetry, with the same ordering of the one-electron levels as for the bisethylene-diamine salts. *Hathaway et al.* (1969a) measured polarized crystal spectra of the $Cu(en)_2X_2$ compounds and established the energy sequence previously proposed on the basis of the powder reflectance data, although the position of the d_{xy} orbital was a little uncertain. *Smith* (1969a) discussed these systems in terms of the angular overlap model with electrostatic corrections.

The *d*—*d* spectra of some copper(II) complexes with other saturated chelating ligands have been discussed. *Hathaway* and *Billing* (1970) have reported the polarized crystal spectrum of $[Cu(1,3\ pn)_2(H_2O)]SO_4$, which is very similar to that of $[Cu(NH_3)_4(H_2O)]SO_4$ where the chromophore is a square pyramidal CuN_4O. The very high energy of the $d_{z^2} \rightarrow d_{x^2-y^2}$ transition in these compounds probably indicates a substantial distortion of the metal from the equatorial plane, as has been observed crystallographically (*Morosin* (1969)) for the tetrammine sulphate monohydrate.

The polarized spectra of $Cu(dien)_2 \cdot Br_2 \cdot H_2O$ and of $Cu(dien)_2(NO_3)_2$ have been reported (*Hathaway et al.* (1969b); *Hathaway et al.* (1970a)). In the latter compound, the chromophore is tetragonally-compressed CuN_6, and the spectrum was assigned in C_2 symmetry. The relative energies of the *d*-orbitals were given as $d_{z^2} > d_{x^2-y^2} > d_{xy} > d_{xz,yz}$. The relative energies of d_{xy} and $d_{xz,yz}$ are very surprising in a tetragonally-compressed system. However, since the real symmetry about the metal is so low in this compound, it may be that extensive mixing of the *d*-orbitals renders simple descriptions of the one-electron levels, and the assignment of bands in terms of these descriptions, rather inappropriate. In $Cu(dien)_2Br_2 \cdot H_2O$, the coordination geometry is a somewhat elongated octahedron, and the *d*-orbital energies were found to follow the usual sequence for ammine complexes. The low energy of the $d_{z^2} \rightarrow d_{x^2-y^2}$ transition was considered to reflect the restricted tetragonal distortion imposed by the tridentate ligands.

3. Complexes with Unsaturated Chelating Amines

The most familiar ligands of this type are 2,2′-dipyridyl (dipy) and 1,10-phenanthroline (phen). The polarized crystal spectra of copper-doped $Zn(dipy)_3Br_2 \cdot 6\,H_2O$ and $Zn(dipy)_3SO_4 \cdot 7\,H_2O$ have been measured by *Palmer* and *Piper* (1966). The low-energy band at 6.4 kK was taken to indicate a large trigonal distortion, although *Jørgensen* (1955) had suggested a restricted tetragonal distortion. *Hathaway* and *Billing* (1970), citing unpublished work from their laboratory, state that the spectrum of $Cu(phen)_3(ClO_4)_2$ can be assigned in D_2 symmetry, with a d_{xy} ground state. The polarized crystal spectra of copper-doped $Zn(phen)_3(NO_3)_2 \cdot 2\,H_2O$ has been reported by *Kokoszka et al.* (1967). Their spectra were very similar to those reported by *Palmer* and *Piper* (1966) for the trisdipyridyl complexes, but *Kokoszka et al.* preferred to suggest a tetragonal Jahn-Teller distortion to explain the optical and magnetic data. It seems that more work remains to be done on these systems; as yet there is little evidence for π-donor or -acceptor behaviour of dipy or phen in these compounds.

Bis-(dipy)copper(II) and bis-(phen)copper(II) compounds appear to be either trigonal bipyramidal or cis-octahedral; these geometries are preferred to square pyramidal or trans-octahedral configurations as a consequence of severe H—H interactions in the latter cases (*McKenzie* (1962)). *Hathaway et al.* (1969c) reported the crystal spectra of the compounds $[Cu(dipy)_2X]Y$, where the chromophore is essentially trigonal bipyramidal. The spectra could be assigned in C_{2v} or D_2 symmetry.

In C_{2v}, the relative energies of the d-orbitals were given as $d_{z^2} > d_{yz} > d_{xz} > d_{xy} > d_{x^2-y^2}$, with the x-axis taken to be the twofold symmetry axis. The role of π-bonding in determining this sequence is uncertain. *Procter et al.* (1969) measured the polarized crystal spectrum of [Cu(dipy)$_2$ NO$_2$]NO$_3$, where the oxygen atoms of the nitrite group complete a cis-octahedral coordination about the metal. The spectrum was assigned in C_{2v} symmetry to yield the one-electron sequence $d_{z^2} > d_{xy} > d_{x^2-y^2} > d_{xz}$. This ordering was in conflict with a point charge crystal field treatment of the one-electron levels, and this was taken to indicate significant π-bonding. However, the significance of this statement is doubtful, especially since the off-diagonal elements between d_{z^2} and $d_{x^2-y^2}$ were neglected.

Billing et al. (1971) reported the polarized crystal spectrum of dichloroaquo(2,9-dimethyl-1,10-phen)copper(II) which can be regarded as a CuN_2Cl_2O chromophore whose geometry is somewhere between a square pyramid and a trigonal bipyramid. The spectrum was assigned assuming C_{2v} symmetry but since the true symmetry is so low it is doubtful whether much of significance can be said about the ligand field splitting.

Bis(dimethylglyoximato)copper(II): Cu(dmg)$_2$ was the subject of an a early single crystal polarized spectroscopic study (*Yamada* and *Tsuchida* (1953)). There is much doubt concerning the assignment of the spectrum since charge transfer bands occur at rather low energies. *Roos* (1967) performed an open shell MO calculation, based on the ZDO-PPP method, on Cu(dmg)$_2$ and was able to give a good account of the electronic spectrum, using the data of *Dryssen* and *Petkovic* (1965). The band observed in chloroform solution at 21 kK was assigned to the d–d transition $d_{z^2} \rightarrow d_{xy}$, but a close examination of the eigenvectors and eigenvalues of the MO's shows that it is hardly possible to write down a d-orbital energy sequence since there are so many ligand orbitals of comparable energy to the d-orbitals. It may be that the basic ideas of ligand field theory are quite inapplicable in this case. *Roos* also discussed the spectra of adducts of Cu(dmg)$_2$ with water and amines, and was able to explain solvent effects on the spectrum.

Many other copper(II) complexes with unsaturated chelating ligands are of great interest, such as copper phthalocyanin, but little is known about the d–d spectra of such molecules; intra-ligand bands, as well as charge transfer bands in some cases, are so intense as to mask the d–d absorption.

4. Complexes with Monodentate Unsaturated Nitrogen Donors

The ligands we discuss in this section are N-bonded nitrite, N-bonded thiocyanate and aromatic heterocyclic bases such as pyridine.

The hexanitrocuprate(II) ion $Cu(NO_2)_6^{4-}$ has attracted a great deal of interest in recent years. *Elliot et al.* (1966) measured the powder reflectance spectra of $K_2BaCu(NO_2)_6$ and $K_2PbCu(NO_2)_6$. They assigned the *d—d* bands on the assumption of a tetragonally-elongated octahedron, with the relative energies $d_{x^2-y^2} > d_{z^2} > d_{xy} > d_{xz,yz}$. However, a neutron diffraction study by *Isaacs* and *Kennard* (1969) showed that the lead salt had a regular octahedral coordination about the copper atom. *Hathaway et al.* (1969d) reported the polarized crystal spectrum of the barium salt (which has a tetragonal unit cell, unlike the cubic lead salt) andconfirmed the results of the earlier powder reflectance studies; a band at about 8 kK certainly indicated a tetragonal chromophore, and the powder reflectance spectrum of the cubic lead salt was very similar to that of the barium salt. From the unit cell data and the 'octahedral' crystallographic parameters, *Hathaway et al.* (1969d) were able to estimate the metal-ligand bond lengths in the tetragonal chromophore as seen by a photon, and discussed the various dynamic effects which could explain the conflict between optical and crystallographic data. *Smith* (1970c) showed, using the angular overlap model, that these bond lengths could be estimated from the spectrum alone, and his results were in good agreement with those of *Hathaway et al.* (1969d).

Thiocyanate complexes of copper(II) have not been studied in much detail. The thiocyanates of $Cu(NH_3)_4^{2+}$ and $Cu(en)_2^{2+}$ have elongated octahedral structures with long Cu—S bonds. Short Cu—N bonds with thiocyanate are uncommon. A good example is $AgCu(NH_3)_2(NCS)_3$, which contains a trigonal bipyramidal CuN_5 chromophore with the ammonia molecules in the axial positions. The polarized crystal spectrum has been measured by *Hathaway et al.* (1970b) and shows much splitting which may be attributed to spin-orbit coupling. The bands were assigned in the double group D_{3h}* to yield the one-electron sequence $d_{z^2} > d_{xz,yz} > d_{xy,x^2-y^2}$. The relative energies of $d_{xz,yz}$ and d_{xy,x^2-y^2} are rather surprising and this was taken by *Hathaway et al.* to indicate strong out-of-plane π-bonding. Thiocyanate may well be a good π-donor but this should be equally effective in the equatorial plane and perpendicular to the equatorial plane, so it is still difficult to rationalize this particular sequence.

Forster and *Goodgame* (1965) have measured the electronic spectra of the complexes $A_2Cu(NCS)_4$ and suggested that the *d—d* bands could be interpreted on the basis of a distorted tetrahedral CuN_4 chromophore. But in $HgCu(NCS)_4$ the chromophore seems to be a distorted octahedron. *Forster* and *Goodgame* (1964) also measured the electronic spectra of $Cu(NCO)_4^{2-}$ and interpreted the results in terms of a distorted tetrahedral chromophore. But more work remains to be done on pseudohalide complexes of copper(II) before we can make any definite statements about the ligand field splittings induced by these ligands.

An important class of copper(II) complexes are those with aromatic heterocyclic bases as ligands. There is the possibility of significant π-bonding in these systems, and the influence of substituents on the aromatic ring is of some interest. There have been many routine measurements of solution and reflectance spectra of such systems, but little quantitative data is available. *Dudley* and *Hathaway* (1970a) have reported the polarized crystal spectrum of dichlorobis(2-methylpyridine)-copper(II) and assigned the d–d bands in C_{2v} symmetry. The relative energies of the d-orbitals were given as $d_{x^2-y^2} > d_{yz} > d_{z^2} > d_{xz}$, with the position of the d_{xy} orbital uncertain. (The x- and y- axes are approximately collinear with the Cu–N and Cu–Cl bonds respectively). This ordering would indicate relatively weak out-of-plane π-bonding involving the picoline ligands.

Yamada and *Miki* (1963) have studied the polarized crystal spectra of the red complexes formed by copper(II) with the succinimidate anion. In the case of $K_2Cu(sucim)_4 \cdot 6\,H_2O$, where a square coplanar CuN_4 chromophore was assumed, the relative energies of the d-orbitals were given as $d_{x^2-y^2} > d_{z^2} > d_{xy} > d_{xz,yz}$. This is the same sequence as is found for the tetrammines and bis(ethylenediamine) complexes so there is no evidence here for π-bonding. *Yamada* and *Miki* (1963) also measured the electronic spectra of $Cu(sucim)_2(amine)_2$ both in the crystalline state and in non-coordinating solvents. In a later study, *Yamada* and *Miki* (1964) measured the crystal spectra of the complexes $Cu(sucim)_2X_2$, where X was pyridine, and various alkyl-substituted pyridines. However, little attempt was made to establish the ligand field splitting.

C. Copper(II)-Oxygen Complexes

1. Aquo complexes

The electronic spectrum of Cu(II) in aqueous solution was investigated by *Bjerrum et al.* (1954) and interpreted in terms of a tetragonally-distorted $Cu(H_2O)_6^{2+}$ ion. It was recognised that the broad band around 12 kK contained at least two d–d bands which were resolved by gaussian analysis.

Holmes and *McClure* (1957) measured the polarized crystal spectrum of $CuSO_4 \cdot 5\,H_2O$ and interpreted the results in terms of a $Cu(H_2O)_4^{2+}$ square coplanar cation with two distant oxygen atoms in the axial positions. The spectrum was resolved into three gaussian components but no assignments were proposed. *Pappalardo* (1961) reported the crystal spectra of $CuSiF_6 \cdot 6\,H_2O$ and copper-doped $ZnSiF_6 \cdot 6\,H_2O$. The spectra were discussed in terms of an octahedral $Cu(H_2O)_6^{2+}$ ion with the added perturbation of spin-orbit coupling. *Roos* (1966) performed a

ZDO–PPP MO calculation on the hexaquocopper(II) ion and showed that the most stable configuration was a tetragonally-distorted octahedron, with axial bond lengths of about 2.5 Å. He successfully calculated the $d-d$ transition energies, the relative energies being $d_{xy} > d_{z^2} > d_{xz,yz} > d_{x^2-y^2}$. It must be noted, however, that these SCF orbital energies are not quite the same as the orbital energies we normally think of in ligand field theory and interelectron repulsion terms have to be included in expressions for the transition energies. With this correction, the ^2E_g state was found to lie higher in energy than the $^2B_{1g}$ state. Note that we are using here the terminology of *Roos* (1966) who chose his coordinate axes to bisect the metal-ligand bonds. It should also be noted that Roos assumed that the twofold axes of the water molecules were collinear with the metal-ligand bonds, and the π-orbitals of the water molecules lay in the equatorial plane. He considered that his calculations could be compared with the experimental data of *Holmes* and *McClure* (1957) for $CuSO_4 \cdot 5\,H_2O$, but this is actually a rather different chromophore from $Cu(H_2O)_6^{2+}$, which appears to prefer a distinctly rhombic environment. It is a most unfortunate feature of the more sophisticated MO calculations on copper(II) systems that the authors have chosen systems where no direct appeal can be made to reliable optical assignments, many of which are now available (*Hathaway* and *Billing* (1970)). Recently, *Clack* and *Farrimond* (1971) have performed an MO calculation on $Cu(H_2O)_6^{2+}$ in the CNDO approximation. Their results predicted that a tetragonally distorted system would be most stable but no correlations with optical data were attempted. Like that of *Roos* (1966), this calculation suffers from the fact that the results are bound to be highly dependent on the assumed orientation of the water molecules. This is a serious problem, since such definite information as we have suggests that hydrogen bonding to oxyanions is at least as important as the requirements of the metal-ligand bond in determining the orientation (*Chidambaram et al.* (1964)).

Billing et al. (1969) have measured the polarized crystal spectrum of metazeunerite, $Cu(UO_2)_2(AsO_4)_2 \cdot 8\,H_2O$, which may be regarded as containing $Cu(H_2O)_4^{2+}$ ions with two more oxygen atoms completing the tetragonally-elongated octahedron. The spectrum was assigned in D_{4h} symmetry to yield the one-electron sequence $d_{x^2-y^2} > d_{xy} > d_{z^2} > d_{xz,yz}$. The reversal of the relative energies of d_{xy} and d_{z^2} compared with the tetrammines may be taken as a hint for appreciable π-bonding in the equatorial plane. Angular overlap calculations by the present author (*Smith* (1970d)), where the σ-bonding parameters were fixed by the position of the $d_{z^2} \rightarrow d_{x^2-y^2}$ transition, indicated that both in-plane and out-of-plane π-bonding were present.

Many other copper(II) complexes containing coordinated water molecules have been studied, and we shall mention some of them elsewhere in the text, under the heading of what is considered to be the principal ligand system present.

2. Oxides

Many complex oxides formally containing the Cu^{2+} ion have been examined spectroscopically. An interesting example is Egyptian Blue, $CaCuSi_4O_{10}$, a rare example of genuinely square coplanar copper(II) chromophore, with no axial ligands. *Clark* and *Burns* (1967) reported its electronic spectrum, with partial polarization data and, by analogy with their previous work on gillespite, the Fe(II) analogue, suggested the one-electron sequence $d_{x^2-y^2} > d_{xy} > d_{xz,yz} > d_{z^2}$. *Smith* (1970a) used this compound as an example of a square coplanar CuO_4 chromophore in order to assess the effects of axial coordination on the $d_{z^2} \rightarrow d_{x^2-y^2}$ transition in a series of tetragonal copper(II)-oxygen systems. Others (*Billing et al.* (1969); *Hathaway* and *Billing* (1970)) have discussed this spectrum in relation to axial coordination and π-bonding.

Reinen and his co-workers (*Reinen* (1968); *Friebel* and *Reinen* (1969); *Propach* and *Reinen* (1969); *Reinen* (1970)) have studied the electronic spectra of Cu(II) in various mixed oxide environments, with a view to relating the spectra to the Jahn-Teller distortion of the lattice by the presence of the d^9 ion. In tetragonal oxide environments, he considers the one-electron sequence to be usually $d_{x^2-y^2} > d_{z^2} > d_{xy} > d_{xz,yz}$, except in special cases such as Egyptian Blue and its analogues. Using crystal field arguments, the spectroscopic data can be used to assess the extent of tetragonal distortion at the octahedral sites occupied by Cu(II). The $d_{z^2} \rightarrow d_{x^2-y^2}$ transition, which is usually observed at very low energies in such systems, is the best index of tetragonality, as also discussed by *Billing* and *Hathaway* (1970). *Schmitz-Dumont* and *Fendel* (1965) have measured the powder reflectance spectra of the system $Cu_xMg_{1-x}O$ and observed a spectrum characteristic of a tetragonally distorted CuO_6 chromophore, although it was not recognised as such at the time. *Schmitz-Dumont* and *Kasper* (1965) also reported the spectra of Cu(II) in mixed oxides with indium, yttrium and lanthanides.

Pappalardo (1961) measured the *d–d* spectrum of Cu(II) doped in Zno, and interpreted the results in terms of a trigonally-distorted tetrahedral chromophore. *Weakliem* (1962) and *Dietz et al.* (1963) have further investigated this system. *Pappalardo* (1961) also measured the spectrum of Cu(II) doped into yttrium garnets; the results indicated that an appreciable proportion of the copper atoms were in tetrahedral sites.

To sum up, we may say that the study of the electronic spectra of copper(II)-containing oxides has yielded much useful information about site preference energies for metal ions, and about Jahn-Teller distortions when a d^9 ion is placed in an octahedral hole. It would be desirable, and certainly possible, to put much of this work on a more quantitative basis since we now have a large body of information which correlates spectra with tetragonality, and it may be possible to actually determine the copper-oxygen distances explicitly by analysis of the spectrum, using the angular overlap model.

3. Carboxylate Complexes

Copper(II) complexes with carboxylates reveal a great variety of stereochemistries and much interesting magnetic behaviour, which has stimulated a good deal of work.

Copper(II) formate tetrahydrate has a rhombic CuO_6 chromophore, with four short-bonded oxygen atoms from formate ions and two longer-bonded oxygens from water molecules, completing what may be regarded as a tetragonally elongated octahedron with a rhombic distortion. The polarized crystal spectrum was measured by *Billing* and *Hathaway* (1968) who assumed approximate D_{4h} symmetry (with the $Cu-OH_2$ vector taken as the z-axis) and found the one-electron energy sequence to be $d_{x^2-y^2} > d_{z^2} > d_{xy} > d_{xz,yz}$. *Smith* (1970 d) discussed the effects of π-bonding involving the equatorial ligands and the influence of the axial ligand atoms on the $d_{z^2} \rightarrow d_{x^2-y^2}$ transition, using the angular overlap model. Dibarium cupric formate tetrahydrate $Ba_2Cu(HCOO)_6 \cdot 4\,H_2O$ also contains a rhombic CuO_6 chromophore, but two of the short bonds are $Cu-OH_2$ bonds while the longest $Cu-O$ bonds are to formate oxygens. The polarized crystal spectrum was assigned in D_{2h} symmetry and the one-electron energies found to be $d_{x^2-y^2} > d_{z^2} > d_{xy} > d_{xz} > d_{yz}$ (*Billing and Hathaway* (1968)). *Smith* (1970d) discussed this system in the angular overlap formalism.

Cupric acetate monohydrate $Cu(CH_3COO)_2 \cdot H_2O$ has been the subject of extensive investigation, mainly as a consequence of its dimeric structure and antiferromagnetic properties. However, the assignment of its electronic spectrum has led to considerable controversy. The crystal spectrum was first measured by *Yamada, Nakahara* and *Tsuchida* (1958) who observed two bands, of different polarizations, at ca. 14 kK and 27 kK. No definite assignments were proposed except to suggest that the 27 kK band was a consequence of metal-metal interaction. *Tonnet, Yamada* and *Ross* (1964) re-examined the spectrum at liquid nitrogen temperature and discussed its assignment by MO and VB models, following the treatment of *Forster* and *Ballhausen* (1962), who postulated a strong σ-bond between the metal d_{z^2} orbitals to explain the antiferromagnetism. *Tonnet et al.* (1964) preferred a δ-bonded scheme (on the basis of ESR results) and assigned the 14 kK band to the $d_{xy} \rightarrow d_{x^2-y^2}$

transition. The assignment of the 27 kK band to the $d_{xz,yz} \rightarrow d_{x^2-y^2}$ transition was considered, but it was concluded that the spectrum as a whole was difficult to interpret. *Reimann et al.* (1965) discovered a new band in the near infra-red at about 11 kK, polarized in the z-direction, like the 27 kK band. They assigned the 14 kK band to the transition $d_{xz,yz} \rightarrow d_{x^2-y^2}$ but could not decide on whether to assign the 27 kK band or the 11 kK band to the $d_{z^2} \rightarrow d_{x^2-y^2}$ transition. Leaving aside the question of the antiferromagnetism, the main spectroscopic problem is the assignment of the 27 kK band. *Boudreaux* (1964) performed a very approximate MO calculation on the system, using the overlap criterion. Assuming that the difference in energy between a bonding and antibonding pair of MO's was proportional to the overlap integral, he assigned the 27 kK band to a '$d_{z^2} \rightarrow d_{z^2}$' transition, i.e. a transition from the bonding MO arising from σ-interaction between the d_{z^2} orbitals of the two copper atoms to the corresponding antibonding MO. *Hansen* and *Ballhausen* (1965) discussed the electronic of the system using a 'coupled chromophore model', in terms of which the 27 kK band was assigned to a transition to a doubly-excited state. In each monomer unit, the d-orbital sequence was taken to be $d_{x^2-y^2} > d_{z^2} > d_{xy} > d_{xz,yz}$ and the transition energies of *Holmes* and *McClure* (1957) for a tetragonal CuO_4 chromophore were assumed.

In the most recent work on this problem, *Dubicki* (1970) argues convincingly that the 27 kK band is in fact a double band arising from ligand to metal charge transfer. The lower energy bands at 14 kK and 11 kK are in fact close to what we might expect for the d–d bands of a square pyramidal CuO_5 chromophore; the water molecule at a distance of 2.20 Å from the metal probably exerts much the same axial field as the water molecules at 2.36 Å in copper formate tetrahydrate and the bands may be assigned along similar lines, the 11 kK band to the $d_{z^2} \rightarrow d_{x^2-y^2}$ transition and the 14 kK band to $d_{xz,yz} \rightarrow d_{x^2-y^2}$ with the weak $d_{xy} \rightarrow d_{x^2-y^2}$ transition as yet unresolved but probably close to 11 kK. Whatever may be the nature of the metal-metal interactions which give rise to antiferromagnetism, there is no evidence that they significantly affect the d-orbital splitting. Weak bands at around 25 kK are seen in other copper(II) oxygen systems, as we shall discuss presently, and their incorrect assignment to d–d transitions has caused much confusion.

A very interesting copper(II) carboxylate system is calcium copper(II) acetate hexahydrate $CaCu(CH_3COO)_4 \cdot 6\,H_2O$. The crystal structure (*Langs* and *Hare* (1967)) reveals a copper(II) atom coordinated by four short bonded (1.97 Å) acetate oxygens in a square somewhat distorted towards a tetrahedron, and four long-bonded acetate oxygens (2.79 Å) completing a distorted dodecahedral arrangement (D_{2d}) about the metal. The polarized crystal spectrum was measured independently by *Billing* *et al.* (1970) and by *Gliemann* and *Morys* (1970). The spectra obtained by the two groups were in close agreement but different assignments were proposed. *Gliemann* and *Morys* (1970) assigned the bands in the double group D_{2d}^* and performed point charge crystal field calculations. They proposed the one-electron sequence $d_{xy} > d_{x^2-y^2} > d_{z^2} > d_{xz,yz}$, whereas *Billing et al.* suggested $d_{xy} > d_{z^2} > d_{xz,yz} > d_{x^2-y^2}$. The latter assignment led to a surprisingly large orbital reduction parameter $k_{\parallel}$, as obtained from the ESR spectrum, which was interpreted by *Billing et al.* in terms of strong out-of-plane π-bonding involving the distant oxygen atoms, with relatively weak in-plane π-bonding. *Mabbs* and *Smail* (1970) presented a crystal field analysis similar to that of *Gliemann* and *Morys* (1970) and, on the basis of the optical data of *Billing et al.* (1970) and ESR measurements, they proposed the same assignment as that of *Gliemann* and *Morys*. The crystal field treatment indicated that the distant oxygen atoms had a considerable influence on the ligand field splitting. However, *Smith* (1971a) published an angular overlap analysis and obtained very good agreement with experiment, with the assignment of *Mabbs* and *Smail* (1970) and *Gliemann* and *Morys* (1970), and concluded that the influence of the axial distant oxygen atoms was relatively small. As we shall see later, the angular overlap and crystal field models often lead to very different assessments of the importance of relatively distant ligand atoms on the *d*-orbital splitting.

A few other copper(II) carboxylate compounds are worthy of mention. *Yamada et al.* (1957, 1958) measured the polarized crystal spectra of copper(II) formate, acetate and propionate. Their spectra of the formate and acetate have been superseded by later work already referred to; the propionate spectrum was very similar to that of the acetate. *Yamada, Nishikawa* and *Tsuchida* (1960) studied the polarized crystal spectra of the mono-, di-, and trichloroacetates, as well as their solution spectra in various solvents. With the 27 kK band taken to be characteristic of the dimeric species, the structures in the solid state and in solution were discussed, although little of interest was discovered about the *d*-orbital splitting. Similar studies have been reported for substituted benzoates of copper(II) (*Yamada, Nishikawa* and *Miki* (1964)).

4. Compounds with Diketones and Related Chelating Ligands

Copper(II) complexes with diketones, or, more precisely, diketoenolato-complexes, have attracted a great deal of attention since the earliest days of the application of ligand field theory to copper(II) compounds. The ideas put forward by *Bjerrum, Ballhausen* and *Jørgensen* (1954) in their classic study of the ammine complexes of copper(II) were further elaborated by *Belford, Calvin* and *Belford* (1957) in their equally classic study of solvent effects on the *d–d* spectra of bis(acetylacetonato)copper(II): Cu $(acac)_2$. The broad envelope of the visible absorption was analysed into three gaussian components, whose behaviour under the influence of axial solvation could be rationalized in terms of a point charge crystal field model. An important feature of the paper was that the authors fully recognised the artificial nature of the model and did not attempt to calculate any crystal field parameters explicitly; they were fitted to experimental transition energies. The authors also noted that the hydrogenlike wave functions used by *Bjerrum, Ballhausen* and *Jørgensen* (1954) were unrealistic compared with the more accurate Hartree-Fock SCF functions. A qualitative MO description of the bonding and ligand field splitting was given. Thus the theoretical content of this paper crystallises much of the development of the ligand field theory in the late 1950's.

A good deal of work has subsequently been done along similar lines, for example by *Graddon* (1960) and *Funck* and *Ortolano* (1968). But the large (and still growing) number of studies of single crystal polarized spectra of $Cu(acac)_2$ and related compounds has been of greater significance.

The pioneering work was done by *Yamada* and *Tsuchida* (1956b). *Ferguson* (1961a) remeasured the polarized crystal spectrum of $Cu(acac)_2$ under better resolution. He assigned the bands in D_{2h} symmetry and obtained the one-electron sequence $d_{yz} > d_{z^2}$ (or $d_{x^2-y^2}$) $> d_{xz} > d_{x^2-y^2}$ (or d_{z^2}) $> d_{xy}$. This ordering is in complete conflict with the crystal field model which would place the d_{xy} orbital highest in energy. *Ferguson* was forced to postulate very strong out-of-plane π-bonding to explain the position of d_{xz}. However, a d_{xz} ground state was also inconsistent with the ESR data of *Maki* and *McGarvey* (1958), which clearly showed a d_{xy} ground state. *Ferguson,* as well as *Maki* and *McGarvey,* had assigned a band at 25 kK to a *d–d* transition; *Graddon* (1961) had similarly assigned a band in bis(ethylacetoacetato)copper(II) in solution. After a redetermination of the crystal structure of $Cu(acac)_2$, *Piper* and *Belford* (1962) carefully examined *Ferguson*'s work and concluded that an unequivocal analysis was not possible, partly due to the fact that the orientation of the molecules in the unit cell was unfavourable, and partly because in D_{2h} symmetry (including the forbidden a_u mode in the vibronic coupling scheme) all *d–d* bands are vibronically allowed in all polarizations. Hence there was no basis for deciding which mode or modes were most effective in promoting vibronic intensity. *Piper* and *Belford* also argued, rather convincingly, that the near u.v. band was not of *d–d* origin. *Ferguson, Belford* and *Piper* (1962) presented a better-resolved spectrum at low temperature but little further in the way of its interpretation. A suggestion by *Dijgraaf* (1965) that Davydov effects might be of importance in the spectrum of $Cu(acac)_2$ was quickly refuted by *Fer-*

guson (1965) and by *Belford* and *Belford* (1965). *Allen* (1966) proposed an assignment in C_{2h} symmetry, with all *d–d* bands assumed to lie under the broad envelope of the visible absorption. His one-electron sequence was $d_{xy} > d_{z^2} > d_{xz,yz} > d_{x^2-y^2}$. However, in other laboratories attempts were being made to find closely similar compounds with more tractable forms of molecular packing in the unit cell, and $Cu(acac)_2$ has received little attention since then, apart from a Wolfsberg-Helmholz semi-empirical MO calculation by *La Mar* (1966). With the inclusion of only oxygen orbitals in the basis set, a $d_{x^2-y^2}$ ground state was predicted (using La Mar's choice of axes), but with the inclusion of the carbon π-orbitals the unpaired electron was in a d_{xz} or d_{yz} orbital, in agreement with Ferguson's original assignment. However, La Mar considered this to be due to defects in the approximate model and did not think the results to be of quantitative significance.

Belford and *Carmichael* (1967) reported the polarized crystal spectrum of $Cu(3\text{-phenyl-acac})_2$, which has a much more favourable crystal structure for polarization studies, as pointed out in a preliminary report by *Basu et al.* (1962). The *d*-orbital energies were given as $d_{xy} > d_{z^2} > d_{x^2-y^2} > d_{xz} > d_{yz}$. Bis(dipivaloylmethanido)copper(II):$Cu(DPM)_2$ also has a suitable crystal structure and the polarized crystal spectrum was studied by *Cotton* and *Wise* (1967). However, the four bands observed in the range 15—20 kK could not be definitely assigned on the basis of the experimental data alone, and the results of a semiempirical MO calculation (*Cotton, Harris* and *Wise* (1967)) were invoked in order to arrive at the relative one-electron energies $d_{xy} > d_{z^2} > d_{xz} > d_{z^2-y^2} > d_{yz}$.

Bis(3-methylpentane-2,4-dionato)copper(II) has only one molecule in the unit cell and is therefore likely to be particularly suitable for polarization studies. *Hathaway et al.* (1970c) have reported its polarized crystal spectrum, assuming D_{2h} symmetry. The one-electron sequence was given as $d_{xy} > d_{x^2-y^2} > d_{xz} > d_{yz} > d_{z^2}$. This sequence seems to fit in well with the body of data available for other copper-oxygen systems, and it may well be that this ordering is general for strictly square coplanar CuO_4 chromophores. Although the picture as a whole is still a little confusing for $Cu(acac)_2$ and its derivatives, it may be thought that this most recent work, performed on a particularly favourable system crystallographically, is the most significant. The most recent solution work, by *Funck* and *Ortolano* (1968), who studied the effects of axial solvation on the *d–d* spectra of bis(hexafluoroacetylacetono)copper(II), leads to the same ordering as that proposed by *Hathaway et al.* (1970c).

Some other copper(II) compounds with ligands of the diketoenolate type are worthy of mention. *Dudley* and *Hathaway* (1970b) have measured the polarized crystal spectrum of bis(aquo)monoacetylacetonatocopper(II) picrate which may be regarded as containing a square pyramidal

CuO_5 chromophore with the diketone oxygens and the water molecules providing the equatorial ligands, while the apical position is occupied by a phenolic oxygen atom, at a distance of 2.76 Å from the metal. The spectrum was assigned in C_{2v} symmetry to yield the partial one-electron sequence $d_{xy} > d_{yz} > d_{xz}$ with the positions of the d_{z^2} and $d_{x^2-y^2}$ orbitals uncertain, but probably below the d_{xz}.

Dudley and *Hathaway* (1970c) also studied the compound 2,2′-dipyridyl bis(hexafluoroacetylacetonato)copper(II) where the coordination geometry is an octahedron with restricted tetragonal distortion. The spectrum was assigned in D_2 symmetry, with the z-axis chosen to be collinear with the long Cu—O bonds, and the x-axis collinear with the C_2 symmetry axis, bisecting the Cu—N bonds. The relative energies were given as $d_{xy} > d_{z^2} > d_{yz} > d_{xz} > d_{x^2-y^2}$, with the d_{z^2} lying rather high in energy in accordance with the restricted tetragonal distortion.

5. Compounds with Chelating Ligands Containing both N and O Donor Atoms

Ferguson (1961b) measured the polarized crystal spectrum of bis(salicylaldiminato)copper(II) and bis(N-methylsalicylaldiminato)copper(II) which may be regarded as containing a planar trans-CuO_2N_2 chromophore. The d-orbital energies were given as $d_{xy} > d_{z^2} > d_{x^2-y^2} > d_{xz} > d_{yz}$ assuming D_{2h} symmetry (a little optimistically, since the true microsymmetry is much closer to C_{2h}).

Hathaway et al. (1969b) have studied the electronic properties of $Cu(H_2edta)(H_2O)$ where the protonated edta molecule acts as a pentadentate ligand with the water molecule completing an octahedral coordination of the metal, with restricted tetragonal distortion. The polarized crystal spectrum was assigned in D_2 symmetry, with a d_{xy} ground state. The relative energies of the d-orbitals were given as $d_{xy} > d_{z^2} > d_{xz} > d_{x^2-y^2} > d_{yz}$, with the z-axis collinear with the long Cu—O bonds and the x- and y-axes bisecting the equatorial metal-ligand bonds.

Many other chromophores of the type CuN_2O_2 have been studied spectroscopically in solution, but little quantitative data have been obtained regarding the ligand field splitting.

D. Fluorocuprates(II)

Relatively little definite information has been obtained on the ligand field splitting in fluorocuprates(II). *Oelkrug* (1971) has recently reviewed the subject of ligand field parameters in tetragonal fluoro-complexes, using both crystal field and angular overlap methods, and the reader is referred to this article for more detail than we give here.

CuF_2 has a distorted rutile structure with a tetragonally-elongated CuF_6 chromophore. Its reflectance spectrum was first measured by *Ludi* and *Feitknecht* (1963) who observed two peaks at 10.4 kK and 11.3 kK, but did not attempt any assignments except to propose a value of 10.4 kK for the parameter 10 Dq in CuF_2. *Oelkrug* (1967) reported three bands in the range 7—12 kK, and interpreted his results by means of point charge crystal field calculations. These indicated the relative energies $d_{x^2-y^2} > d_{z^2} > d_{xy} > d_{xz,yz}$. The assignment of the band at 8.8 kK to the transition $d_{xy} \rightarrow d_{x^2-y^2}$ led to an empirical value of the radial integral $\langle r^4 \rangle$; by adjusting the exponent of a Slater-type orbital for the $3d$-function to yield this value, the other radial parameter $\langle r^2 \rangle$ was obtained and the other two transition energies could be calculated, in good agreement with experiment. *Schmitz-Dumont* and *Grimm* (1967) also measured the powder reflectance spectrum of CuF_2 and reported only two bands, although there was some evidence of a third at low energy. These workers also investigated the mixed phases $Cu_xMg_{1-x}F_2$, where a three band structure was evident, and proposed assignments similar to those of *Oelkrug* (1967).

Trifluorocuprates(II) and tetrafluorocuprates(II) of the alkali metals also contain tetragonal CuF_6 chromophores. The tetragonal distortion is reported to be an axial elongation, except in the case of K_2CuF_4 where *Knox* (1959) reported an axially compressed octahedral chromophore. *Oelkrug* (1967) and *Schmitz-Dumont* and *Grimm* (1967) measured the powder reflectance spectra of these, and reported spectra very similar to that of CuF_2. *Oelkrug* (1967, 1971) has found it very difficult to explain the spectrum of K_2CuF_4 on the basis of the crystallographic parameters reported by *Knox* (1959). The crystal field model indicates the one-electron sequence $d_{z^2} > d_{x^2-y^2} > d_{xz,yz} > d_{xy}$ with the $d_{x^2-y^2} \rightarrow d_{z^2}$ transition at very low energy, ca. 3—4 kK. The angular overlap model leads to the same conlcusion. Yet the spectra of Na_2CuF_4 and K_2CuF_4 are virtually identical. *Babel* (1967) has expressed some doubts about the reported axially-compressed geometry about the copper(II) atom in K_2CuF_4 on purely crystallographic grounds. A redetermination of the crystal structure is clearly desirable; preliminary e.s.r. data indicate axially-elongated octahedra (*Friebel* and *Reinen* (1969)) however. In order to obtain more definite assignments of the spectra of fluorocuprates(II), single crystal polarized spectra would be desirable, and *Ferguson* (1970) has reported some preliminary results for $KCuF_3$, K_2CuF_4, and Cu(II) doped into $KZnF_3$. His spectra of $KCuF_3$ and $KCu_xZn_{1-x}F_3$ were rather different, indicating that although the geometry about the copper atom in the doped crystals is certainly distorted, it is not quite the same as in pure $KCuF_3$. In $KCuF_3$, *Ferguson* observed four d–d bands, in the range 7–13 kK; this may indicate either an appreciable rhombic distortion or the effects of spin-orbit coupling. *Ferguson* also looked at K_2CuF_4; his room-temperature single crystal spectrum was similar to the powder reflectance spectra of *Oelkrug* (1967) and *Schmitz-Dumont* and *Grimm* (1967).

Ba_2CuF_6 is an interesting example of a discrete hexafluorocuprate(II) and its structure (V. *Schnering* (1967)) indicates an axial compression. The powder reflectance spectrum (*Oelkrug* (1971)) looks very much like those of the tetrafluorocuprates(II) and trifluorocuprates(II), and *Oelkrug* expressed doubts about the reported axial compression.

Ferguson (1970) has given preliminary single crystal data for $CuF_2 \cdot 2\,H_2O$, containing a rhombic CuO_2F_4 chromophore but little detail could be seen and no interpretation was offered.

Many papers have presented theoretical studies of fluoro-complexes of the transition metals, and fluorocuprates(II) have not escaped. *Lohr* and *Lipscomb* (1963) performed semi-empirical MO calculations on CuF_6^{4-}, with a view to determining the most stable configuration of the ion. Two minima in the total electron energy were found, corresponding to slightly elongated and slightly compressed octahedra, apparently in good agreement with the experimental evidence. The calculated d–d transition energies are also in good agreement with the powder reflectance spectra of *Oelkrug* (1967, 1971).

Johansen and *Ballhausen* (1966) reported the results of an extended Wolfsberg-Helmholz MO calculation of CuF_6^{4-}. A regularly octahedral configuration was assumed. The calculation gave no information about d-orbital splittings in real CuF_6 chromophores.

A rather more sophisticated calculation, based on the CNDO model, has been reported by *Allen* and *Clack* (1970) for CuF_6^{4-}. As regards the most stable configuration of the ion, the CNDO results were in agreement with the much simpler calculations of *Lohr* and *Lipscomb* (1963). No attempt was made to compare the results with optical data; the relative energies of the highest occupied SCF MO's for the axially-compressed configuration would, in ligand field language, lead to the one-electron sequence $d_{z^2} > d_{x^2-y^2} > d_{xz,yz} > d_{xy}$, in agreement with qualitative crystal field or molecular orbital considerations, but it must be remembered that even in a one-hole d^9 system these SCF orbital energies are not the same as the one-electron energies we normally consider in ligand field theory.

Basch et al. (1969) have performed a non-empirical ab initio SCF–MO calculation on the CuF_2 triatomic molecule, with some interesting results. A linear configuration was found to be most stable for the ground state, and for the first two excited states. The ground state was predicted to be $^2\Sigma_g^+$, in agreement with simple crystal field considerations; however, the lowest-lying excited state was found to be $^2\Delta_g$ and not $^2\Pi_g$ as we might expect. Moreover, the first $^2\Pi_g$ excited state appears to arise from promotion of an electron from the second highest occupied π_g orbital to the σ_g^+ orbital. The authors noted that one-electron energy differences do not determine the ordering of electronic states even in this d^9 system, in conflict with all the usual assumptions of ligand field theory. However, the calculated transition energies of ca. 11 kK and ca. 14 kK for the transitions $^2\Sigma_g^+ \rightarrow {}^2\Delta_g$ and $^2\Sigma_g^+ \rightarrow {}^2\Pi_g$ respectively are not in agreement with the experimental data of *Kasai et al.* (1966) who studied the ESR spectrum of CuF_2 trapped in an argon matrix at 4 °K. They were unable to find any optical absorption above 11 kK or below 5 kK and concluded that the *d—d* bonds must lie in the range 5—11 kK, although it is just possible that they were too weak to be observed. *Smith* (1971b) showed by angular overlap calculations that the *d—d* transitions of CuF_2 (calculated from parameters deduced from the analysis of other fluorocuprate(II) systems) may indeed lie in this range.

E. Chlorocuprates(II)

Much more spectroscopic data have been accumulated for chlorocuprates(II) than for any other halocuprate(II). The chlorocuprates are noted for the wide variety of stereochemistries encountered, and it will be convenient to subdivide this section accordingly.

1. Tetragonal Chlorocuprates(II)

Anhydrous $CuCl_2$ may be regarded as containing a $CuCl_6$ chromophore with a pronounced tetragonal elongation. The powder reflectance spectrum has been reported by *Ludi* and *Feitknecht* (1963) and by *Hatfield* and *Piper* (1964) but yields rather little information. Both groups reported only a single broad peak at about 12 kK. *Hatfield* and *Piper* (1964) performed point charge crystal field calculations on a number of chlorocuprates(II), including $CuCl_2$ and fitted all the observed absorption bands to two crystal field parameters, taking account of the different bond lengths in the series of compounds. These calculations predicted that the $d_{xz,yz} \rightarrow d_{x^2-y^2}$ and $d_{xy} \rightarrow d_{x^2-y^2}$ transitions should occur fairly close together at ca. 11 kK but the $d_{z^2} \rightarrow d_{x^2-y^2}$ transition was predicted to lie at a much lower energy, ca. 6 kK, although no bands were observed in this region. *Day* (1964) performed similar crystal field calculations, with much the same results. *Ros* and *Schuit* (1966) reported extensive semi-empirical MO calculations on chlorocuprates(II), using a fairly sophisticated version of the Wolfsberg-Helmholz model, but they assumed an idealised cubic structure for $CuCl_2$ with equal Cu—Cl bond lengths of 2.22 Å; since the structure is in fact highly distorted, with four Cu—Cl distances of 2.30 Å and two of 2.95 Å, their results are of little significance.

The trichlorocuprates(II) are of considerable interest since they usually contain the dimeric ion $Cu_2Cl_6^{2-}$. The copper atoms have a distorted octahedral environment with two distant axial chlorine atoms. Powder reflectance spectra of trichlorocuprates(II) have been measured by *Hatfield* and *Piper* (1964), *Day* (1964) and *Willett* and *Liles* (1967). The spectra are similar to that of $CuCl_2$, with the main absorption maximum at about 12 kK and a shoulder at slightly lower energy. The crystal field calculations of *Hatfield* and *Piper* (1964) and *Day* (1964) led to the assignment of these bands to the $d-d$ transitions $d_{xz,yz} \rightarrow d_{x^2-y^2}$ and $d_{xy} \rightarrow d_{x^2-y^2}$ respectively, with the $d_{z^2} \rightarrow d_{x^2-y^2}$ transition predicted to occur much lower in energy, around 4 kK, but unobserved. *Willett* and *Liles* (1967) also discussed the band seen at 19 kK, and assigned it to a ligand-to-metal charge transfer band.

Compounds of the type A_2CuCl_4 either contain a tetragonally-elongated $CuCl_6$ chromophore or a distorted tetrahedral $CuCl_4^{2-}$ ion; the latter systems will be discussed later. The best example of square coplanar $CuCl_4^{2-}$ is in $Pt(NH_3)_4CuCl_4$. *Hatfield* and *Piper* (1964) measured the powder reflectance spectrum of this compound and observed three $d-d$ bands in the range 10—15 kK, which were assigned on the basis of the one-electron sequence $d_{x^2-y^2} > d_{z^2} > d_{xy} > d_{xz,yz}$ which was indicated by their crystal field calculations. It is unfortunate that no single crystal polarized spectra have been measured since $Pt(NH_3)_4CuCl_4$ (which is isomorphous with Magnus' green salt $Pt(NH_3)_4PtCl_4$) has an ideal crystal structure for such measurements, with the square coplanar ions stacked along the tetragonal *c*-axis. The present author has tried unsuccessfully to grow suitable crystals for this purpose. Thus the assignment of *Hatfield* and *Piper* (1964) remains uncertain. Angular overlap calculations (*Smith* (1969b)) suggest that the d_{z^2} orbital lies below d_{xy}, assuming that π-overlap is important.

Willett, Liles and *Michelson* (1967) reported the powder reflectance spectra of several tetrachlorocuprates(II) containing the $CuCl_6$ tetragonal chromophore, and observed a broad band at about 13 kK with a shoulder at 10—11 kK. The latter was assigned to the $d_{z^2} \rightarrow d_{x^2-y^2}$ transition while the other two $d-d$ transitions were considered to lie under the broad envelope of the 13 kK band. However, they pointed out that better agreement with the ESR data was obtained by assigning the shoulder to the transition $d_{xy} \rightarrow d_{x^2-y^2}$; the significance of this argument is doubtful (*Smith* (1970a)). At any rate, *Willett et al.* (1967) were satisfied that all three $d-d$ bands lay under the near infra-red band in the region of 10—13 kK. However, the crystal field parameters of *Hatfield* and *Piper* (1964) and *Day* (1964) would place the $d_{z^2} \rightarrow d_{x^2-y^2}$ transition at about 5 kK.

Furlani et al. (1967a) measured the crystal spectrum of $(CH_3NH_3)_2CuCl_4$ at room temperature and at 77 °K. At the lower temperature, three bands were seen in the range 10—14 kK, and these were assigned with the orbital sequence $d_{x^2-y^2} > d_{z^2} > d_{xy} > d_{xz,yz}$. The spectrum was carefully examined for bands at lower energy, but none was observed. Thus *Furlani et al.* stated that all *d*—*d* transitions were in the range 10—14 kK, in conflict with the predictions of crystal field theory.

Ros and *Schuit* (1966) included square coplanar $CuCl_4^{2-}$ in their series of semi-empirical MO calculations on chlorocuprates(II) to which we have already referred. They obtained the energy sequence $d_{x^2-y^2} > d_{xy} > d_{xz,yz} > d_{z^2}$, with an overall splitting of about 9 kK, in fair agreement with experiment. The less sophisticated MO model of *Lohr* and *Lipscomb* (1963) gave the ordering $d_{x^2-y^2} > d_{z^2} > d_{xy} > d_{xz,yz}$ without consideration of $3d$—$4s$ mixing; inclusion of such mixing in the calculations placed the d_{z^2} orbital lowest in energy, as found by *Ros* and *Schuit* (1966). *Demyunck* and *Viellard* (1970) have reported an ab initio SCF—MO calculation on square coplanar $CuCl_4^{2-}$, and found the relative energies of the SCF orbitals to correspond to the sequence $d_{x^2-y^2} > d_{z^2} > d_{xz,yz} > d_{xy}$.

Angular overlap calculations on tetragonal chlorocuprates give rather different results depending on the stress placed on π-bonding, and on the extent to which distant ligands are considered. *Day* and *Jørgensen* (1964) considered only σ-bonding, and applied their Ξ^2 model to a $CuCl_6$ chromophore with equatorial bond lengths of 2.28 Å and axial bond lengths of 2.65 Å. This placed the d_{z^2} orbital only 4 kK below the $d_{x^2-y^2}$ orbital. (However, there seems to be an error in Table 2 of this paper; the transition energy $d_{z^2} \rightarrow d_{x^2-y^2}$ should be 8.0 kK). The more refined angular overlap treatment of *Smith* (1969b) placed the d_{z^2} orbital higher than d_{xy} if covalent bonding involving the axial ligands was included, but lower if these ligands were deemed to be only electrostatically bound. *Howald* and *Keeton* (1966) performed 'point bond' calculations (analogous to the angular overlap method) on halocuprates, considering both σ- and π-covalence and obtained good agreement with experiment as regards transition energies, but did not publish their assignments in detail.

Before concluding this discussion of tetragonal chlorocuprates(II) we must mention the compound $(dienH_3)CuCl_5$, which contains an elongated $CuCl_6$ chromophore (*Zaslow* and *Ferguson* (1967)). The powder reflectance spectrum was measured by *Allen* and *Hush* (1967) along with other pentachlorocuprates(II), and was found to be very similar to other tetragonal chlorocuprates; these optical and crystallographic studies have not prevented the assignment of its vibrational spectrum in D_{3h} symmetry, however (*Boorman et al.* (1969)).

It now seems reasonably clear that all three $d-d$ bands in tetragonal chlorocuprates(II) occur in the range 10—14 kK. The angular overlap model, as well as more sophisticated MO treatments, also give this result. The crystal field model, however, consistently places the d_{z^2} orbital too high in energy. We shall discuss the significance of this later. Since there is still no definite evidence concerning the relative energies of the *d*-orbitals, we cannot offer any further comparisons of the relative merits of the rival theories.

2. The Tetrachlorocuprate Ion (D_{2d})

A number of compounds of the type A_2CuCl_4 contain discrete $CuCl_4^{2-}$ ions, which are flattened tetrahedra of approximately D_{2d} symmetry. The interpretation of the ligand field spectra of these compounds has been the subject of a good deal of work. Preliminary single crystal spectra, measured by *Helmholz* and *Kruh* (1952) and *Yamada* and *Tsuchida* (1954), yielded little information about the $d-d$ bands except to show the existence of weak $d-d$ absorption below 10 kK. *Felsenfeld* (1956) attempted to explain the tetragonal distortion of the tetrahedron by calculating the balance between the crystal field stabilization energy and inter-anion repulsions, using a point charge crystal field model with rather crude Slater-type wave functions. He found an energy minimum remarkably close to the experimentally-determined equilibrium geometry; the relative energies of the *d*-orbitals were found to be $d_{xy} > d_{x^2-y^2} > d_{xz,yz} > d_{z^2}$, with the $d-d$ transitions predicted to occur in the range 2–6 kK. *Ballhausen* (1962) was of the opinion that the $d-d$ absorption should lie at around 6 kK. *Furlani* and *Morpurgo* (1963) measured the spectrum of $CuCl_4^{2-}$ in non-aqueous solvents, and assigned a broad band around 9 kK to the $d-d$ absorption. They performed point charge crystal field calculations which indicated the energy sequence $d_{xy} > d_{x^2-y^2} > d_{xz,yz} > d_{z^2}$, in agreement with *Felsenfeld* (1956).

Morosin and *Lawson* (1964a,b) measured the polarized crystal spectra of Cs_2CuCl_4 and $(NMe_4)_2CuCl_4$, and found bands at about 9 kK, 6 kK and 4 kK. They invoked the model of Felsenfeld which led to the rejection of the 9 kK band as not of $d-d$ origin and proposed that the lowest energy $d-d$ band, considered to be $d_{x^2-y^2} \rightarrow d_{xy}$, must lie around 2.5 kK. However, the semi-empirical MO calculations of *Lohr* and *Lipscomb* (1963), which led to an energy minimum close to the observed geometry, gave the sequence $d_{xy} > d_{xz,yz} > d_{x^2-y^2} > d_{z^2}$, with the $d-d$ bands predicted to appear in the range 4—9 kK. *Day* (1964) obtained the same result with a point charge crystal field calculation where the $d-d$ spectra of several chlorocuprates(II) were fitted to two freely-chosen parameters. *Hatfield* and *Piper* (1964) measured the single crystal spectrum of Cs_2CuCl_4 doped in Cs_2ZnCl_4 and assigned the bands at 8.3 kK and 4.5 kK to the transitions $d_{z^2} \rightarrow d_{xy}$ and $d_{xz,yz} \rightarrow d_{xy}$ with the position of the $d_{x^2-y^2}$ orbital uncertain. The definitive work came with the measurement by *Ferguson* (1964) of the polarized crystal spectrum of Cs_2CuCl_4 coated into Cs_2ZnCl_4. His data clearly indicated the sequence $d_{xy} > d_{xz,yz} > d_{x^2-y^2} > d_{z^2}$ and the transition energies were in good agreement with the calculated energies of *Day* (1964) and *Lohr* and *Lipscomb* (1963). The 9 kK band was assigned to the $d-d$ transition $d_{z^2} \rightarrow d_{xy}$, contrary to the interpretation by *Morosin* and *Lawson* (1964a,b). *Ferguson* also discussed the effects of spin-orbit coupling, which apparently splits the ^2E state, and of the lowered site symmetry (C_s) of the chromophore. *Furlani et al.* (1967b) reported the polarized crystal spectrum of bis-(trimethylbenzylammonium) tetrachlorocuprate(II) and assigned the bands in the same manner as *Ferguson* (1964). They discussed the short-comings of the crystal field model in the interpretation of their results, and suggested that a point dipole model (rather than point charge) was more appropriate. However, this ignores the fact that *Day* (1964) successfully predicted the relative energies of the d-orbitals by means of the point charge model before the definitive assignment of *Ferguson* (1964) was published. However, *Furlani et al.* (1967b) did not take the radial parameters $\langle r^2 \rangle$ and $\langle r^4 \rangle$ as independently disposable parameters but considered only the variation of the orbital exponent of the Slater-type wave function.

Day and *Jørgensen* (1964) applied the Ξ^2 angular overlap model to halocuprates(II), considering only the effects of σ-bonding. Their calculated transition energies for tetrachlorocuprate(II) (using one freely-chosen parameter to fit the spectra of all chlorocuprates(II)) were in good agreement with the experimental data of *Ferguson* (1964) but with the wrong ordering of the d-orbitals; the Ξ^2 model reverses the relative energies of the d_{z^2} and $d_{x^2-y^2}$ orbitals. This flaw no doubt reflects the consequences of neglecting π-overlap in chlorocuprates(II). *Smith* (1970e) included π-bonding and electrostatic effects in the angular overlap model, and obtained good agreement with experiment using parameters deduced empirically in the discussion of other chlorocuprates(II) (*Smith* (1969b)). *Ros* and *Schuit* (1966) performed semi-empirical MO calculations on the undistorted $CuCl_4^{2-}$ ion, but these have little relevance to the distorted system as found in practice.

To summarise the situation, we can say that the d-orbital splitting in the D_{2d} ion $CuCl_4^{2-}$ is well-documented experimentally, and there is now general agreement that the assignment of *Ferguson* (1964) is correct. There is less agreement concerning the origin of the splitting and the applicability of the various theoretical models. *Ferguson* (1964, 1970) was impressed by the apparent failure of the crystal field model to explain his results while the MO calculations of *Lohr* and *Lipscomb* (1963) led to transition energies in excellent agreement with his spectrum. He considered that $3d-4p$ mixing was important, and its consideration was necessary in order to obtain the correct splitting. However, the angular overlap calculations of *Smith* (1970*e*) gave equally good agreement without the need to invoke appreciable $3d-4p$ mixing, and crystal field calculations also give good results if two freely-chosen parameters are used (*Day* (1964)). Thus it seems that Ferguson's objections to the simple ligand field models can be overruled.

3. The Pentachlorocuprate Ion (D_{3h})

The trigonal bipyramidal ion $CuCl_5^{3-}$ occurs in $Co(NH_3)_6CuCl_5$, $Cr(NH_3)_6CuCl_5$ and probably in $Rh(NH_3)_6CuCl_5$ and $Ir(NH_3)_6CuC^{l}{}_5$. *Day* (1964) measured the single crystal spectrum of $Co(NH_3)_6CuCl_5$ and observed only one band, at 9.5 kK. This was assigned to $d_{xz,yx} \rightarrow d_{z^2}$. Crystal field calculations placed the $d_{xy,x^2-y^2} \rightarrow d_{z^2}$ transition at about 6.5 kK, where no band was observed. *Hatfield* and *Pieper* (1964) measured mull spectra at liquid nitrogen temperature, and were able to resolve two bands, at 10.4 kK and 8.2 kK. Their crystal field treatment led to similar results to those of *Day* (1964). Since the two bands were of comparable intensity, they thought it possible that the two bands could be the spin-orbit components of the $d_{xz,yz} \rightarrow d_{z^2}$ transition (which is dipole-forbidden in D_{3h}) with the $d_{xy,x^2-y^2} \rightarrow d_{z^2}$ transition at lower energy. Alternatively, it was suggested that spin-orbit mixing was so strong as to break down the D_{3h} selection rules. The former possibility is extremely unlikely, since the E″ state is split by only $k\lambda$ under the perturbation of spin-orbit coupling, and this cannot be sufficient to account for the observed splitting of ca. 2 kK. *Allen* and *Hush* (1967) published low temperature powder reflectance spectra of pentachlorocuprates(II) and confirmed the presence of two bands in the near infra-red. These were assigned to the two $d-d$ bands with the energy sequence $d_{z^2} > d_{xy,x^2-y^2} > d_{xz,yz}$, as predicted by crystal field theory (*Day* (1964); *Hatfield* and *Piper* (1964); *Krishnamurthy* and *Schaap* (1969, 1970)). *Day* and *Jørgensen* (1964) applied the Ξ^2 model, considering only σ-bonding, to $CuCl_5^{3-}$ and predicted that the two transition energies should be in the ration 22/13, rather larger than the experimental ratio of ca. 1.2—1.3 (the calculated transition energy for $d_{xy,x^2-y^2} \rightarrow d_{z^2}$ in Table 2 of this paper appears to be in error, and should read 5.7 kK).

A Wolfsberg-Helmholz semi-empirical MO calculation on $CuCl_5^{3-}$ was performed by *Hatfield et al.* (1965). The results suggested that the 9 kK band was partly due to ligand-to-metal charge transfer, as well as the $d_{xz,yz} \rightarrow d_{z^2}$ transition. The $d_{xy,x^2-y^2} \rightarrow d_{z^2}$ transition was predicted to occur around 6 kK. However, it is doubtful whether these calculations are of much significance.

Becker et al. (1968) have discussed in detail the effects of spin-orbit coupling in trigonal bipyramidal d^9 systems, but without explicit reference to any experimental optical data. Substituting what appeared to be reasonable values of the crystal field parameters into their equations, the $E'(d_{xy,x^2-y^2})$ state was predicted to split by 1.05 kK and the $E''(d_{xz,xy})$ state by 0.44 kK. These are not very different from the values that might be predicted by a simple first-order treatment, assuming k (the orbital reduction factor) to be around 0.6. Single crystal spectra (Smith (1968)) show some evidence of splitting of the low-energy band. An angular overlap treatment (Smith (1969b)) taking into account σ- and π-covalence and electrostatic effects, led to excellent agreement with experiment for both d–d transition energies. The inclusion of π-bonding is particularly important in order to obtain the correct separation of the E' and E'' states.

Some aspects of the ligand field theory of $CuCl_5^{3-}$ have been discussed by *Furlani* (1968) and *Ciampolini* (1969) in their reviews on five-coordinate complexes.

4. Molecular $CuCl_2$

The triatomic molecule $CuCl_2$ can be studied in the gas phase and is believed to be linear, in common with $ZnCl_2$ and $FeCl_2$ whose structures have been established by electron diffraction. The electronic absorption spectrum of gaseous $CuCl_2$ was first measured by *Hougen et al.* (1961) who observed a relatively weak band at 9 kK and a stronger one at 18 kK. These were interpreted in terms of crystal field theory and assigned to the transitions $d_{xz,yz} \rightarrow d_{z^2}$ and $d_{xy,x^2-y^2} \rightarrow d_{z^2}$ respectively. The crystal field parameters were thought to be the reasonably consistent with those appropriate to other chlorocuprates(II). However, in a later study by *DeKock* and *Gruen* (1966) the higher energy band was shown to be too intense for a d–d transition ($\varepsilon \sim 3000$). *Jørgensen* (1964) had reached the same conclusion on the basis of the earlier work of *Hougen et al.* (1961). Thus *DeKock* and *Gruen* assigned the 9 kK band to the $d_{xy,x^2-y^2} \rightarrow d_{z^2}$ transition and the higher energy band, which they observed at 19 kK, to a charge transfer transition. The $d_{xz,yz} \rightarrow d_{z^2}$ transition was thought to lie below 4 kK and could not be observed with the instrumentation available. A crystal field treatment similar to that of *Hougen et al.* (1961) was given. *Smith* (1970f, 1971b) has challenged this assignment, on the basis of an angular overlap treatment. Making reasonable assumptions regarding the Cu—Cl bond length in $CuCl_2$, he suggested that the $d_{xz,yz} \rightarrow d_{z^2}$ transition must lie around 9 kK, with the $d_{xy,x^2-y^2} \rightarrow d_{z^2}$ transition rather higher in energy, obscured by the charge transfer absorption at 19 kK. The same result can be obtained by crystal field theory; taking the radial parameters which were successfully used by *Day* (1964) and *Hatfield* and *Piper* (1964) to rationalize the d–d spectra of chlorocuprates, and assuming a Cu—Cl distance of 2.1 Å, both *d*—*d* transitions are predicted to occur above 10 kK. It therefore seems most unlikely that the $d_{xz,yz} \rightarrow d_{z^2}$ transition can lie below 4 kK and it must be assigned to the 9 kK band.

Lohr (1968) published a semi-empirical MO calculation on $CuCl_2$, with a view to obtaining wave functions for the calculation of vibronic phenomena. His calculated *d*—*d* transitions were in fair agreement with the assignment of *DeKock* and *Gruen* (1966) but he used a Cu—Cl bond length of 2.37 Å which is certainly too large.

F. Bromocuprates(II)

Rather little is known about the *d*—*d* spectra of bromocuprates(II). *Ludi* and *Feitknecht* (1963) reported the powder reflectance spectrum of $CuBr_2$ and observed a single broad band around 12 kK. *Day* (1964) gave a point charge crystal field treatment of $CuBr_2$, and *Day* and *Jørgensen* (1964) performed an angular overlap analysis, using the Ξ^2 model. Although it seems very likely that all three *d*—*d* bands lie under the envelope of the 12 kK band, this calculation placed the $d_{z^2} \rightarrow d_{x^2-y^2}$ transition at about 7 kK (the figure of 3.9 kK in Table 2 of this paper seems to be an error).

The tetrabromocuprate(II) ion in Cs_2CuBr_4 has the same distorted tetrahedral structure as $CuCl_4^{2-}$ in the caesium salt, although the degree of flattening of the tetrahedron is somewhat greater in the tetrabromocuprate(II). The electronic spectrum was first measured by *Karipides* and *Piper* (1962) who reported two oppositely-polarized bands at 8.0 kK and 7.57 kK in mixed $Cs_2Zn(Cu)Br_4$ crystals at 77 °K. These were assigned to the transitions $d_{z^2} \rightarrow d_{xy}$ (8.0 kK) and $d_{x^2-y^2} \rightarrow d_{xy}$ (7.57 kK). Point charge crystal field calculations were performed and values of the tetragonal parameters Ds and Dt were obtained. These predicted that the $d_{xz,yz} \rightarrow d_{xy}$ transition should appear around 4.5 kK. Spin-orbit coupling was taken into account. However, *Morosin* and *Lawson* (1964a,b) remeasured the spectrum and concluded that the peaks observed by *Karipides* and *Piper* were due to water in inadequately dried samples. *Morosin* and *Lawson* reported bands at 8.5 kK, 5.4 kK and 4.5 kK, and assigned these on the basis of their crystal field calculation to which we have already referred in the section on tetrachlorocuprates(II). These gave the relative energies $d_{xy} > d_{x^2-y^2} > d_{xz,yz} > d_{z^2}$ but the subsequent work of *Ferguson* (1964) on tetrachlorocuprate(II) suggests that this sequence is wrong. The point charge crystal field treatment of *Day* (1964) gave the relative energies $d_{xy} > d_{xz,yz} > d_{x^2-y^2} > d_{z^2}$ for $CuBr_4^{2-}$, as in $CuCl_4^{2-}$. His calculated *d*—*d* transition energies are in good agreement with those observed by *Morosin* and *Lawson* (1964). The Ξ^2 angular overlap treatment of *Day* and *Jørgensen* (1964) also produced good agreement with the experimental results of *Morosin* and *Lawson* (1964a,b) although *Day* and *Jørgensen* were apparently unaware of this work, and compared their calculations with the experimental results of *Karipides* and *Piper* (1962). Note, however, that the Ξ^2 model reverses the relative energies of the d_{z^2} and $d_{x^2-y^2}$ orbitals, as in the case of $CuCl_4^{2-}$. This no doubt arises from the neglect of π-bonding. Thus the available experimental and theoretical results for $CuBr_4^{2-}$ present a confusing picture, but, in the opinion of the present author, the *d*-orbital sequence is almost certainly the same as in $CuCl_4^{2-}$ and the bands reported by *Morosin* and *Lawson* (1964a,b) can be satisfactorily assigned on this basis.

The pentabromocuprate(II) ion occurs in $Co(NH_3)_6CuBr_5$ and $Cr(NH_3)_6CuBr_5$. *Raymond* (1970) has confirmed that $CuBr_5^{3-}$ in $Cr(NH_3)_6CuBr_5$ is trigonal bipyramidal. The powder reflectence spectrum of *Day* (1964) showed a peak at 8.7 kK, quite unlike the spectrum of a tetragonal bromocuprate(II) like $CuBr_2$ but similar to $CuCl_5^{3-}$. A better-resolved spectrum has been published by *Allen* and *El-Sharkaway* (1970) who observed bands at 10.3 kK and 8.6 kK. The spectrum was very similar to that of $CuCl_5^{3-}$ and the bands were assigned to $d_{xy,x^2-y^2} \rightarrow d_{z^2}$ (8.6 kK) and $d_{xz,yz} \rightarrow d_{z^2}$ (10.3 kK). It is of great interest to note that these bands occur at slightly higher energy than in pentachlorocuprates, suggesting that the relative positions of Cl and Br in the spectrochemical series are reversed in Cu(II) systems. There is evidence that fluorine produces the smallest splittings of all the halogens in copper(II) systems, contrary to the generally-accepted sequence in the spectrochemical series. However, it is doubtful whether the spectrochemical series has much meaning in copper(II) chemistry, with the remarkable variations in bond lengths and stereochemistries encountered.

III. Theoretical Models for Computing Ligand Field Splitting

A. The Crystal Field Model

The success of the crystal field model in dealing with cubic complexes is well-established; however, its general applicability to grossly non-cubic complexes is less well-documented since, until quite recently, there was a lack of definitive spectroscopic assignments which could be correlated with reasonably precise crystallographic data.

The calculation of the one-electron levels in the crystal field model requires the construction of a crystal field potential which describes the field produced by the ligands, regarded as point charges or point dipoles.

This may be regarded as a perturbation to the initially degenerate set of d-orbitals, and the matrix elements are evaluated. There are two distinct ways in which the potential may be constructed and the one-electron levels formulated.

1. The potential may be written down as the well-known cubic potential with an additional perturbation to take account of, say, a tetragonal or trigonal distortion (*Ballhausen* (1962)). The evaluation of the matrix elements under this additional perturbation leads to the d-orbital energies relative to the cubic energies of the t_2 and e orbitals. This additional perturbation requires two parameters, as well as the cubic parameter Dq. For tetragonally distorted systems, *Moffitt* and *Ballhausen* (1956) introduced the notation Ds and Dt for these parameters. Thus the d-orbital energies for a tetragonal complex can be written as:

$$
\begin{aligned}
E(d_{x^2-y^2}) &= 6\,\mathrm{Dq} + 2\,\mathrm{Ds} - \mathrm{Dt} \\
E(d_{z^2} &= 6\,\mathrm{Dq} - 2\,\mathrm{Ds} - 6\,\mathrm{Dt} \\
E(d_{xy}) &= -4\,\mathrm{Dq} + 2\,\mathrm{Ds} - \mathrm{Dt} \\
E(_{xz,yz}) &= -4\,\mathrm{Dq} - \mathrm{Ds} + 4\,\mathrm{Dt}
\end{aligned}
$$

Note that a baricentre rule holds for the Ds terms, but not for the Dt terms. Thus the splitting of the cubic e and t_2 orbitals on applying a tetragonal distortion does not obey a centre of gravity rule, although many authors have assumed that it does in discussions of tetragonal distortions in copper(II) complexes[1]).

Similar expressions can be obtained for a trigonal system; *Liehr* (1960) introduced the notation Dσ and Dτ for the two additional parameters in this case, but *Piper* and *Carlin* (1960) suggested that the symbols Ds and Dt should be retained in all cases.

The advantage of this method is that the same equations as given above for the tetragonal case are applicable to all systems which can be regarded as distorted octahedral or tetrahedra while preserving a C_4 (or S_4) axis. Thus the same set of equations will be applicable for $CuCl_6^{4-}$ (elongated or compressed octahedron), $CuCl_5^{3-}$ (square pyramidal) and $CuCl_4^{2-}$ (square coplanar or flattened tetrahedron). However, there are disadvantages as well, and this method has not been so widely applied in the discussion of copper(II) systems as it has in tetragonal Cr(III),

---

1) As Prof. D. Reinen has kindly informed the author in a private communication, the baricentre rule can be shown to hold for small tetragonal distortions, if allowance is made for the contraction in the equatorial bond length as axial ligands are withdrawn (Reinen (1968)).

Co(III) and Ni(II). There are three freely-chosen parameters — explicit calculations of Dq, Ds and Dt are seldom attempted. However, there are rarely as many as three experimental spectroscopic transition energies to which these may be fitted. Moreover, the theory gives little guidance as to the relative magnitudes or even the signs of the parameters. In copper(II) compounds (as opposed to tetragonal Cr(III), Co(III), etc.) there are such gross variations in bond lengths and angles from one compound to another related one that there is little to be gained by comparing the variation of the parameters in a series of compounds, and little hope that the parameters are transferable from one system to another. In copper(II) systems it is often unprofitable to regard the ligand field as being slightly distorted from cubic symmetry, and it is usually better to adopt method (2) discussed below.

2. In this method, the crystal field potential is developed with explicit regard for each ligand atom and its position relative to the metal. The construction of the potential has been discussed in detail elsewhere (*Companion* and *Komarynsky* (1964); *Krishnamurthy* and *Schaap* (1969, 1970); *Kettle* (1969)). The matrix elements can be expressed in terms of the radial parameters a_0, a_2 and a_4 where

$$a_n = q\langle r^n\rangle / \mathrm{R}^{n+1}$$

q is the magnitude of the negative point charge deemed to reside on the ligand, $\langle r^n\rangle$ is the mean value of the n th power of the distance of a d-electron from the nucleus, and R is the metal-ligand distance. If the ligands are divided into classes of equivalent atoms, the a_n parameters will take distinct values for each equivalent set. The terms in a_0 (the spherically symmetric term) are the same for all d-orbitals, so that the relative energies of the d-orbitals may be expressed in terms of the two parameters a_2 and a_4. Many authors use the notation Cp and Dq, where $\mathrm{Cp} = (2/7)a_2$ and $\mathrm{Dq} = (1/6)a_4$. For an MX_n chromophore where all the ligands X are equivalent, a single value for each of a_2 and a_4 will suffice; if the X atoms are identical but not equivalent, with different M—X interatomic distances, the problem can still be reduced to one containing only two parameters by invoking the dependence of a_2 and a_4 on R^{-3} and R^{-5} respectively. Where completely different ligand atoms are found in the chromophore, it is necessary to use distinct values of a_2 and a_4 for each type of ligand atom.

The methods of calculating the one-electron energies in terms of the a_n parameters have been discussed in detail (*Companion* and *Komarynsky* (1964); *Krishnamurthy* and *Schaap* (1969, 1970); *Hartmann* and *Koenig* (1961)). Here we shall follow the treatment of *Companion* and *Komaryns-*

ky. The crystal field potential can be set up in a general form and the matrix elements H_{ij} (where $H_{ij} = \langle \phi_i | V | \phi_j \rangle$) are expressible in terms of the functions D_{lm} and G_{lm}, which are the sums over all point charges of the ligand position functions G^i_{lm} and D^i_{lm}.

$$D_{lm} = \sum_i D^i_{lm}$$

$$G_{lm} = \sum_i G^i_{lm}$$

In Table 1 we give the matrix elements H_{ij} in terms of D_{lm} and G_{lm}. The position functions D^i_{ml} and G^i_{lm} are given in Table 2. These tables should enable the reader to calculate the relative one-electron energies of the *d*-orbitals in any complex of known or assumed geometry, without the need to construct the crystal field potential. Note that in low symmetries (e.g. D_{2h}, C_{2v}, D_{3d}) where different *d*-orbitals (or sets of d-orbitals) are mixed by the ligand field and non-zero off-diagonal matrix elements H_{ij} $(i \neq j)$ arise, it is necessary to solve the secular equations.

The parameters a_2 and a_4 are simply related to the parameters Dq, Ds and Dt as discussed above for the case of a general tetragonal potential:

$$\begin{aligned} Dq &= (1/6)a^e_4 \\ Ds &= (2/7)\,(a^e_2 - a^a_2) \\ Dt &= ((2/21)\,(a^e_4 - a^a_4) \end{aligned}$$

Table 1. *Matrix elements of the crystal field potential in terms of the position functions D_{lm} and G_{lm}*

$\langle d_{x^2-y^2} \vert V \vert d_{x^2-y^2} \rangle$	$= D_{00} - (1/7)D_{20} + (1/56)D_{40} + (5/24)D_{44}$
$\langle d_{z^2} \vert V \vert d_{z^2} \rangle$	$= D_{00} + (1/7)D_{20} + (3/28)D_{40}$
$\langle d_{xy} \vert V \vert d_{xy} \rangle$	$= D_{00} - (1/7)D_{20} + (1/56)D_{40} - (5/24)D_{44}$
$\langle d_{xz} \vert V \vert d_{xz} \rangle$	$= D_{00} + (1/14(D_{20} - (1/14)D_{40} + (3/14)D_{22} + (5/42)D_{42}$
$\langle d_{yz} \vert V \vert d_{yz} \rangle$	$= D_{00} + (1/14)D_{20} - (1/14)D_{40} - (3/14)D_{22} - (5/42)D_{42}$
$\langle d_{x^2-y^2} \vert V \vert d_{z^2} \rangle$	$= -(\sqrt{3}/7)D_{22} + 5\,(\sqrt{3}/84)D_{42}$
$\langle d_{x^2-y^2} \vert V \vert d_{xy} \rangle$	$= (5/24)G_{44}$
$\langle d_{x^2-y^2} \vert V \vert d_{xz} \rangle$	$= (3/7)D_{21} - (5/28)D_{41} + (5/12)D_{43}$
$\langle d_{x^2-y^2} \vert V \vert d_{yz} \rangle$	$= -(3/7)G_{21} + (5/28)G_{41} + (5/12)G_{43}$
$\langle d_{z^2} \vert V \vert d_{xy} \rangle$	$= -(\sqrt{3}/7)G_{22} + 5\,(\sqrt{3}/84)G_{42}$
$\langle d_{z^2} \vert V \vert d_{xz} \rangle$	$= (\sqrt{3}/7)D_{21} + 5\,(\sqrt{3}/14)D_{41}$
$\langle d_{z^2} \vert V \vert d_{yz} \rangle$	$= (\sqrt{3}/7)G_{21} + 5\,(\sqrt{3}/14)G_{41}$
$\langle d_{xy} \vert V \vert d_{xz} \rangle$	$= (3/7)G_{21} - (5/28)G_{41} + (5/12)G_{43}$
$\langle d_{xy} \vert V \vert d_{yz} \rangle$	$= (3/7)D_{21} - (5/28)D_{41} - (5/12)D_{43}$
$\langle d_{xz} \vert V \vert d_{yz} \rangle$	$= (3/14)G_{22} + (5/42)G_{42}$

Table 2. *Ligand position functions D^i_{lm} and G^i_{lm}*

$D^i_{00} = a^i_0$	
$D^i_{20} = a^i_2$	$(3 \cos^2 \theta_i - 1)$
$D^i_{40} = a^i_4$	$(35/3 \cos^4 \theta_i - 10 \cos^2 \theta_i + 1)$
$D^i_{21} = a^i_2$	$\sin \theta_i \cos \theta_i \cos \phi_i$
$D^i_{22} = a^i_2$	$\sin^2 \theta_i \cos 2 \phi_i$
$D^i_{41} = a^i_4$	$\sin \theta_i \cos \theta_i (7/3 \cos^2 \theta_i - 1) \cos \phi_i$
$D^i_{42} = a^i_4$	$\sin^2 \theta_i (7 \cos^2 \theta_i - 1) \cos 2 \phi_i$
$D^i_{43} = a^i_4$	$\sin^3 \theta_i \cos \theta_i \cos 3 \phi_i$
$D^i_{44} = a^i_4$	$\sin^4 \theta_i \cos 4 \phi_i$
$G^i_{21} = a^i_2$	$\sin \theta_i \cos \theta_i \sin 2 \phi_i$
$G^i_{22} = a^i_2$	$\sin^2 \theta_i \sin 2 \phi_i$
$G^i_{41} = a^i_4$	$\sin \theta_i \cos \theta_i (7/3 \cos^2 \theta_i - 1) \sin \phi_i$
$G^i_{42} = a^i_4$	$\sin^2 \theta_i (7 \cos^2 \theta_i - 1) \sin 2 \phi_i$
$G^i_{43} = a^i_4$	$\sin^3 \theta_i \cos \theta_i \sin 3 \phi_i$
$G^i_{44} = a^i_4$	$\sin^4 \theta_i \sin 4 \theta_i$

θ is the angle which the metal-ligand vector makes with the principal symmetry axis, usually denoted z, and ϕ is the angle between the projection of the metal-ligand vector in the xy plane and the x-axis.

The superscripts a and e refer to the axial and equatorial ligands respectively. In the limit of a square coplanar system, Dq = (7/4) Dt, as discussed by *Martin et al.* (1962), and the problem is reduced to one of only two parameters. In the general case of a tetragonal system, Ds and Dt (and Dσ and Dτ in the trigonal case) are to be regarded as composite parameters. But it is usually best, in the case of copper(II) systems, to consider the axial and equatorial contributions to the splitting explicitly and separately for the reason already given.

The parameters a_2 and a_4 (or Cp and Dq) are seldom calculated explicitly from first principles. We can calculate $\langle r^2 \rangle$ and $\langle r^4 \rangle$ from Hartree-Fock wave functions (*Richardson et al.* (1962); *Clementi* (1965)). For Cu^{2+}, $\langle r^2 \rangle$ is 1.07 (a.u.)^2 and $\langle r^4 \rangle$ 2.76 (a.u.)^4. For Cu^+, these are increased to 1.26 (a.u.)^2 and 4.38 (a.u.)^4 respectively. For a series of chlorocuprates(II) and bromocuprates(II), *Day* (1964) noted that the Hartree-Fock values of $\langle r^2 \rangle$ and $\langle r^4 \rangle$ gave transition energies which were much too small. In order to fit the observed transition energies it was necessary to use values of $\langle r^2 \rangle$ and $\langle r^4 \rangle$ which were respectively three and twenty times too large compared with the Hartree-Fock values compiled by *Freeman* and *Watson* (1965). Some workers have proposed that, although the absolute magnitudes of the radial parameters $\langle r^2 \rangle$ and $\langle r^4 \rangle$ must be regarded as disposable parameters, some guidance regarding their relative magnitudes may be obtained from the Hartree-Fock value of their ratio. Such a proposal has little to recommend itself from the logical point of view since the absolute values of the radial parameters are at such variance from those determined on an empirical basis. It seems that in order to fit experimental data, widely different values of the ratio $\langle r^2 \rangle / \langle r^4 \rangle$ must be invoked. For chlorocuprates(II), from the calculations of *Day* (1964), *Hatfield* and *Piper* (1964), and *Krishnamurthy* and *Schaap* (1969, 1970), the ratio a_2/a_4 has to be close to unity for reasonable agreement, in contrast to the Hartree-Fock value of ca. 7. for Cu^{2+} and ca. 6 for Cu^{+}. In the case of calcium copper(II) acetate hexahydrate, crystal field calculations (*Mabbs* and *Smail* (1970) *Gliemann* and *Morys* (1970)) demand that $a_2/a_4 =$ ca. 2.5, compared with the Hartree-Fock value of 5—6. As pointed out by the present author (*Smith* (1971a)) a single value of the ratio a_2/a_4 will not suffice for all copper(II)-oxygen systems. This may be explained by the fact that the possibilities for π-bonding in copper(II) — oxygen compounds are widely variable, depending on the involvement of the oxygen π-orbitals in bonding within the ligand molecule and on the orientation of the ligand molecules with respect to the metal. One frequently encounters the statement that the relative energies of the *d*-orbitals as indicated by experimental data differ from those predicted by crystal field theory, and that this may be taken as an indication of π-bonding. This kind of argument is rather unsatisfactory, since the crystal field theory, properly applied, should take account (albeit implicitly) of all the interactions which give rise to the ligand field splitting; one is entitled to enquire why σ-bonding is not taken into account, if π-bonding is so important in determining the energy sequence. There is indeed a considerable body of evidence which suggests that covalence is the major contributor to the ligand field splitting. According to the MO model of *Sugano* and *Shulman* (1963) the octahedral splitting parameter in $KNiF_3$ includes positive contributions from both crystal field effects and covalence, the latter being much larger. More sophisticated interpretations of 10 Dq (*Watson* and *Freeman* (1964); *Offenhartz* (1969, 1970)) paint rather a different picture from that of *Shulman* and *Sugano*, but agree to the extent that covalent bonding is the primary source of the splitting. On a qualitative level, even rather elementary text books stress the fact that the spectrochemical series cannot be rationalized on the basis of an electrostatic model, although a covalent model, considering the π-donor and π-

acceptor characteristics of the various types of ligands, does afford some explanation of the relative positions of the ligands. Many workers who have applied the crystal field model to non-cubic copper(II) systems have made it clear that the model provides a useful and simple means of taking implicit account of the effects of covalence, but others have sought to invest the crystal field parameters a_2 and a_4 with real physical significance, and have tried to explain the magnitude of the ratio a_2/a_4 and its apparent variation from one system to another by crystal field arguments. These arguments are not very convincing.

A further defect of the crystal field model which is particularly serious in dealing with copper(II) systems is that the dependence of a_2 and a_4 on R^{-3} and R^{-5} respectively can lead to large errors when substantial variations of R have to be considered. The influence of distant ligands appears to be grossly exaggerated in the crystal field model. This is well-illustrated by the crystal field calculations on tetragonal chlorocuprates(II) (*Day* (1964); *Hatfield* and *Piper* (1964); *Furlani et al.* (1967a)). The crystal field model gives such weight to the distant chlorines (ca. 2.7—3.0 Å from the metal) that the d_{z^2} orbital is driven much too high in energy and the position of the $d_{z^2} \rightarrow d_{x^2-y^2}$ transition is wrongly predicted to be about 4—6 kK. This is also true for tetragonal copper(II) — oxygen systems (*Smith* (1971a,c)). It seems that over a relatively small range of internuclear distances close to what might be regarded as 'normal' for the metal and ligand concerned the crystal field model seems to give quite a good account of the variation in the splitting parameters with internuclear distance (*Drickamer* (1967); *Smith* (1969c)); but over a greater range of internuclear distances, the crystal field model predicts excessively high values of a_2 and a_4 (particularly of the former) for relatively distant ligand atoms.

We may summarise the present status of the crystal field model in its application to copper(II) systems as follows;

(i) It is very simple to calculate the relative energies of the *d*-orbitals in terms of the crystal field parameters a_2 and a_4.

(ii) The dependence of these parameters on internuclear distance is easy to determine.

(iii) The value of the ratio a_2/a_4 required to fit the experimental data is usually very different from that predicted from Hartree-Fock wave functions and varies widely and in an unpredictable manner from one system to another.

(iv) The absolute magnitudes of a_2 and a_4 required to fit experimental data bear no resemblance whatever to the Hartree-Fock values.

(v) The radial dependence of the parameters a_2 and a_4 as predicted by crystal field theory leads to excessively large contributions to the splitting from relatively distant ligand atoms.

A refinement of the point charge crystal field model has recently been suggested in the form of the 'diffuse charge' model of *Bonacic* and *Randic* (1968, 1971). Here the source of the electrostatic perturbation is not just a single point charge located at the ligand nucleus but a large number of charges (up to 2635) uniformly distributed over 18 concentric shells centred on the ligand atom. The model was applied to chlorocuprates(II), using realistic wave functions. The results obtained were no better than those which a simple point charge model might have afforded, although no attempt was made to compare the results with experimental spectroscopic data. It is of interest to note that as the number of charges used in the calculation was increased, the calculated splittings first began to deviate quite strongly from the simple point charge values, but then with the largest number of charges (2635) the diffuse charge model gave results comparable to those of the point charge model. Some measure of rationalization of the spectrochemical series is possible with the diffuse charge model; but otherwise it seems to offer few advantages.

Thus, although the crystal field model has proved to be quite useful in discussions of ligand field splittings in copper(II) systems, it suffers from some serious defects and a comparable empirical model which takes explicit account of σ- and π-covalence has obvious appeal. We must therefore turn to the angular overlap model.

B. The Angular Overlap Model

The basic ideas behind the angular overlap model (AOM) have appeared in a variety of guises over the years. In its earliest form (*Yamatera* (1958); *McClure* (1961)) it was restricted to the discussion of the spectra of systems such as $Co^{III}A_4B_2$. The basic assumption was that the ligand field splitting arose from covalent bonding, and that, applying second-order perturbation theory, the extent to which a *d*-orbital is destabilized by virtue of becoming weakly antibonding is proportional to the square of the group overlap integral between the *d*-orbital and the appropriate symmetry-adapted combination of ligand orbitals. In the model used by *Yamatera* and *McClure*, the *d*-orbital energies were expressed in terms of the destabilization parameters $\delta\sigma$ and $\delta\pi$. These ideas were placed on a more formal and general basis by *Jørgensen et al.* (1963) who were concerned with the *f*–*f* spectra of lanthanide ions in various environments. Their method is often known as the Ξ^2 model. Only σ-

bonding was considered originally; the group overlap integral G_{mx} was divided into a radial part G^*_{mx} and an angular part Ξ:

$$G_{mx} = G^*_{mx} \cdot \Xi$$

G^*_{mx} is proportional to the diatomic overlap integral S_{mx}, and hence depends on the metal and ligand wave functions and on the metal-ligand distance, but not on the molecular geometry. Ξ is a simple function of the geometry. Applying second order perturbation theory, the energy of a *d*-orbital is proportional to the square of the appropriate value of Ξ, the proportionality constant being denoted σ^*. Thus the one-electron energies are expressible in terms of a single parameter σ^*. *Day* and *Jørgensen* (1964) applied the model to a number of chlorocuprates(II) and bromocuprates(II) with apparent success, as we have previously discussed in Sect. II E. However, this model ignores π-bonding which we might expect to be of some importance in chlorocuprates(II) and bromocuprates(II). *Day* and *Jørgensen* (1964) used the variation of the diatomic overlap integral S_{mx} in order to determine the dependence of σ^* on the internuclear distance; however, for the metal they used a ls-type function which they felt adequately represented the tail of the 3*d*-function in the overlap region. Simple Slater exponents were used for the ligand wave functions.

Chlorocuprates(II) and bromocuprates(II) were discussed by *Howald* and *Keeton* (1966) in terms of the point bond model, which is closely related to the angular overlap model and takes account of both σ- and π-bonding. They were able to successfully interpret the charge transfer spectra as well as *d*—*d* spectra.

The angular overlap model was further developed in more general form by *Schäffer* and *Jørgensen* (1965) and by *Schäffer* (1966). The ligand field which perturbs the metal wave functions is represented by an operator V whose form need not be explicitly discussed. The matrix elements V_{mx} are deemed to be proportional to the group overlap integral G_{mx} and the *d*-orbital energies are expressible in terms of the parameters e_σ and e_π. These are related to the parameters σ^* and π^* (which are also widely used) by the relations $e_\sigma = 5\sigma^*$ and $e_\sigma = 5\pi^*$. *Schäffer* and *Jørgensen* (1965) give tabulations of ligand field splitting parameters in terms of e_σ and e_π for several fairly regular geometries. In the more general case where the symmetry may be quite low and the ligand atoms not all equivalent, it is best to proceed as follows:

1. Write down the position of each ligand atom in terms of the polar coordinates (R, θ, ϕ) (as defined in Table 2).

2. Each ligand atom is deemed to have available for bonding a σ-orbital, directed along the metal-ligand axis. Where appropriate, two π-orbitals may also be considered, a $\pi(\perp)$ orbital which is symmetric with respect to reflection in a plane containing the z-axis, and a $\pi(\|)$ orbital which is antisymmetric with respect to such a reflection.

3. The ligand group orbitals are written down as symmetry-adapted linear combinations of ligand orbitals. The group orbitals may be classified as σ, $\pi(\perp)$ and $\pi(\|)$.

4. The metal orbitals are classified according to symmetry; the destabilization energy of a d-orbital can be written as:

$$E(\phi_d) = \sum_i \frac{\langle\phi_d|V|\phi_L^i\rangle \langle\phi_L^i|V|\phi_d\rangle}{E(\phi_d) - E(\phi_L^i)}$$

$$= a \sum_i k_{mi}^2 \{\sigma\} + b \sum_i k_{mi}^2 \{\pi(\perp)\} + c \sum_j k_{mi}^2 \{\pi(\|)\},$$

$$\text{where } G_{mi}(\lambda) = k_{mi} S_{mi}(\lambda)$$

The constants a, b and c are simply the parameters e_σ, $e_\pi(\perp)$ and $e_\pi(\|)$, which can be written as $5\,\sigma^*$, $5\,\pi^*(\perp)$ and $5\,\pi^*(\|)$ in the alternative nomenclature. The squared group overlap integral coefficients k_{mx}^2 are given by the expressions in Tables 3–5.

In low symmetries such as D_{2h}, D_{3d} and C_{2v} where two (or more) sets of d-orbitals belong to the same irreducible representation it is necessary to solve a secular determinant. The off-diagonal elements will be given by:

$$\sum_i \langle\phi_d|V|\phi_L^i\rangle \langle\phi_L^i|V|\phi_d'\rangle$$

if ϕ_d and ϕ_d' are the d-orbitals which mix.

Thus in the case of an MX_n chromophore where all the ligands are equivalent, the relative energies of the d-orbitals can be written down in terms of the three parameters σ^*, $\pi^*(\perp)$ and $\pi^*(\|)$. If the ligand valence orbitals σ, $\pi(\perp)$ and $\pi(\|)$ can be regarded as approximately equivalent, for example in the case of a halide complex, ignoring $s-p$ hybridisation of the σ-orbitals, then we can assume that $\pi^*(\perp) = \pi^*(\|)$ and that $\sigma^*/\pi^* = (S_{mx}^\sigma/S_{mx}^\pi)^2$. Thus, considering only the ligand np-orbitals as being of importance, the problem can be reduced to one of only a single freely-chosen parameter by evaluating the diatomic overlap integrals S_{mx}^σ and S_{mx}^π. Computer programs for the evaluation of these overlap integrals are freely available in most laboratories where theoreti-

Table 3. *The squared group overlap integral coefficients* $k^2_{mx}\{\sigma\}$

Metal orbital	$k^2_{mx}\{\sigma\}$
$d_{x^2-y^2}$	$(3/4)\sin^4\theta\,(\sum_N c_n \cos 2\phi_n)^2$
d_{z^2}	$N\,(\cos^2\theta - \frac{1}{2}\sin^2\theta)^2$
d_{xy}	$(3/4)\sin^4\theta\,(\sum_N c_n \sin 2\phi_n)^2$
d_{xz}	$(3/4)\sin^2 2\theta\,(\sum_N c_n \cos\phi_n)^2$
d_{yz}	$(3/4)\sin^2 2\theta\,(\sum_N c_n \sin\phi_n)^2$

Note: N is the number of ligand atoms involved in the appropriate group orbital; c_n is the coefficient of the *n th* ligand orbital in the group orbital.

Table 4. *The squared group overlap integral coefficients* $k^2_{mx}\{\pi(\perp)\}$

Metal orbital	$k^2\{\pi(\perp)\}$
$d_{x^2-y^2}$	$(1/4)\sin^2 2\theta\,(\sum_N c_n \cos 2\phi_n)^2$
d_{z^2}	$N(3/4)\sin^2 2\theta$
d_{xy}	$(1/4)\sin^2 2\theta\,(\sum_N c_n \sin 2\phi_n)^2$
d_{xz}	$\cos^2 2\theta\,(\sum_N c_n \cos\phi_n)^2$
d_{yz}	$\cos^2 2\theta\,(\sum_N c_n \sin\phi_n)^2$

Table 5. *The squared group overlap integral coefficients* $k^2_{mx}\{\pi(\parallel)\}$

Metal orbital	$k^2\{\pi(\parallel)\}$
$d_{x^2-y^2}$	$\sin^2\theta\,(\sum_N c_n \sin 2\phi_n)^2$
d_{xy}	$\sin^2\theta\,(\sum_N c_n \cos 2\phi_n)^2$
d_{xz}	$\cos^2\theta\,(\sum_N c_n \sin\phi_n)^2$
d_{yz}	$\cos^2\theta\,(\sum_N c_n \cos\phi_n)^2$

cal chemistry is pursued. Dr. *A.D.C. Towl* has kindly made such a program available to the present author, who is willing to compute any overlap integrals on request. Analytic Hartree-Fock radial wave functions for copper are given by *Watson* (1960), *Richardson et al.* (1962) and *Clementi* (1965), who gives wave functions for all important ligand atoms as well. Numerical tabulations of overlap integrals involving $3d$-orbitals are given by *Jaffe* (1953), *Jaffe* and *Doak* (1953) and *Craig et al.* (1954).

A group orbital must be a linear combination of equivalent ligand atomic orbitals; if there is more than one group of identical ligand atoms differing in their distance from the metal, the relative values of σ^* and π^* can be estimated by assuming their proportionality to $(S^{\sigma}_{mx})^2$ and $(S^{\pi}_{mx})^2$ respectively. Thus in a halogen complex only one parameter suffices to express the relative energies of the d-orbitals, irrespective of the molecular geometry.

In many cases, the ligand atom will be bound to another atom or atoms in the ligand molecule; if the symmetry of the ligand is less than spherical, the three ligand np-orbitals can no longer be regarded as approximately equivalent. It may be possible to set $\pi^*(\perp)$ or $\pi^*(\|)$ (or both) equal to zero if an inspection of the ligand molecule suggests that the relevant ligand orbitals are not available for bonding to the metal and are preoccupied in holding the ligand molecule together. Thus it is generally assumed that in ammine complexes, only σ-bonding need be considered.

The angular overlap model offers some distinct advantages over the crystal field model. The effects of σ- and π-covalent bonding are explicitly considered, consistent with the view that these make the greatest contribution to the ligand field splitting. While two (or three) parameters are required for each type of ligand atom present, the relative magnitudes of these can be estimated in a rational manner as described above. This cannot be said for the crystal field model. There has been much discussion (*Krishnamurthy* and *Schaap* (1969, 1970); *Garner* and *Mabbs* (1970); *Mabbs* and *Smail* (1970); *Companion* and *Komarynsky* (1964); *Kibler* (1970, 1971)) as to what is the 'best value' for the ratio a_2/a_4 to use in crystal field calculations on copper(II) systems, but there is no general agreement on the matter. Thus *Day* (1964) and *Hatfield* and *Piper* (1964), together with *Krishnamurthy* and *Schaap* (1969, 1970), found that $a_2/a_4 = 1$ gave good results for chlorocuprates(II); *Mabbs* and *Smail* (1970) and *Gliemann* and *Morys* (1970) preferred values of a_2/a_4 of about 1.7 for copper-oxygen systems while *Kibler* (1970, 1971) argues on theoretical and experimental grounds for $a_2/a_4 = (5/9)$.

Another aspect of the relative merits of the angular overlap and crystal field models concerns the radial dependence of the parameters. This is, of course, a matter of particular importance in copper(II) chemistry. As we have already noted, the crystal field model gives a good account of the apparent dependence of the parameters a_2 and a_4 on the metal-ligand distance over a fairly small range close to the 'normal' bonded distance. However, in dealing with more distant ligands such as are usually encountered in tetragonal copper(II) compounds we find that the crystal field model gives undue weight to these ligands in the calculation of the relative energies of the d-orbitals. Indeed, as pointed out by *Garner* and *Mabbs* (1970), in the point charge model ligand atoms at two or three times the 'normal' distance from the metal can significantly affect the relative energies of the d-orbitals. If this were really true, it would be necessary to include the effects of atoms well outside the primary coordination sphere in crystal field calculations. However, there is little evidence to suggest that d-orbital splittings are at all sensitive to such interactions. On the other hand, the angular overlap model gives a good account of the variation of the splitting parameters with internuclear distance, over both small and great variations of R (*Smith* (1969c, 1970d, 1971a)).

As regards the angular dependence of the splitting parameters, the crystal field and angular overlap methods give comparable results, though these are usually quite sensitive to the assumed values of the ratios σ^*/π^* and a_2/a_4.

The angular overlap model is a relatively crude method which appears to yield results at least as good as those afforded by the crystal field model. As with all simple empirical models, the AOM depends on many approximations and assumptions which cannot be expected to be even approximately correct. Thus, for example, the parameter σ^* is assumed to depend only on the identity of the metal and the ligand, and on the internuclear distance; it is independent of the stoichiometry or stereochemistry. The theoretical basis for assuming the proportionality of the AOM matrix elements to overlap integrals is closely related to the Wolfsberg-Helmholz approximation for the off-diagonal matrix elements of the one-electron operator:

$$H_{mx} = kG_{mx}(H_m + H_x)/2$$

It is evident from the form of this equation that σ^* should be strongly dependent on H_m and H_x which are steep functions of the charge distribution. Now, the charge distribution (and hence the values of H_m and H_x) cannot be the same in, say, $CuCl_4^{2-}$, as in $CuCl_5^{3-}$; yet we assume that a given value of σ^* (adjusted to take account of the different internuclear distances concerned) can fit the d–d spectra of both compounds. It is of interest to note that in the well-documented cubic systems it is readily apparent that the ligand field splitting indeed depends only on the identity of the metal and of the ligands, and on the internuclear distances, and not on the charge distribution. Thus NiF_2, $KNiF_3$ and K_2NiF_4 all contain the octahedral $Ni^{II}F_6$ chromophore with closely similar Ni—F distances, and have almost identical spectra despite the fact that the charge distribution cannot be the same in all. The success of the concept of inorganic chromophores, (*Jørgensen* (1963)) depends on the constancy of σ^* and π^* for a given metal ion surrounded by given ligands at a fixed internuclear distance, irrespective of the formal charges which may be placed on the metal, the ligands, and the complex as a whole.

Another possible source of error in the angular overlap method is the neglect of $3d$—$4s$ and $3d$—$4p$ mixing in certain symmetries. Thus *Ferguson* (1964) considered that the relative energies of the d-orbitals in $CuCl_4^{2-}$ (D_{2d}) were strongly dependent on d—p mixing since the crystal field model apparently failed in this case. However, as we have already pointed out, both the crystal field and angular overlap models are remarkably successful in rationalizing the d—d spectra of chlorocuprates in various stereochemistries. There is no cogent experimental evidence for the need to consider d—s and d—p mixing, although semi-empirical MO calculations (*Roos* (1966); *Lohr* and *Lipscomb* (1963)) indicate that the relative one-electron energies are significantly affected.

In most angular overlap treatments it has been common practice to neglect ligand-ligand overlap. Strictly speaking, the normalisation constant which precedes the expression for a ligand group orbital must contain such terms. There are indeed circumstances where the inclusion of these ligand-ligand overlap integrals may significantly affect the results. Consider, for example, a square coplanar complex. The b_{1g} group orbital is usually given the normalisation constant $\frac{1}{2}$; including ligand-ligand overlap this becomes:

$$\tfrac{1}{2}\,[1 - 2\,S_1 - S_2]^{-\frac{1}{2}}$$

where S_1 is the overlap integral between two σ-orbitals on ligands *cis* to one another, and S_2 is the corresponding overlap integral between two *trans* σ-orbitals. Numerical values of these overlap integrals have been given (*Smith* (1970a)). Including ligand-ligand overlap, the normalisation term for the a_{1g} group orbital is

$$\tfrac{1}{2}\,[1 + 2\,S_1 + S_2]^{-\frac{1}{2}}$$

If the ligand orbitals are deemed to be nitrogen $2p$-orbitals, with a metal-ligand distance of 2.0 Å, the normalisation constant for the b_{1g}

group orbital is 0.53, while that for the a_{1g} group orbital is 0.47. Carrying through the usual angular overlap procedure to find the relative energies of the d_{z^2} (a_{1g}) and $d_{x^2-y^2}$ (b_{1g}) orbitals, we find:

$$E(d_{z^2}) = 4.4\ \sigma^* \text{ and } E(d_{x^2-y^2}) = 16.8\ \sigma^*$$

compared with the well-known values of $E(d_{z^2}) = 5\ \sigma^*$ and $E(d_{x^2-y^2}) = 15\ \sigma^*$ which are obtained if ligand-ligand overlap is neglected. With oxygen or fluorine ligands, the difference would be much less since their more compact valence orbitals overlap less strongly with one another; with the more diffuse orbitals of chlorine and sulphur, ligand-ligand overlap will certainly make significant differences to angular overlap calculations. The present author is currently investigating this and related points in more detail.

In this author's opinion, the angular overlap method offers the best empirical way of dealing with ligand field splittings in copper(II) compounds, although further investigation of its theoretical basis would be desirable. We should bear in mind that the purely electrostatic contribution to the ligand field splitting, although small compared with that due to covalence, is significant and if realistic wave functions are used in a point charge model, the inclusion of these electrostatic terms can significantly affect the relative energies of the d-orbitals. We shall discuss this point in more detail in Sct. IV when the model used by the present author will be discussed in more detail.

C. Semi-Empirical MO Calculations

In the last ten years or so a great deal of effort has been devoted to the development of semi-empirical LCAO—MO calculations for transition metal complexes. We shall not discuss these in great detail since this article is more concerned with simpler empirical models which the experimentalist may wish to employ on a routine basis.

By a semi-empirical MO calculation we mean one in which some attempt is made to solve the Schroedinger Equation for the system and obtain explicit LCAO wave functions, without actually evaluating all the necessary integrals from first principles. Some integrals are obtained by direct appeal to experimental data, others are approximated or neglected altogether. A considerable variety of levels of approximation are to be found in the literature. In the simplest models, a one-electron Hamiltonian is considered; the diagonal matrix elements H_m and H_x for the metal and ligand functions respectively are assumed to be obtainable from ionization potential data, though they depend very much on the assumed charge distribution. The off-diagonal elements H_{mx} are found by appeal to the Wolfsberg-Helmholz approximation:

$$H_{mx} = \boldsymbol{k}\, G_{mx}(H_m + H_x)/2$$

(Sometimes the geometric mean of H_m and H_x has been used). $\boldsymbol{k}$ is sometimes taken as a disposable parameter to be fitted to spectroscopic data, or is arbitrarily set equal to ca. 2. The secular equations can then be set up on the basis of some assumed charge distribution and solved to yield the one-electron eigenvalues and eigenfunctions. From these eigenfunctions a new charge distribution is obtained, usually by means of a Mulliken population analysis, and the secular equations solved again. The process is repeated until the output charge distribution is the same as the input, and self-consistency is achieved. This, in essence, is the SCCC method (*Basch et al.* (1966)) (SCCC = Self Consistent Charge and Configuration). Rather more sophisticated models have also been employed, analogous to the semi-empirical MO methods which have been successful in organic chemistry. In these, some attempt is made to solve the Roothaan SCF equations, taking explicit account of two-electron interactions, while making simplifications. Thus *Roos* (1966) has applied the ZDO—PPP method to $Cu(NH_3)_6^{2+}$ and $Cu(H_2O)_6^{2+}$, while CuF_6^{4-} and $Cu(H_2O)_6^{2+}$ have been treated by the CNDO method (*Clack* and *Farrimond* (1971); *Allen* and *Clack* (1970)). These models enquire more closely the nature of the Hamiltonian although fairly gross approximations are still made.

It is difficult to say how useful these calculations have been in the interpretation of the ligand field splitting in copper(II) complexes. The calculations of *Lohr* and *Lipscomb* (1963), using the Wolfsberg-Helmholz method, were remarkably successful in dealing with not only the *d—d* spectra but also the stereochemistry of chlorocuprates(II) and fluorocuprates(II); in particular, they were able to predict the correct *d—d* spectrum of the $CuCl_4^{2-}$ (D_{2d}) ion before the definitive spectroscopic work of *Ferguson* (1964) was published. Most semi-empirical MO calculations which set out to rationalize the *d—d* spectra of copper(II) compounds have apparently been reasonably successful but one cannot help noting that the results are very sensitive to the nature of the approximations and assumptions made. It seems unlikely that these methods can really give a reliable account of the *d*-orbital splitting in rather asymmetric complexes. Semi-empirical MO calculations appear to offer few advantages over simpler empirical models, except where it is necessary to obtain eigenvectors, or information about charge transfer states. It is pertinent to note that the least sophisticated (*Lohr* and *Lipscomb* (1963)) and most sophisticated (*Roos* (1966); *Allen* and *Clack* (1970); *Clack* and *Farrimond* (1971)) models appear to give equally good results for the *d*-orbital energies and equilibrium geometries of copper(II) systems.

It has been strongly argued (*Dahl* and *Ballhausen* (1968)) that semi-empirical MO calculations should not seek to calculate spectroscopic parameters such as the octahedral 10 Dq, since such quantities are not given directly by the difference between the orbital energies obtained from Wolfsberg-Helmholz calculations. This arises from the fact that the calculation of virtual orbital energies involves the interaction of an electron with itself. When an electron is excited from a singly-occupied orbital ϕ_i to a virtual orbital ϕ_j, the excitation energy involves an interelectron repulsion term $-J_{ij}$ (if the excited state is a triplet) or $2\,K_{ij}-J_{ij}$ (if the excited state is a singlet). These are very large terms — much larger than 10 Dq — and cannot be readily calculated. However, *Offenhartz* (1969, 1970) has shown that 10 Dq for octahedral systems can be expressed as the difference between two orbital energies provided that these orbital energies are eigenvalues of the proper open-shell Roothaan-Fock operator. A different operator should be used for each state, but Offenhartz assumed an 'average of configuration' value of 10 Dq.

Two remarks seem worth noting in the context of this article. First, the definition and calculation of the octahedral splitting parameter 10 Dq by semi-empirical methods has been a source of much difficulty and argument. It seems unlikely that these methods can give a meaningful account of the ligand field in grossly noncubic systems of the type which dominate the chemistry of copper(II), where we are often concerned with the relative energies of *d–d* states differing by a fraction of an electron volt — chickenfeed to theoretical chemists. Second, the objections which have been raised to the interpretation of spectroscopic parameters in terms of MO orbital energies are relevant to our discussion to the extent that the simple 'one-positron' description usually invoked in the case of copper(II) systems may not be entirely adequate since the different spacial extension of the various *d*-orbitals may lead to appreciably different interelectron repulsion parameters in the excited state.

It must be admitted that any semi-empirical MO model which seeks to provide a basis for the interpretation of *d–d* spectra of non-cubic complexes must involve drastic approximations. It seems that we are faced with two alternatives: either we stick to simpler empirical models (such as the crystal field or angular overlap methods) or we make a more serious attempt to solve the problem exactly. We therefore turn to non-empirical *ab initio* SCF calculations which are now becoming feasible for transition metal compounds.

D. *Ab initio* SCF–MO Calculations

Ab initio calculations make no use of experimental data, except perhaps the parameters which determine the molecular geometry. All electrons are included in the calculation and all molecular integrals are explicitly calculated, starting from the Roothaan SCF equations (*Roothaan* (1960); *Richards* and *Horsely* (1970)). Such calculations on transition metal compounds involve very extensive computations but *ab initio* calculations on copper(II) compounds are beginning to appear, and many more will undoubtedly be performed in the near future. One might hope that calculations of this kind will provide the definitive theoretical treatment of spectroscopic data. However, this is far from true. '*Ab initio*' does not mean 'accurate'. In simple diatomic molecules of the first period, electronic transition energies are reasonably well accounted for by *ab initio* calculations, but the calculated transition energies can still be in error by several thousand wavenumbers. The Roothaan-Hartree-Fock SCF method does not take account of electron correlation and therefore gives a relatively poor account of interelectron repulsion. Since however interelectron repulsion is explicitly considered in constructing the wave functions, the *ab initio* methods are likely to lead to very different results compared with the simpler semi-empirical calculations usually classified as 'extended Wolfsberg-Helmholz'; but it is worth noting that SCF calculations on the free transition metal atoms and ions lead to interelectron repulsion parameters which may be some 30—50% too large. It is therefore not very likely that they can deal with interelectron repulsion in molecules very convincingly and we are left with some reservations as to whether the calculation of spectroscopic parameters on the basis of *ab initio* calculations will assist us in the rationalization of the ligand field splitting in non-cubic copper(II) systems. The results so far are not very encouraging. We have already referred to the calculations of *Basch et al.* (1969) on molecular CuF_2. It is not absolutely certain that their calculated transition energies are in disagreement with the observed spectrum of *Kasai et al.* (1966) but their results are somewhat disturbing from the point of view of conventional ligand field arguments, as we have already discussed. *Demnynck* and *Viellard* (1970), who performed an ab initio calculation on square coplanar $CuCl_4^{2-}$, did not attempt any correlations with optical data; as discussed by *Dahl* and *Ballhausen* (1968) this is a tricky business in the SCF framework.

No doubt further *ab initio* SCF calculations on copper(II) systems will appear in the near future, and the results are awaited with great interest. They should provide much useful information about bonding and electron distribution, but it is less likely that they will be as useful in the routine interpretation of optical data as the simpler empirical methods.

IV. Application of the Angular Overlap Model with Crystal Field Corrections for Copper(II) Complexes

A. Critique of Model

In the previous Sect. III we have summarised the essential features of the various theoretical models which may be used to interpret the ligand field spectra of copper(II) compounds. In the opinion of this author, the energy differences involved are so small that a perturbation approach seems best, rather than attempt to calculate (or estimate) all the interaction integrals explicitly. It is clear that optical transition energies are expressible as the algebraic sum of several quantities all of which are very much larger than the experimental transition energies. To try to determine whether, say, the $^2A_1$ state in $CuCl_4^{2-}$ (D_{2d}) lies higher in energy than the $^2B_1$ state by means of a rigorous MO calculation is rather like trying to decide whether the captain of a ship is heavier than his first officer by weighing the whole ship first with only the captain on board, then with only the first officer on board, and taking the difference.

The model advocated here and in this author's recent publications supposes that the splitting of the *d*-orbitals is largely a consequence of covalent bonding, but that purely crystal field effects are not negligible. The affects of covalence are most easily dealt with by means of the angular overlap model, while the crystal field effects are deemed to be the consequence of a potential due to the presence of negative point charges located at the ligand atoms. The latter are evaluated using Hartree-Fock wave functions. We have already discussed our reasons for believing that covalent bonding is the major effect; however, we should also try to justify our contention that crystal field effects must be included as well. The most sophisticated electrostatic calculations (*Freeman* and *Watson* (1960)) indicated that the octahedral splitting parameter 10 Dq was small but positive, and not quite negligible. Thus for $KNiF_3$ the crystal field model of *Freeman* and *Watson* (1960) gave 10 Dq as 1.1 kK (*Watson* and *Freeman* (1964); *Sugano* and *Shulman* (1963)) and these authors considered that this had to be regarded as a genuine contribution to the experimental 10 Dq. A simple point charge calculation of 10 Dq in $KNiF_3$ gives a value of 1.4 kK (using the Hartree-Fock value of $\langle r^4 \rangle$) so that the crude point charge crystal field model is apparently in good agreement with the more sophisticated Freeman-Watson model. The latter, incidentally, leads to an expression for 10 Dq which is the summation of several large terms which almost cancel out, so that the apparent success and applicability of the point charge model in assessing the electrostatic contribution is fortuitous.

We therefore have sound theoretical grounds for including crystal field effects, and reasonable grounds for representing these by the point charge model. There is also experimental evidence that crystal field effects significantly affect the d-orbital sequence. Thus in tetragonal $Cu(NH_3)_4X_2$ systems, where we may reasonably neglect π-bonding, the angular overlap model predicts that the d_{xy} and $d_{xz,yz}$ orbitals should remain accidentally degenerate, since they are σ-nonbonding. *Kibler* (1970, 1971) has discussed the significance of these accidental degeneracies in the angular overlap model. There is experimental evidence, however, both from optical and ESR spectra, that the d_{xy} and $d_{xz,yz}$ orbitals may be appreciably split (*Hathaway* and *Tomlinson* (1970); *Hathaway* and *Billing* (1970)). The only consistent description of the bonding which is in agreement with the optical and ESR data requires that π-bonding is negligible but that the d_{xy} and $d_{xz,yz}$ orbitals are split by about 2 kK. This, as shown by *Smith* (1969a), can be readily accounted for by applying an electrostatic perturbation of point charges (with $q \sim 0.5$) located at the ligand atoms.

There are two other possible perturbations which might lead to splitting of the d_{xy} and $d_{xz,yz}$ orbitals in tetragonal systems. First, the distant axial ligands may provide π-type orbitals which could overlap with the $d_{xz,yz}$ orbitals. However, this would have the effect of placing $d_{xz,yz}$ above d_{xy}, contrary to the experimental evidence, and in any case π-interaction with distant ligands is very weak (*Smith* (1970b)). A second conceivable source of splitting might be spin-orbit coupling. In tetragonal systems, this has the effect of splitting the degeneracy of the E state ($d_{xz,yz}$) and mixing one of the components with d_{xy}, the other with d_{z^2}. Let us consider a σ-bonded system where the d_{z^2} orbital lies a few kK above the accidentally-degenerate d_{xy} and $d_{xz,yz}$ levels. Ignoring for the meanwhile the d_{z^2} orbital, the effects of spin-orbit coupling can be described by the matrix:

	d^+_{xy}	d^-_{xz}	d^-_{yz}
d^+_{xy}	E_{xy}	$-i\,\zeta/2$	$\zeta/2$
d^-_{xz}	$i\,\zeta/2$	E_{xz}	$i\,\zeta/2$
d^-_{yz}	$\zeta/2$	$-i\,\zeta/2$	E_{yz}

If we arbitrarily set $E_{xz} = E_{yz} = 0$, and $E_{xy} = n\,\zeta$, the one-electron energies are then given by the roots of the equation:

$$E^3 - n\,\zeta\,E^2 - 3\,(\zeta^2/4)E - (1-n)\,(\zeta^3/4) = 0$$

It is easy to find that the roots of this equation are $-\zeta/2$ and $(\zeta/2)(n+\tfrac{1}{2}) \pm (\zeta/2)\,[n^2 - n + (9/4)]^{\frac{1}{2}}$. In the absence of any other perturbation

($n=0$) spin-orbit coupling produces a splitting of $(3/2)\zeta$ which will be about 0.9 kK in a copper(II) tetrammine complex. In order to explain the observed splitting, it is necessary to postulate that E_{xy}—$E_{xz,yz}$ is about 1.5 kK at least. It may also be noted that the proximity of the d_{xy} orbital to the $d_{xz,yz}$ orbitals has the effect of considerably reducing the extent to which the spin-orbit components of the $d_{xz,yz}$ level are split. In the point group D_4^*, the $d_{xz,yz}$ functions are split into components belonging to $e_{\frac{1}{2}}$ and $e_{\frac{3}{2}}$. In the absence of any second-order mixing these are split by ζ, with the $e_{\frac{1}{2}}$ level lying lowest in energy. The d_{xy} orbital transforms as $e_{\frac{3}{2}}$ in D_4^* so that the mixing of the two $e_{\frac{3}{2}}$ orbitals reduces the splitting of the two lowest one-electron functions. This is shown in Fig. 1. Thus in σ-bonded tetragonal copper(II) systems spin-orbit coupling does not manifest itself to any appreciable extent at a qualitative level.

In systems where π-bonding may be appreciable a very different picture may emerge. The d_{z^2} orbital may now lie very close to the $d_{xz,yz}$ orbitals, with the d_{xy} lying somewhat higher in energy (Fig. 2). We must

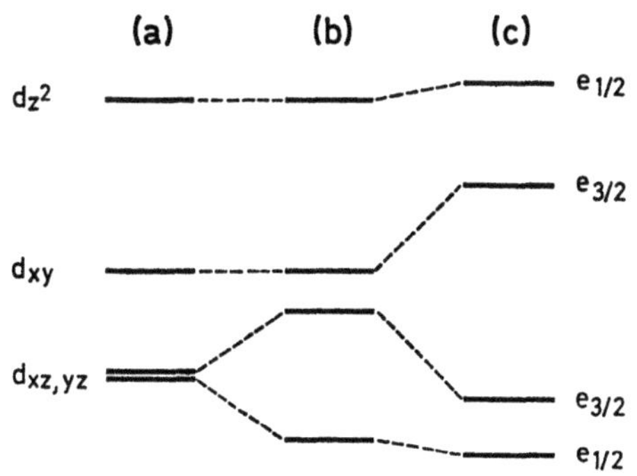

Fig. 1. Effect of spin-orbit coupling on the d-orbitals when d_{xy} lies close to $d_{xz,yz}$; (a) no spin-orbit coupling (b) first-order perturbation (c) second order perturbation

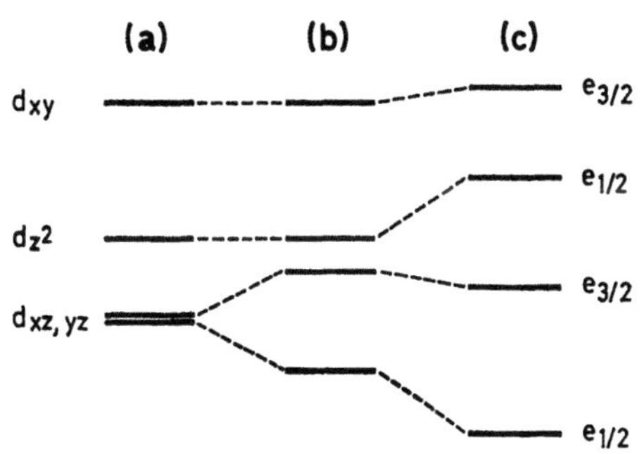

Fig. 2. Effect of spin-orbit coupling on the d-orbitals when d_{z^2} lies close to $d_{xz,yz}$; (a) no spin-orbit coupling (b) first-order perturbation (c) second-order perturbation

now consider the matrix:

	$d^+_{z^2}$	d^-_{xz}	d^-_{yz}
$d^+_{z^2}$	E_{z^2}	$-(\sqrt{3}/2)\zeta$	$i(\sqrt{3}/2)\zeta$
d^-_{xz}	$-(\sqrt{3}/2)\zeta$	E_{xz}	$i\,\zeta/2$
d^-_{yz}	$-i(\sqrt{3}/2)\zeta$	$-i\,\zeta/2$	E_{yz}

If we set $E_{xz}=E_{yz}=0$, and $E_{z^2}=n\ \zeta$, the one-electron energies are given by: $\zeta/2$ and $(\zeta/2)\ (n-\frac{1}{2})\pm(\zeta/2)\ [n^2+n+25/4]^{\frac{1}{2}}$. When $n=0$, these reduce to ζ, $\zeta/2$ and $-3\ \zeta/2$. Thus if the d_{z^2} orbital lies close to the $d_{xz,yz}$ orbitals the latter may be split by up to $2\ \zeta$. A well-authenticated example experimentally occurs in calcium copper(II) acetate hexahydrate (*Billing et al.* (1970); *Mabbs* and *Smail* (1970); *Gliemann* and *Morys* (1970); *Smith* (1971a)). In this D_{2d} CuO_8 chromophore, the ^2E state is apparently split by 1.4 kK in the two different polarizations. Taking the orbital reduction factor to be 0.8, $2\ \zeta$ is about 1.3 kK. In systems where the $d_{xz,yz}$ orbitals are sufficiently far removed in energy from other d-orbitals and the second order matrix elements are small, the E state is seen to be split by ζ, *e.g.* in Cs_2CuCl_4 (*Ferguson* (1964)).

Having discussed the role of spin-orbit coupling in the interpretation of copper(II) d—d spectra, we now proceed to discuss other features of our model in more detail. The effects of covalence are dealt with by means of the angular overlap method, and the one-electron energies can be written down in terms of the covalence parameters σ^*, $\pi^*(\perp)$ and $\pi^*(\parallel)$, as described in IIIB. The relative magnitudes of the σ^* and π^* parameters can be estimated from the respective overlap integrals, with $\pi^*/\sigma^* = (S^{\pi}_{mx}/S^{\sigma}_{mx})^2$ in cases where $\pi^*(\perp)$ and $\pi^*(\parallel)$ are deemed to be equal, as in halide complexes. With less simple ligands, $\pi^*(\perp)$ and $\pi^*(\parallel)$ have to be determined empirically. We propose that σ^* depends only on the identity of the metal and ligand atoms and on the internuclear distance so that we need only determine the constant $\sigma^*/(S^{\sigma}_{mx})^2$ for each of the common ligand atoms N, O, F, Cl and Br. For the halides, the single parameter $\pi^*/(S^{\pi}_{mx})^2$ can likewise be specified. However, for copper-oxygen and copper-nitrogen complexes where π-bonding may be of importance, the parameters $\pi^*(\perp)$ and $\pi^*(\parallel)$ must be taken separately but their radial variation may also be deemed to be proportional to $(S^{\pi}_{mx})^2$. For closely related ligand systems, and for complexes where the orientations of these ligands with respect to the metal are similar, we may hope that the ratio $\pi^*/(S^{\pi}_{mx})^2$ is transferable. It may be possible in some cases to exclude one or other of the π^* parameters if an inspection of the ligand atom precludes the likelihood of appreciable in-plane or out-of-plane π-bonding.

The crystal field perturbation is then applied. The matrix elements are found as described in III A and the electrostatic corrections to the one-electron levels found. These are expressible in terms of the parameters a_n. The radial integrals $\langle r^n \rangle$ are calculated from Hartree-Fock wave functions. Two points now require further consideration. First, the values of the $\langle r^n \rangle$ (and hence the a_n) depend on the fractional charge on the metal. Simple methods of estimating this were discussed by this author (*Smith* (1969a)), but these were rather crude and neglected the Madelung forces which tend to enhance the charge separation between the metal and ligand atoms (*Jørgensen et al.* (1967)). This author has usually taken the formal charge on the metal to be $+0.5$, in accordance with many semi-empirical MO calculations, but, as pointed out by *Jørgensen et al.* (1967), this may be an underestimate. It may be best to use a value of $+1$, which leads to $\langle r^2 \rangle = 1.26$ (a.u.)^2 and $\langle r^4 \rangle = 4.38$ (a.u.)^4. The second problem is the value of the point charge q deemed to reside on each ligand atom. In the tetrammines, a value of $q = 0.5$ gave good results, and we have advocated this value for neutral nitrogen and oxygen complexes where stoichiometry gives no guidance. For halide complexes, the fractional charges on the metal and ligand are fixed stoichiometrically, and if we identify q with the fractional negative charge in the ligand atoms, its value can be determined from the assumed charge on the metal and the stoichiometry. Values thus determined lead to good results. Thus in any event we can determine the electrostatic parameters a_n without being forced to make any ad hoc assumptions in order to fit the facts in each particular case.

The d–d transition energies can then be written down in terms of the parameters σ^*, $\pi^*(\perp)$, $\pi^*(\parallel)$, a_2 and a_4. There is now a large body of experimental data on which to base empirical values of the covalence parameters. We shall now give tabulations of these parameters as a function of the internuclear distance for the most important ligand atoms encountered in copper(II) compounds. From these tables it should be possible to predict the d–d spectrum of any copper(II) chromophore of known or assumed geometry.

B. Nitrogen Ligands

We may classify most nitrogen-donor ligands into one of three categories:

(i) Purely σ-donors, such as ammonia or ethylenediamine; the only covalence parameter required will be σ^*.

(ii) Unsaturated N-donor ligands where the nitrogen atom may have one π-orbital available for bonding to the metal, but uses a second p-orbital in bonding within the ligand molecule to such an extent that we may ignore its interaction with the metal. In this case one π^* parameter is required, but the ratio π^*/σ^* not necessarily given by $(S^{\pi}_{mx}/S^{\sigma}_{mx})^2$.

(iii) Ligands such as NCS^- where two π-orbitals on the N-atom are possibly available for bonding to the metal, and these are equivalent, so that only one π^* parameter is required.

Table 6. *Splitting parameters* σ^*, a_2 *and* a_4 *and* $(S^{\pi}_{mx})^2$ *as functions of internuclear distance R for nitrogen bonded to copper(II)*

R (Å)	σ^* (kK)	$(S^{\pi}_{mx})^2$	a_2 (kK)	a_4(kK)
1.90	1.34	0.00394	2.98	0.80
1.95	1.22	0.00325	2.77	0.70
2.00	1.10	0.00264	2.56	0.62
2.05	0.99	0.00218	2.37	0.54
2.10	0.89	0.00176	2.22	0.48
2.15	0.79	0.00144	2.09	0.44
2.20	0.69	0.00116	1.95	0.39
2.25	0.61	0.00095	1.82	0.35
2.30	0.53	0.00076	1.71	0.31
2.35	0.46	0.00062	1.59	0.28
2.40	0.40	0.00049	1.49	0.25
2.45	0.35	0.00040	1.40	0.22
2.50	0.30	0.00032	1.32	0.20
2.55	0.25	0.00026	1.25	0.18
2.60	0.21	0.00020	1.19	0.17
2.70	0.15	0.00013	1.05	0.14
2.80	0.11	0.00008	0.95	0.12
2.90	0.08	0.00005	0.85	0.10
3.00	0.05	0.00003	0.77	0.08

Table 6 contains a set of recommended values of σ^* over a suitable range of internuclear distances. We also give in Table 6 the appropriate values of $(S^{\pi}_{mx})^2$ so that the radial dependence of π^* may be gauged, without introducing any assumption regarding the relative magnitudes of σ^* and π^*. As in all the tabulations of quantities related to overlap integrals given here, only overlap between copper $3d$-orbitals (for Cu^+) and neutral ligand np-orbitals are considered. We also give values of the crystal field parameters a_2 and a_4, assuming a charge of $+1$ on the metal and $q=0.5$. These values give good results for tetragonal copper ammine complexes, except that the d_{z^2} orbital is placed a little too high in energy; the results can be much improved by taking into account ligand-ligand overlap in the group orbitals, and it is always possible that $3d_{z^2}$—$4s$ mixing is far from negligible. This of course is always neglected in the angular overlap approach.

C. Oxygen Ligands

Oxygen-donor ligands are the most troublesome to treat theoretically. A case where the ligand $\pi(\perp)$ and $\pi(\|)$ orbitals can be regarded as even approximately equivalent is very rare in copper(II) chemistry, even in oxides where the structure may be dominated by, for example, SiO_4 tetrahedra. The hydroxide ion, where we may expect to encounter a pair of equivalent π-orbitals on the oxygen, is usually found as a bridging ligand in copper(II) compounds, and in more complex oxygen ligand systems, such as carboxylates, the orientation of the ligand groups with respect to the metal can be highly variable. Water is particularly difficult to treat since it is not possible to make any sensible proposals regarding the relative amounts of in-plane or out-of-plane π-bonding without information about the orientation of the molecules — and this information is obviously difficult to obtain (*Smith* (1970d)). Thus the π-bonding situation in copper(II)-oxygen systems must to some extent be treated empirically in each individual case, although it is sometimes possible to transfer parameters from one system to another where the bonding situation looks similar. The best guidance we can offer is that $\pi^*(\perp)$ always seems to be somewhat larger than $\pi^*(\|)$, and both values are a little larger than would be expected on the basis of the assumption that $\pi^*/\sigma^* = (S^{\pi}_{mx}/S^{\sigma}_{mx})^2$. It does appear, however, that a single value of the ratio $\sigma^*/(S^{\sigma}_{mx})^2$ suffices for all copper(II)-oxygen systems, since *Smith* (1970d) found that the energy of the $d_{z^2} \rightarrow d_{x^2-y^2}$ transition (which does not depend on π-bonding) could be very well fitted to a single parameter in a series of tetragonal copper(II)-oxygen systems with very different types of ligand molecules, and varying tetragonality. In Table 7 we give recommended values of σ^* over a range of internuclear distances, together with values of $(S^{\pi}_{mx})^2$. The electrostatic parameters can be taken from Table 6.

D. Fluoride

Rather little definite spectroscopic data is available for fluorocuprates(II), as we discussed in IID, so we have a less satisfactory experimental basis for our tabulation of covalence parameters. From the spectra reported by *Oelkrug* (1967, 1971) and *Schmitz-Dumont* and *Grimm* (1967) we have arrived at a suitable value of $\sigma^*/(S^{\sigma}_{mx})^2$. Values of σ^* are given in Table 8, together with values of π^*, assuming $\pi^*/\sigma^* = (S^{\pi}_{mx}/S^{\sigma}_{mx})^2$. We do not give values of the crystal field parameters since these will depend on the stoichiometry. These may be readily found for any system as we have already discussed.

Table 7. *Angular overlap parameters σ^* and $(S^{\pi}_{mx})^2$ as functions of the internuclear distance R for oxygen bonded to copper(II)*

R(Å)	σ^*(kK)	$(S^{\pi}_{mx})^2$
1.90	1.52	0.00286
1.95	1.36	0.00233
2.00	1.20	0.00185
2.05	1.06	0.00147
2.10	0.93	0.00112
2.15	0.81	0.00094
2.20	0.70	0.00076
2.25	0.60	0.00062
2.30	0.51	0.00048
2.35	0.44	0.00039
2.40	0.37	0.00030
2.45	0.32	0.00024
2.50	0.26	0.00019
2.55	0.22	0.00015
2.60	0.19	0.00012
2.70	0.13	0.00007
2.80	0.09	0.00004
2.90	0.06	0.00003
3.00	0.04	0.00002

Table 8. *Angular overlap parameters σ^* and π^* as functions of the internuclear distance R for copper(II) fluorine polyhedra*

R(Å)	σ^*(kK)	π^*(kK)
1.80	1.08	0.33
1.85	0.95	0.27
1.90	0.84	0.22
1.95	0.74	0.18
2.00	0.64	0.14
2.05	0.55	0.11
2.10	0.47	0.08
2.15	0.40	0.06
2.20	0.34	00.5
2.25	0.30	0.04
2.30	0.25	0.03
2.35	0.21	0.02
2.40	0.18	0.02
2.50	0.11	0.01
2.60	0.08	0.01
2.70	0.05	0.00
2.80	0.03	0.00

E. Chloride

Again, we can assume that the ligand atoms present an equivalent set of *p*-orbitals for bonding to the metal. Values of σ^*, based on the results of calculations on chlorocuprates(II) of various geometries, are collected in Table 9, together with the appropriate values of π^*.

F. Bromide

Rather little data is available on which to base a set of parameters for bromocuprates(II). As we pointed out in Sect. II F, there is still much confusion in the literature regarding the spectrum of $CuBr_4^{2-}$. The parameters tabulated in Table 10 are based on the electronic spectrum of $CuBr_5^{3-}$, reported by *Allen* and *El-Sharkaway* (1970), and the crystallographic parameters of *Raymond* (1970).

Table 9. *Angular overlap parameters σ^* and π^* as functions of the internuclear distance R of copper (II) — chlorine polyhedra*

R(Å)	σ^*(kK)	π^*(kK)
2.20	1.24	0.35
2.25	1.12	0.29
2.30	1.00	0.25
2.35	0.89	0.20
2.40	0.79	0.16
2.45	0.70	0.14
2.50	0.62	0.11
2.55	0.54	0.09
2.60	0.47	0.07
2.65	0.41	0.06
2.70	0.35	0.05
2.75	0.30	0.04
2.80	0.26	0.03
2.85	0.22	0.02
2.90	0.19	0.02
2.95	0.16	0.02
3.00	0.14	0.01
3.10	0.10	0.01
3.20	0.07	0.01
3.30	0.05	0.00

Table 10. *Angular overlap parameters σ^* and π^* as functions of the internuclear distance R for copper (II) — bromide polyhedra*

R(Å)	σ^*(kK)	π^*(kK)
2.30	1.35	0.38
2.35	1.23	0.33
2.40	1.12	0.29
2.45	1.00	0.25
2.50	0.89	0.21
2.55	0.80	0.18
2.60	0.71	0.15
2.65	0.63	0.12
2.70	0.55	0.10
2.75	0.48	0.09
2.80	0.43	0.07
2.85	0.38	0.06
2.90	0.33	0.05
2.95	0.29	0.04
3.00	0.25	0.03
3.10	0.19	0.02
3.20	0.14	0.01
3.30	0.10	0.01

V. References

Allen, H. C., (1966): Ground state of bis(acetylacetonate)copper(II). J. Chem. Phys. *45*, 553.

Allen, G. C., Clack, D. W. (1970): Molecular orbital calculations on transition metal complexes. Part I. Hexafluorometallate ions. J. Chem. Soc. (A) 2668.

— *El-Sharkaway, G. A. M.* (1970): Electronic spectrum of the pentabromocuprate ion. Inorg. Nucl. Chem. Letters *6*, 281.

— *Hush, N. S.* (1967): Reflectance spectra and electronic states of the pentachlorocuprate ion in a number of crystal lattices. Inorg. Chem. *6*, 4.

Babel, D. (1967): Structural chemistry of the octahedral fluorocomplexes of the transition metals. Struct. Bonding *3*, 1.

Baglio, J. A., Vaughan, P. A. (1970): The structures of the tetramminecopper(II) dihalocuprates(I). Part II. Structures of $Cu(NH_3)_4 \cdot (CuBr_2)_2$ and $Cu(NH_3)_4 \cdot (CuCl_2)_2 \cdot H_2O$. J. Inorg. Nucl. Chem. *32*, 803.

Ballhausen, C. J. (1954): Studies of Absorption Spectra. II. Theory of copper(II) spectra. Kgl. Danske Videnskab. Selskab., Mat.-Fys. Medd. 29.

— (1962): Introduction to ligand field theory. New York: McGraw-Hill.

Basch, H., Viste, A., Gray, H. B. (1966): Molecular orbital theory for octahedral and tetrahedral complexes. J. Chem. Phys. *44*, 10.

— *Hollister, C., Moskowitz, J. W.* (1969): Geometry and electronic states of CuF_2. Chem. Phys. Letters *4*, 79.

Basu, G., Belford, R. L., Dickerson, R. E. (1962): Observation of four *d—d* transitions in a copper chelate. Inorg. Chem. *1*, 438.

Becker, C. A. L., Meek, D. W., Dunn, T. M. (1968): Crystal Field and Spin-Orbit treatment of d^1 and d^9 trigonal bipyramidal complexes. J. Phys. Chem. *72*, 3588.

Belford, R. L., Belford, G. (1965): Group-theoretic criterion for selection rules in the vibronic mechanism and applications to copper acetylacetonate. Theoret. Chim. Acta *3*, 465.

— *Calvin, M., Belford, G.*(1957): Bonding in copper(II) chelates: solvent effects on their visible solution spectra. J. Chem. Phys. *26*, 1165.

— *Carmichael, J. W.* (1967): Bis(3-phenyl-2,4-pentanedionato) copper(II). II. Polarized crystal spectrum. J. Chem. Phys. *46*, 4515.

Billing, D. E., Dudley, R. J., Hathaway, B. J., Tomlinson, A. A. G. (1971): Single crystal electronic and e. s. r. spectrum of dichloroaquo(2,9-dimethyl-1,10-phenanthroline)copper(II). J. Chem. Soc. (A) 691.

— *Hathaway, B. J., Nicholls, P.* (1969): Polarizations of the copper(II) and uranyl electronic spectra of meta-zeunerite single crystals. J. Chem. Soc. (A) 316.

— — — (1970): Single crystal electronic and e.s.r. spectra of calcium copper(II) acetate hexahydrate. J. Chem. Soc. (A) 1877.

— — (1968): Crystal field energy levels of dibarium copper(II) formate tetrahydrate and copper(II) formate tetrahydrate. J. Chem. Soc. (A) 1516.

Bjerrum, J. (1932): Untersuchungen über Kupferammoniakverbindungen. II. Die Komplexitätskonstante des Pentamminkuprikomplexes und die Absorptionsspektren der Amminkupriionen. Kgl. Danske Videnskab. Selskab Mat.-Fys. Medd. *11*, 3.

— *Ballhausen, C. J., Jørgensen, C. K.* (1954): Studies of Absorption spectra. I. Results of calculations on the spectra and configuration of copper(II) ions. Acta Chem. Scand. *8*, 1275.

Bonacic, V., Randic, M. (1968): Ligand field splitting for a diffuse charge model. Theoret. Chim. Acta *10*, 337.

— — (1971): Ligand field splitting in square planar, trigonal bipyramidal and octahedral stereochemistries by the diffuse charge model. J. Mol. Struct. *7*, 9.

Boorman, P. M., Craig, P. J., Swaddle, T. W. (1969): Far infrared and Raman spectra of the pentachlorocuprate(II) ion. J. Chem. Soc. (A) 2970.

Boudreaux, E. A. (1964): The δ-bond in binuclear complexes of copper(II); an approximate molecular orbital treatment. Inorg. Chem. *3*, 506.

Bukovska, M., Porai-Koshits, M. A. (1960): Crystal structure of $Cu(NH_3)_5NH_4(ClO_4)_3$. Kristallografiya *5*, 140.

Chidambaram, R., Sequeira, A., Sikka, S. K. (1964): Neutron diffraction study of the structure of potassium oxalate monohydrate: lone pair coordination of hydrogen bonded water molecules in crystals. J. Chem. Phys. *41*, 3616.

Ciampolini, M. (1969): Spectra of 3*d* five-coordinate complexes. Structure and Bonding *6*, 52.

Clack, D. W., Farrimond, M. S. (1971): Molecular orbital calculations on transition metal complexes. Part II. Hexa-aquometal(II) and hexa-aquometal (III) ions. J. Chem. Soc. (A) 299.

Clark, M. G., Burns, R. G. (1967): Electronic spectra of Cu^{2+} and Fe^{2+} square planar coordinated by oxygen in $BaXSi_4O_{10}$. J. Chem. Soc. (A) 1034.

Clementi, E. (1965): Tables of atomic wave functions. Supplement to IBM J. Res. Develop. *9*, 2.

Companion, A. L., Komarynsky, M. A. (1964): Crystal field splitting diagrams. J. Chem. Educ. *41*, 257.

Cotton, F. A., Harris, C. B., Wise, J. J. (1967): Extended Huckel calculations of the molecular orbitals in bis(β-ketoenolate) complexes of Cu^{II} and Ni^{II}. Inorg. Chem. *6*, 909.

— *Wise, J. J.* (1967): Assignment of the electronic absorption spectrum of bis(β-diketoenolate) complexes of Cu(II) and Ni(II). Inorg. Chem. *6*, 917.

Craig, D. P., Maccoll, A., Nyholm, R. S., Orgel, L. E., Sutton, L. E. (1954). Chemical bonds involving *d*-orbitals. J. Chem. Soc. 332, 354.

Dahl, J. P., Ballhausen, C. J. (1968): Molecular orbital theories of inorganic complexes. Advan. Quantum Chem. *4*, 170.

Day, P. (1964): Point charge crystal field calculations for cupric halides. Proc. Chem. Soc. (London) 18.

— *Jørgensen, C. K.* (1964): A simple molecular orbital model for transition metal complexes. J. Chem. Soc. 6226.

DeKock, C. W., Gruen, D. M. (1966): Electronic absorption spectra of the gaseous 3*d*-transition metal dichlorides. J. Chem. Phys. *44*, 4387.

Demuynck, J., Veillard, A. (1970): Ab initio calculation of the electronic structure of the $CuCl_4^{2-}$ ion. Chem. Phys. Letters *6*, 204.

Dietz, R. E., Kaminura, H., Sturge, M. D., Yariv, A. (1963): Electronic structure of Cu impurities in ZnO. Phys. Rev. *132*, 1558.

Dijkgraaf, C. (1965): Interpretation of the polarized absorption spectrum of copper acetylacetonate. Theoret. Chim. Acta *3*, 38.

Distler, T., Vaughan, P. A. (1967): The crystal structure of the hexamminecopper(II) halides. Inorg. Chem. *6*, 126.

Drickamer, H. G. (1967): Effect of the interatomic distance on the ligand field in NiO. J. Chem. Phys. *47*, 1880.

Dyrssen, D., Petkovic, D. (1965): Formation of adducts with bis-(dimethylglyoximato)copper(II). Acta Chem. Scand. *19*, 653.

Dubicki, L. (1970): The electronic spectrum of binuclear copper acetate. Proceedings of the XIII International Conference on Coordination Chemistry; ed. W. Kakolowicz, Polish Academy of Sciences, Vol. 1, p. 129.

Dudley, R. J., Hathaway, B. J. (1970a): Single crystal electronic and ESR spectrum of dichlorobis(2-methyl pyridine)copper(II). J. Chem. Soc. (A) 2799.

— — (1970b): Single crystal electronic and ESR spectra of bis(aquo)monoacetylacetonatocopper(II) picrate. J. Chem. Soc. (A) 1725.

— — (1970c): Single crystal electronic and ESR spectra of 2,2′-dipyridylbis(hexafluoroacetylacetonato)copper(II). J. Chem. Soc. (A) 2794.

Elliott, H., Hathaway, B. J. (1966): The hexammine complexes of the copper(II) ion. Inorg. Chem. *5*, 885.

— — *Slade, R. C.* (1966): The properties of the hexanitrocomplexes of divalent Fe, Co, Ni and Cu. Inorg. Chem. *5*, 669.

Felsenfeld, G. (1956): Structure of the chlorocuprate ion. Proc. Roy. Soc. (London) *A 236*, 506.

Ferguson, J. (1961a): Crystal spectra of metal coordination compounds. IV. Bis-acetylacetonatocopper(II). J. Chem. Phys. *34*, 1609.

— (1961b): Crystal spectra of metal coordination compounds. VI. Bis-salicylaldiminatocopper(II) and bis-N-methylsalicylaldiminatocopper(II). J. Chem. Phys. *35*, 1612.

— (1964): Electronic absorption spectrum and structure of $CuCl_4^{2-}$. J. Chem. Phys. *40*, 3406.

— (1965): Davydov splitting and the absorption spectra of transition metal ion crystals. Theoret. Chim. Acta *3*, 287.
— (1970): Spectroscopy of the 3*d* metals. Prog. Inorg. Chem. *12*, 159.
— *Belford, R. L., Piper, T. S.* (1962): Crystal spectrum of copper acetylacetone. J. Chem. Phys. *37*, 1659.
Forster, D., Ballhausen, C. J. (1962): Binuclear complexes. I. The electronic structure of copper acetate. Acta Chem. Scand. *16*, 1385.
— *Goodgame, D. M. L.* (1964): Preparation, electronic spectra and magnetic properties of some transition metal isocyanato complexes. J. Chem. Soc. 2790.
— — (1965): Isothiocyanato-complexes of Ni(II) and Cu(II). Inorg. Chem. *4*, 823.
Freeman, A. J., Watson, R. E. (1960): Calculation of the crystal field strength in chrome alum. Phys. Rev. *120*, 1254.
— — (1965): Hyperfine interactions in magnetic materials. Magnetism, Vol. II A, p. 167; eds. *Rado, G. T., Suhl, H.* New York: Academic Press.
Friebel, C., Reinen, D. (1969): ESR-Untersuchungen zum Jahn-Teller-Effekt des Cu^{2+}-Ions in oxidischen Perowskiten. Z. Naturforsch. *24*, 1518.
Funck, L. L., Ortolano, T. R. (1968): The effects of axial ligation on the ligand field spectra of copper(II) β-diketonates. Inorg. Chem. *7*, 567.
Furlani, C., Morpurgo, G. (1963): Properties and electronic structures of tetrahalogenocuprates(II). Theoret. Chim. Acta *1*, 102.
— *Sgamellotti, A., Magrini, F., Cordischi, O.* (1967a): ESR and optical spectrum of $(CH_3NH_3)_2CuCl_4$. J. Mol. Spectra. *24*, 270.
— *Cervone, E., Calzona, F., Baldanza, B.* (1967b): Crystal spectrum of bis-(trimethylbenzylammonium)tetrachlorocuprate(II). Theoret. Chim. Acta *7*, 375.
— (1968): Ligand field interpretation of some cases of pentacoordination. Coord. Chem. Rev. *3*, 141.
Garner, C. D., Mabbs, F. E. (1970): Studies in eight-coordination. Part I. Crystal field energies in the D_{2d} point group. J. Chem. Soc. (A) 1711.
Gliemann, G., Morys, P. (1970): Kristallspektren und elektronische Struktur von oktacoordiniertem Cu(II) in Kalziumkupferacetat-6-Hydrat. Z. Phys. Chem. (Leipzig) *243*, 281.
Graddon, D. P. (1960): Absorption spectra of complex salts. III. Cupric ethylacetoacetate. J. Inorg. Nucl. Chem. *14*, 161.
— (1961): Absorption spectra of complex salts. IV. Cupric alkanoates. J. Inorg. Nucl. Chem. *17*, 222.
Hansen, A. E., Ballhausen, C. J. (1965): Electronic structure of copper acetate monohydrate based on a coupled chromophore model. Trans. Faraday Soc. *6*, 631.
Hartmann, H., Koenig, E. (1961): Matrixelemente des Ligandfeldpotentials in Komplexverbindungen der Übergangsmetalle. Z. Physik. Chem. (Frankfurt) *28*, 425.
Hatfield, W. E., Piper, T. S. (1964): Spectral and magnetic properties of chlorocuprates(II). Inorg. Chem *3*, 841.
— *Whyman, R.* (1969): Copper complexes. Transition Metal Chem. *5*, 147.
— *Bedon, H. D., Horner, S. M.* (1965): Molecular orbital theory for the pentachlorocuprate(II) ion. Inorg. Chem. *4*, 1181.
Hathaway, B. J., Billing, D. E. (1970): The electronic properties and stereochemistry of mononuclear complexes of the copper(II) ion. Coord. Chem. Rev. *5*, 143.
— *Stephens, F. S.* (1970): The electronic properties and crystal structure of $Na_4Cu(NH_3)_4[Cu(S_2O_3)_2]_2 \cdot L$ ($L = H_2O$ or NH_3). J. Chem. Soc. (A) 884.
— *Tomlinson, A. A. G.* (1970): Copper(II)-ammonia complexes. Coord. Chem. Rev. *5*, 1.

— *Billing, D. E., Nicholls, P., Procter, I. M.* (1969a): Crystal field energy levels of some bis-ethylenediaminecopper(II) complexes. J. Chem. Soc. (A) 319.

— *Bew, M. J., Billing, D. E., Dudley, R. J., Nicholls, P.* (1969b): Electronic and e.s.r. spectra of dihydrogenethylenediaminetetraacetatoaquocopper(II) and bis(diethylenetriamine)copper(II) bromide monohydrate. J. Chem. Soc. (A) 2312.

— *Procter, I. M., Slade, R. C., Tomlinson, A. A. G.* (1969c): Electronic properties and stereochemistry of the copper(II) ion; Part VI. Bis(dipyridyl)copper(II) complexes. J. Chem. Soc. (A) 2219.

— *Dudley, R. J., Nicholls, P.* (1969d): Electronic properties and stereochemistry of $K_2BaCu(NO_2)_6$. J. Chem. Soc. (A) 1845.

— *Bew, M. J., Billing, D. E.* (1970a): Single crystal and ESR spectra of bis(diethylenetriamine)copper(II) nitrate. J. Chem. Soc. (A) 1090.

Hathaway B. J., Billing, D. E., Dudley, R. J., Fereday, R. J., Tomlinson, A. A. G. (1970b): Electronic and ESR spectra of diamminocopper(II) trithiocyanatoargentate(I) and iodobis(2,2′-dipyridyl)copper(II) iodide. J. Chem. Soc. (A) 806.

— — — (1970c): Single crystal electronic and e.s.r. spectra of bis(3-methylpentane-2,4-dionato(copper(II). J. Chem. Soc. (A) 1420.

Helmholz, L., Kruh, R. F. (1952): The crystal structure of caesium chlorocuprate, Cs_2CuCl_4, and the spectrum of the chlorocuprate ion,. J. Am. Chem. Soc. *74*, 1176.

Holmes, O. G., McClure, D. S. (1957): Optical spectra of hydrated ions of the transition metals. J. Chem. Phys. *26*. 1686.

Hougen, J. T., Leroi, G. E., James, T. C. (1961): Application of ligand field theory to the electronic spectra of gaseous $CuCl_2$, $NiCl_2$ and $CoCl_2$. J. Chem. Phys. *34*, 1670.

Howald, R. A., Keeton, D. P. (1966): Charge transfer spectra and structures of the copper(II) halide complexes. Spectrochim. Acta *22*, 1211.

Hush, N. S., Hobbs, R. J. M. (1968): Absorption spectra of crystals containing ions of the transition metals. Prog. Inorg. Chem. *10*, 259.

Ilse, F. E., Hartmann, H. (1951). Termsysteme electrostatischer Komplexionen der Übergangsmetalle mit einen *d*-Elektron. Z. Physik. Chem. (Leipzig) *197*, 239.

Isaacs, N. W., Kennard, C. H. L. (1969): A neutron diffraction study of potassium lead hexanitrocuprate(II), an example of a regular octahedral copper(II) complex. J. Chem. Soc. (A) 386.

Jaffe, H. H. (1953): Some overlap integrals involving *d*-orbitals. J. Chem. Phys. *21*, 258.

— *Doak, G. O.* (1953): On the stability of metallo-organic compounds. J. Chem. Phys. *21*, 196.

Johansen, H., Ballhausen, C. J. (1966): An extended Wolfsberg-Helmholz calculation on CuF_6^{4-}. Mol. Phys. *10*, 178.

Johrdahl, O. M. (1934): Effect of crystalline electric fields on the paramagnetic susceptibilities of cupric salts. Phys. Rev. *45*, 87.

Jørgensen, C. K. (1955): Comparative crystal field study of some ligands and the lowest singlet state of paramagnetic nickel(II) complexes. Acta Chem. Scand. *9*, 1362.

— (1963): Inorganic Complexes. New York: Academic Press.

— (1964): Molecular orbital description of diatomic molecules containing a transition metal atom. Mol. Phys. *7*, 417.

— *Pappalardo, R., Schmidtke, H.-H.* (1963): Do the 'ligand field' parameters in lanthanides represent weak covalent bonding? J. Chem. Phys. *39*, 1422.

— *Horner, S. M., Hatfield, W. E., Tyree, S. Y.* (1967): Influence of Madelung (interatomic coulomb) energy on Wolfsberg-Helmholz calculations. Intern. J. Quantum Chem. *1*, 191.

Karipides, A. G., Piper, T. S. (1962): Crystal spectrum of Cs_2CuBr_4. Inorg. Chem. *1*, 971.

Kasai, P. H., Whittle, E. B., Weltner, W. (1966): ESR of $Cu(NO_3)_2$ and CuF_2 molecules oriented in neon and argon matrices at 4 °K. J. Chem. Phys. *44*, 2581.

Kettle, S. F. A. (1969): Crystal field potentials. J. Chem. Educ. *46*, 339.

Kibler, M. R. (1970): Comparison between the point charge electrostatic model and the Ξ^2 model. Chem. Phys. Letters *7*, 83.

— (1971): Explanation of some casual degeneracies introduced by the Ξ^2 model. Chem. Phys. Letters *8*, 142.

Knox, K. (1959): Structure of K_2CuF_4; a new type of distortion for octahedral copper(II) complexes. J. Chem. Phys. *30*, 991.

Kokoszka, G. F., Reimann, C. W., Allen, H. C., Gordon, G. (1967): Optical and magnetic measurements on single crystals of copper-doped tris(1,10phenanthroline) zinc(II) nitrate trihydrate. Inorg. Chem. *6*, 1657.

Krishnamurthy, R., Schaap, W. B. (1969): Computing ligand field potentials and relative energies of *d*-orbitals; a simple, general approach. J. Chem. Educ. *46*, 799.

— — (1970): Computing ligand field potentials and relative energies of *d*-orbitals; theory. J. Chem. Educ. *47*, 433.

Langs, D. A., Hare, C. R. (1967): The crystal structure of calcium copper acetate hexahydrate and its isomorph calcium cadmium acetate hexahydrate. Chem. Commun. 890.

La Mar, G. N. (1966): Molecular orbital calculations on bis(acetylacetonato)copper(II). Acta Chem. Scand. *20*, 1359.

Lever, A. B. P. (1968): The electronic spectra of tetragonal metal complexes: analysis and significance. Coord. Chem. Rev. *3*, 119.

Liehr, A. D. (1960): The reciprocation of electrostatic and electromagnetic forces in ligand field theory. J. Phys. Chem. *64*, 43.

Lohr, L. L. (1968): Molecular orbital study of vibronic interactions in the $CuCl_2$ molecule. Inorg. Chem. *7*, 2093.

— *Lipscomb, W. N.* (1963): An LCAO—MO study of static distortions in transition metal complexes. Inorg. Chem. *2*, 911.

Ludi, A., Feitknecht, W. (1963): Lichtabsorption und Struktur von Kristallverbindungen der Übergangsmetalle. II. Lichtabsorption und Struktur der Halogenide, des Hydroxides und der Hydroxidhalogenide von Kupfer. Helv. Chim. Acta. *46*, 2238.

Mabbs, F. E., Smail, W. R. (1970): Studies in eight-coordination. II. The e.s.r. of, and crystal field in, calcium copper acetate hexahydrate. J. Chem. Soc. (A) 1710.

McClure, D. S. (1961): Electronic structure of transition metal complex ions. Advances in the Chemistry of coordination compounds, p. 498; ed. *Kirshner, S.* New York: McMillan and Co.

Martin, D. S., Fenske, R. F., Ruedenberg, K. (1962): Ligand field theory of square planar platinum(II) complexes. Inorg. Chem. *1*, 441.

McGarvey, B. R. (1966): Electron spin resonance of transition metal complexes. Transition Metal Chem. *3*, 89.

McKenzie, E. D. (1962): A physico-chemical study of transition metal complexes of 1,10-phenanthroline and related ligands. Ph. D. Thesis, University of New South Wales.

Maki, A. H., McGarvey, B. R. (1958): Electron spin resonance of transition metal chelates. I. Copper bis-acetylacetonate. J. Chem. Phys. *29*, 31.
Mellor, J. W. (1923): A comprehensive treatise on inorganic and theoretical chemistry, Vol. 3, p. 173. London: Longmans, Green and Co.
Moffitt, W., Ballhausen, C. J. (1956): Quantum theory. Ann. Rev. Phys. Chem. *7*, 107.
Morosin, B. (1969): The crystal structures of copper tetrammine complexes. A. $Cu(NH_3)_4SO_4 \cdot H_2O$ and $Cu(NH_3)_4SeO_4$. Acta Cryst. *B 25*, 19.
— *Lawson, K.* (1964a): The configuration and electronic absorption spectra of tetrachloro-and tetrabromocuprate(II) ions. J. Mol. Sepctry. *12*, 98.
— — (1964b): The configuration and electronic spectra of tetrachloro- and tetrabromocuprate(II) ions. II. Additional comments and erratum. J. Mol. Spectry. *14*, 397.
Oelkrug, D. (1967): Optische Absorption and Struktur des CuF_6^{4-}. Z. Physik. Chem. (Frankfurt) *56*, 325.
— (1971): Absorption spectra and ligand field parameters of tetragonal 3*d*-transition metal fluorides. Struct. Bonding *9*, 1.
Offenhartz, P. O. D. (1969): On the calculation of 10 Dq. J. Am. Chem. Soc. *91*, 5699.
— (1970): On the calculation of 10 Dq. II. Approximations in the open-shell framework. J. Am. Chem. Soc. *92*, 2599.
Orgel, L. E. (1952): The effects of crystal fields on the properties of transition metal ions. J. Chem. Soc. 4756.
Palmer, R. A., Piper, T. S. (1966): 2,2'-bipyridine complexes. I. Polarized crystal spectra of tris (2,2'-bipyridine)copper(II),-nickel(II), -iron(II) and ruthenium(II). Inorg. Chem *5*, 864.
Pappalardo, R. (1961): Absorption spectra of Cu^{2+} in different crystal coordinations. J. Mol. Spectry. *6*, 554.
Piper, T. S., Belford, R. L. (1962): The ground state of copper acetylacetonate. Mol. Phys. *5*, 169.
— *Carlin, R. L.* (1960): Axial crystal fields in the ionic model. J. Chem. Phys. *33*, 1208.
Procter, I. M., Hathaway, B. J., Nicholls, P. (1968): Electronic properties and stereochemistry of the copper(II) ion. Part I. Bis-ethylenediamine complexes. J. Chem .Soc. (A) 1678.
— — *Billing, D. E., Dudley, R. J., Nicholls, P.* (1969): Electronic energy levels and stereochemistry of the cis-distorted octahedral complex nitratobis(2,2'-bipyridyl) copper(II) nitrate. J. Chem. Soc. (A) 1192.
Propach, V., Reinen, D. (1969): Über den Jahn-Teller-Effekt des Cu^{2+} Ions in oxidischen Festkörpern. II. Untersuchungen an Kristallgittern vom Trirutil-, Niobit- und Perowskittyp. Z. Anorg. Allgem. Chem. *369*, 278.
Raymond, K. N. (1970): Changes in geometry of five-coordinate transition metal complexes: structure and bonding. Proceedings of the XIIIth. International Conference on Coordination Chemistry, Vol. II, p. 94; ed. W. Kakolowicz, Polish Academy of Sciences.
Reimann, C. W., Kokoszca, G. F., Allen, G. (1965): Additional observations on the electronic spectrum of copper(II) acetate monohydrate. Inorg. Chem. *4*, 1082.
Reinen, D. (1968): Über den Jahn-Teller-Effekt des Cu^{2+}-Ions in oxidischen Festkörpern. I. In oktaedrischer Koordination. Z. Naturforsch. *23*. 521.
— (1970): Kationenverteilung zweiwertiger 3*d*-Ionen in oxidischen Spinell-, Granat- und anderen Strukturen. Struct. Bonding *7*, 114.
Richards, W. G., Horsley, J. A. (1970): Ab initio molecular orbital calculations for chemists. Oxford: Clarendon Press.

Richardson, J. W., Nieupoort, W. C., Powell, R. R., Edgell, W. F. (1962): Approximate radial functions for first-row transition metal atoms and ions I. Inner-shell $3d$ and $4s$ atomic orbitals. J. Chem. Phys. *36*, 1057.
Roothaan, C. C. J. (1960): Self-consistent field theory for open shells of electronic system. Rev. Mod. Phys. *32*, 179.
Roos, B. (1966): A semi-empirical open shell SCF—MO method for transition metal complexes. Acta Chem. Scand. *20*, 1673.
— (1967): A theoretical investigation of the electronic structure of the normal and excited states of copper dimethylglyoxime and its adducts with water and amines. Acta Chem. Scand. *21*, 1855.
Ros, P., Schuit, G. C. A. (1966): Molecular orbital calculations on copper-chloride complexes. Theoret. Chim. Acta *4*, 1.
Schaffer, C. E. (1966): A ligand field approach to orthoaxial complexes. Perturbation treatment using zero-order functions with octahedral basis. Theoret. Chim. Acta *4*, 166.
— *Jørgensen, C. K.* (1965): The angular overlap model, an attempt to revive the ligand field approaches. Mol. Phys. *9*, 401.
Schmitz-Dumont, O., Fendel, H. (1965): Farbe und Konstitution bei anorganischen Feststoffen. 9 Mitt. Die Lichtabsorption des zweiwertigen Kupfers in System $Cu_xMg_{1-x}O$. Monatsh. Chem. *96*, 496.
— *Grimm, D.* (1967): Farbe und Konstitution bei anorganischen Feststoffen. XVI. Die Lichtabsorption des zweiwertigen Kupfers in binären und ternären Fluoriden. Z. Anorg. Allgem. Chem. *355*, 280.
— *Kasper, H.* (1965): Farbe und Konstitution bei anorganischen Feststoffen. 10. Mitt. Die Lichtabsorption des zweiwertigen Kupfers in Kupferindiumoxid und analogen Verbindungen mit Yttrium und einigen Lanthaniden. Monatsh. Chem. *96*, 505.
von Schnering, H. G. (1967): Chemische und magnetische Eigenschaften der Barium-fluorometallate(II) Ba_2MF_6 mit $M = Zn$, Cu, Ni, Co, Fe. Z. Anorg. Allgem. Chem. *353*, 1.
Shupack, S. I. (1967): Molecular orbital theory for metal ammine complexes. I. $Cu(NH_3)_4^{2+}$, $Cu(NH_3)_6^{2+}$ and $Ni(NH_3)_6^{2+}$. Inorg. Chim. Acta *1*, 435.
Smith, D. W. (1966): Polarized crystal spectrum of pentaamminecupric ammonium perchlorate. Inorg. Chem. *5*, 2236.
— (1968): Single crystal spectrophotometry of inorganic complexes. D. phil. Thesis, Oxford University.
— (1969a): Ligand field splittings in noncubic complexes. Part I. Copper-nitrogen systems. J. Chem. Soc. (A) 1708.
— (1969b): Ligand field splittings in noncubic complexes. Part II. Copper(II)-chlorine systems. J. Chem. Soc. (A) 2529.
— (1969c): Dependence of 10 Dq on the metal-ligand distance in NiO. J. Chem. Phys. *50*, 2784.
— (1970a): Relationship between e.s.r. g-values and covalent bonding in tetragonal copper(II) compounds. J. Chem. Soc. (A) 3108.
— (1970b): Overlap considerations relevant to the geometry of copper(II) complexes. J. Chem. Soc. (A) 1498.
— (1970c): De structura electronica et stereochimica ionis $Cu(NO_2)_6^{4-}$. Theoret. Chim. Acta. *17*, 98.
— (1970d): Ligand field splittings in noncubic complexes. Part III. Copper(II)-Oxygen systems. J. Chem. Soc. (A) 176.
— (1970e): Ligand field splittings in noncubic complexes Part IV. The tetrachlorocuprate(II) ion in distorted tetrahedral symmetry. J. Chem. Soc. (A) 2900.

— (1970f): On the assignment of the electronic spectrum of gaseous $CuCl_2$. Chem. Phys. Letters. *6*, 83.
— (1971a): Ligand field splittings in noncubic complexes. Part VI. Comparison of the crystal field and angular overlap models for eight-coordinate D_{2d} systems. J. Chem. Soc. (A) 1024.
— (1971b): Ligand field splittings in noncubic complexes. Part V. Analysis of the *d—d* spectra of the gaseous dichlorides of the first transition series. Inorg. Chim. Acta. *5*, 231.
— (1971c): Ligand field splittings in noncubic complexes. Part VII. Calcium copper (II) acetate hexahydrate. J. Chem. Soc. (A) 1209.
Sugano, S., Shulman, R. G. (1963): Covalency effects in $KNiF_3$, III. Theoretical studies. Phys. Rev. *130*, 517.
Tomlinson, A. A. G., Hathaway, B. J. (1968a): Electronic properties and stereochemistry of the copper(II) ion. Part III. Some pentammine complexes. J. Chem. Soc. (A) 1905.
— — (1968b): Electronic properties and stereochemistry of the copper(II) ion. Part IV. Some diammine complexes. J. Chem. Soc. (A) 2578.
— — (1968c): Electronic properties and stereochemistry of the copper(II) ion. Part II. Monoamine adducts of bisethylenediaminecopper(II) complexes. J. Chem. Soc. (A) 1685.
— — *Billing, D. E., Nicholls, P.* (1969): Electronic properties and stereochemistry of the copper(II) ion. Part V. Tetrammine complexes. J. Chem. Soc. (A) 65.
Tonnet, M. C., Yamada, S., Ross, I. G. (1964): The metal-metal bond in binuclear copper(II) acetate. Part III. Electronic spectrum and *g*-factors. Trans. Faraday Soc. *60*, 840.
Watson, R. E. (1960): Iron series Hartree-Fock calcultions. Phys. Rev. *118*, 1036.
— *Freeman, A. J.* (1964): Covalency in crystal field theory — $KNiF_3$. Phys. Rev. *134*, A 1526.
Weakliem, H. A. (1962): Optical spectra of Ni^{2+}, Co^{2+} and Cu^{2+} in tetrahedral sites in crystals. J. Chem. Phys. *36*, 2117.
Willett, R. D., Liles, O. L. (1967): The electronic absorption spectrum and bonding in the $Cu_2Cl_6^{2-}$ dimer. Inorg. Chem. *6*, 1666.
— — *Michelson, C.* (1967): The electronic absorption spectra of monomeric copper (II) chloride species and the ESR spectrum of the a square planar $CuCl_4^{2-}$ ion. Inorg. Chem. *6*, 1885.
Yamada, S., Tsuchida, R. (1953): Direct interaction between metals atoms in the crystals of bis(dimethylglyoxime)nickel(II) and platinum(II). J. Am. Chem. Soc. *75*, 6351.
— — (1954): Spectrochemical study of microscopic crystals. VI. Light absorption of non-planar quadricoordinate complexes. Bull. Chem. Soc. Japan *27*, 436.
— — (1956a): Spectrochemical study of microscopic crystals. X. The absorption spectra and configuration of bis-ethylenediamine-copper(II) complexes. Bull. Chem. Soc. Japan *29*, 289.
— — (1956b): Spectrochemical study of microscopic crystals. XII. Polarized absorption spectra of cupric complexes with tropolone and related compounds. Bull. Chem. Soc. Japan *29*, 694.
— *Nakamura, H., Tsuchida, R.* (1957): Spectrochemical study of microscopic crystals. XVI. Spectra of cupric formate, acetate and propionate. Bull. Chem. Soc. Japan *30*, 953.
— — — (1958): Spectrochemical study of microscopic crystals. XVII. Structures of cupric salts with monocarboxylic fatty acids. Bull. Chem. Soc. Japan *31*, 303

— *Nishikawa, H., Tsuchida, R.* (1960): Spectrochemical study of microscopic crystals. XXII. The structures of cupric mono-, di- and trichloroacetate. Bull. Chem. Soc. Japan *33*, 1278.
— — *Miki, S.* (1964): Spectrochemical study of microscopic crystals. XXIV. The structure and light absorption of copper(II) complexes of benzoic acid and substituted benzoic acids. Bull. Chem. Soc. Japan *37*, 576.
— *Miki*, S. (1963): Bis-(succinimidato)copper(II) complexes with pyridine, picolines and lutidines. Bull. Chem. Soc. Japan *37*, 1044.
— — (1964): Spectrochemical study of microscopic crystals. XXIII. The structure and light absorption of the red cupric complexes with succinimide. Bull. Chem. Soc. Japan *36*, 680.
Yamatera, H. (1958): On the absorption spectrum of hexamminecobalt(III) and related complexes. II. Theoretical study of the shifting and splitting of the first and second bands due to the substitution of ligands. Bull. Chem. Soc. Japan *31*, 95.
Zaslow, B., Ferguson, G. L. (1967): Crystal structure of bis(2-ammonioethyl)-ammonium monochloride tetrachlorocuprate(II). Chem. Commun. 822.

Received August 30, 1971.

Phenomenological Approach to Cation-Solvent Interactions

U. Mayer and V. Gutmann

Institut für Anorganische Chemie, Technische Hochschule, Wien, Austria

Table of Contents

1. Characterization of Ion Solvation by Thermodynamic Quantities 113

2. "Donicity" of Donor Solvents, Definition and Relation to Solvating Properties Towards Cations 116
   a) Steric Factors 116
   b) Dependence of Relative Donor Strengths on the Nature of the EPA .. 116
   c) Type of Bonding between EPD and EPA 118

3. Polarographic Reduction of Solvated Metal Cations in EPD Solvents 118

4. Complex Formation of Co^{2+} with Halide and Pseudohalide Ions 126

5. Ion-Pair Equilibria 129
   a) General 129
   b) UV Spectroscopy 129
   c) IR Investigations 130
   d) NMR Investigations 132

6. Quantum-Mechanical Considerations 133

7. Functional Approach to Chemical Interactions 134

8. Conclusion 137

9. References 138

1. Characterization of Ion Solvation by Thermodynamic Quantities

The solvation of ions in different solvents is described most satisfactorily in terms of single-ion free energies, enthalpies and entropies of solvation. These quantities are related by the well-known Gibbs-Helmholtz equation.

$$\Delta G_{sv} = \Delta H_{sv} - T\Delta S_{sv}$$

Thermodynamic considerations are applicable only to electro-neutral combinations of cations and anions. Single-ion values can be obtained

(a) by extrapolation procedures,

(b) by making use of certain assumptions concerning the change in enthalpy or free energy on transfer of certain ions from one solvent to another and

(c) by theoretical calculations.

a) Extrapolation procedures for the evaluation of single-ion enthalpies of solvation are all based on the Born equation or modifications thereof and consequently require an exact knowledge of ionic radii (*1, 2*). Various sets of ionic radii have been proposed by different authors, which naturally produce different values of single-ion solvation enthalpies (*2*).

The values given by *Pauling* (*3*) and *Goldschmidt* (*4*) have recently been criticized by *Gourrary* and *Adrian* (*5a*) who proposed a new set of data based on the electron-density diagram of solid NaCl.

Absolute values of free energies of hydration were made accessible by the work of *Randles* (*1*) who determined the absolute free energy of hydration of the K^+ ion from measurements of Volta potentials. Since the absolute entropy of H^+ is fairly well established (*2*), absolute values of enthalpy, free energy and entropy can be calculated for the hydration of ions. No comparable measurements have been carried out in non-aqueous solvents. Application of extrapolation procedures to nonaqueous solvents is frequently restricted by the low solubility of salts in these solvents and lack of experimental data.

b) Values of single-ion enthalpies and free energies may be obtained on the assumption that changes in enthalpy or free energy on transfer from one solvent to another will be equal for certain pairs of bulky ions with highly dispersed charges or certain pairs consisting of an ion and a neutral molecule, for example $[AsPh_4]^+$ $[BPh_4]^-$; $[NBu_4]^+$ $[BBu_4]^-$; I_2, I_3^- and $[BPh_4]^-$, CPh_4· (*5b—8*). Although this approach has provided valuable information about the solvating properties of protic and aprotic solvents (*9, 10*), it is qualitative in nature and data are available for only a very limited number of solvents (*11*).

c) Theoretical calculations of single-ion enthalpies or free energies are usually based on electrostatic models. Quite recently efforts have been made to calculate hydration energies by the use of quantum-mechanical models (see Sect. 6). The simplest approach is provided by the Born equation (*12*) which relates the free enthalpy of solvation ΔG_{sv} to the radius r of the ion and the dielectric constant ε of the solvent:

$$\Delta G_{sv} = -\frac{N(ze)^2}{2r}\left(1-\frac{1}{\varepsilon}\right) \quad (1)$$

The corresponding enthalpy change (*13*) is given by Eq. (2).

$$\Delta H_{sv} = -\frac{N(ze)^2}{2r}\left\{1 - \frac{1}{\varepsilon} - \frac{T}{\varepsilon}\left(\frac{\delta \ln \varepsilon}{\delta T}\right)_P\right\} \tag{2}$$

It has long been recognized that the Born equation cannot account for differences in the solvation energy of ions of the same size. Various modifications of the Born equation have been proposed in order to account for the effect of partial dielectric saturation in the near vicinity of ions and or to allow for differences in the size of ions in solution and in a crystal lattice (*14—18*). The modified equations cannot explain differences in the solvating power of different aprotic solvents as will be shown later.

Modern theories of ion solvation are based on models (*19—21*) which take into account the interactions between the ions and solvent dipoles and multipoles as well as the energetics of structural changes imposed on the solvent molecules under the influence of the ion. *Morf* and *Simon* (*22*) proposed an electrostatic model for calculating free energies of hydration, with which very good agreement between calculated and experimental data was obtained for 27 cations. It is, however, doubtful whether this model can be applied with equal success to nonaqueous solvents. Charge distribution and steric requirements are much more difficult to evaluate for organic solvent molecules than for simple molecules such as water or ammonia. In fact, formula (3)

$$E_{pot} = -\frac{Q \cdot \mu}{r^2} \tag{3}$$

which is frequently used to describe ion dipole interactions (Q = charge of the ion) is valid only if the distance r is large compared to the distance ℓ (Fig. 1a) between the centres of positive and negative charges of the dipole solvent molecule S_1.

Potential electrostatic energies of ion-dipole interactions may differ considerably for molecules of the same dipole moment but of different

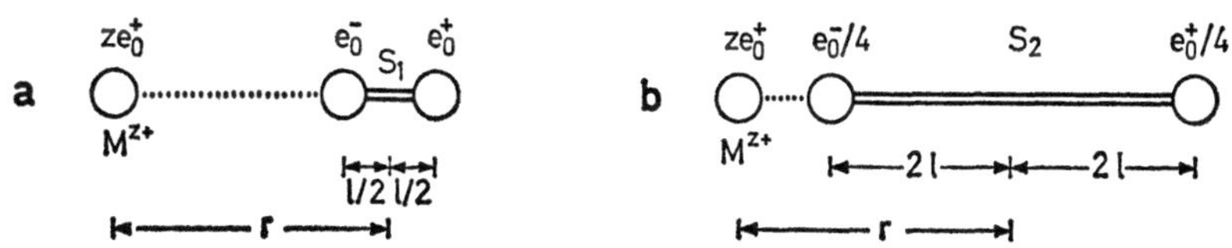

Fig. 1. Ion dipole interactions, the dipoles having the same dipole moment μ but different charge distribution.

a) $\mu S_1 = e_0 \cdot \ell$, $E_{pot}(S_1) = -z.e_0^2/(r - \ell/2) + z.e_0^2/(r + \ell/2)$.

b) $\mu S_2 = (e_0/4)\, 4\ell = e_0 \cdot \ell$, $E_{pot}(S_2) = -z.(e_0^2/4)/(r - 2\ell) + z.(e_0^2/4)/(r + 2\ell)$

geometrical arrangement of charges (Fig. 1). If one assumes, for example, $r = 3\ell$, one obtains $E_{pot}(S_1) = 0.114 . z.e_o^2/\ell$ and $E_{pot}(S_2) = 0.200 . z.e_o^2/\ell$, i.e. differences of nearly 100% in potential electrostatic energies, although both molecules have the same dipole moment μ.

In view of the lack of thermodynamic data, an empirical approach to cation solvation is presented in the following sections.

2. "Donicity" of Donor Solvents, Definition and Relation to Solvating Properties Towards Cations

The ability of a solvent to function as an electron-pair donor (EPD) towards antimony(V) chloride is represented by the ΔH value of the reaction of the EPD with $SbCl_5$ in a dilute solution of 1,2-dichloroethane (*23, 24*):

$$\text{EPD} + SbCl_5 \longrightarrow \text{EPD.}SbCl_5; \quad -\Delta H_{EPD.SbCl_5} \equiv \text{DN}$$

The negative ΔH value of this reaction is called "donor number" or "donicity" (*23*). Donicities (DN) and dielectric constants (ε) for various EPD solvents are listed in Table 1. Systematic studies of interactions between EPD molecules and various electron-pair acceptors (EPA) reveal that the observed EPD strengths towards $SbCl_5$ follow essentially the same order as towards a considerable number of other EPA molecules, such as iodine, phenol, $SbCl_3$, $Sn(CH_3)_3Cl$ and $VO(acac)_2$ (*24*). Since not only neutral molecules but also cations are capable of acting as EPA's (*97*), one might suppose there to be a relationship between the donicity of the EPD and solvation enthalpies for a given cation, as long as the following factors do not come into play:

a) Steric Factors

Only 1:1 adducts are formed by the reaction of $SbCl_5$ with EPD molecules. Cations in solution, however, have high coordination numbers so that the magnitude of solvation enthalpies also depends on the coordination number and steric factors. For example, Co^{2+} is coordinated in most cases by six solvent molecules (octahedral coordination) whereas tetrahedral coordination is observed towards bulky EPD molecules such as HMPA and $(CH_3)_3NO$ (*25–28*).

b) Dependence of Relative Donor Strengths on the Nature of the EPA

A distinction is made between hard and soft EPD and EPA entities (*29*). A hard EPA will react preferentially with a hard EPD and a soft EPA with

Table 1. *Donicities DN and dielectric constants ε of various EPD-solvents*

Solvent	Symbol	DN	ε
1,2-Dichloroethane		—	10.1
Sulphuryl chloride		0.1	10.0
benzene		0.1	2.3
Thionyl chloride		0.4	9.2
Acetyl chloride		0.7	15.8
Tetrachloroethylene carbonate		0.8	9.2
Benzoyl fluoride	(BF)	2.0	22.7
Benzoyl chloride		2.3	23.0
Nitromethane	(NM)	2.7	35.9
Dichloroethylene carbonate	(DEC)	3.2	31.6
Nitrobenzene	(NB)	4.4	34.8
Acetic anhydride		10.5	20.7
Phosphorus oxychloride		11.7	14.0
Benzonitrile	(BN)	11.9	25.2
Selenium oxychloride		12.2	46.0
Acetonitrile	(AN)	14.1	38.0
Sulpholane		14.8	42.0
Propanediol-1,2-carbonate	(PDC)	15.1	69.0
Benzyl cyanide		15.1	18.4
Ethylene sulphite	(ES)	15.3	41.0
iso-Butyronitrile		15.4	20.4
Propionitrile		16.1	27.7
Ethylene carbonate	(EC)	16.4	89.1
Phenylphosphonic difluoride		16.4	27.9
Methylacetate		16.5	6.7
n-Butyronitrile		16.6	20.3
Acetone		17.0	20.7
Ethyl acetate		17.1	6.0
Water		18.0	81.0
Phenylphosphonic dichloride		18.5	26.0
Diethyl ether		19.2	4.3
Tetrahydrofurane		20.0	7.6
Diphenylphosphonic chloride		22.4	—
Trimethyl phosphate	(TMP)	23.0	20.6
Tributyl phosphate	(TBP)	23.7	6.8
Dimethylformamide	(DMF)	26.6	36.1
N-methyl-ε-caprolactame	(NMC)	27.1	—
N-methyl-2-pyrrolidone	(NMP)	27.3	—
N,N-Dimethylacetamide	(DMA)	27.8	38.9
Dimethyl sulphoxide	(DMSO)	29.8	45.0
N,N-Diethylformamide		30.9	—
N,N-Diethylacetamide		32.2	—
Pyridine	(py)	33.1	12.3
Hexamethylphosphoricamide	(HMPA)	38.8	30.0

a soft EPD (*29*). Thus, if relative donor strengths are determined for both hard and soft EPD molecules towards both hard and soft EPA's their relative donor strengths may be reversed. For example, an EPD sulfide (soft) will coordinate more strongly towards iodine (soft) than the corresponding oxide, which is hard (*30*). The reverse is true for the hard EPA, phenol (*31*). Thus only for EPD molecules of the same type essentially the same order of EPD strengths is expected towards both hard and soft EPA molecules (*24*). In fact, most of the commonly used EPD solvents are hard, containing either oxygen or nitrogen as the coordinating atom.

c) Type of Bonding between EPD and EPA

Adduct formation of $SbCl_5$ with an EPD involves the formation of a covalent bond. The order of relative EPD strengths as measured towards $SbCl_5$ may be expected to be valid only for EPD-EPA interactions where coordinate bonds of similar type are formed.

Interactions between alkali metal ions and solvent molecules are considered to be essentially electrostatic in nature. Although covalent contributions to the solvate bonds cannot be completely neglected (see Sect. 5d and 6), one would not expect the donicities to be generally applicable to the solvation of alkali metal ions. This problem will be discussed in detail in the following section.

Covalent contributions should be considerably higher for solvation of soft univalent cations such as Tl^+ or Ag^+. The same applies to solvation of transition metal ions, where even stronger covalent bonds are formed. Hence, one would expect there to be a relationship between solvent donicities and solvation enthalpies of a given metal ion.

In the following sections the results of various experimental investigations will be presented.

3. Polarographic Reduction of Solvated Metal Cations in EPD Solvents

The relationship between the free energy of solvation of a cation M^{z+} and its polarographic half-wave potential may be derived from the energy cycle shown in Fig. 2. ΔG_{red} denotes the free energy of reduction of the solvated metal ion to the metal amalgam M_{am}, ΔG_{sv} the free energy of solvation, ΔG_{subl} the free energy of sublimation, ΔG_{am} the free energy of amalgamation and ΣI the sum of ionization potentials. ΔG_{red} is related to the standard electrode potential E by the equation:

$$\Delta G_{red} = -z.F.E \qquad (4)$$

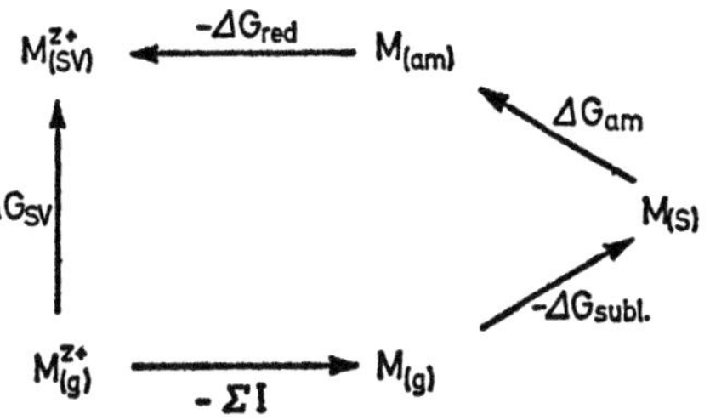

Fig. 2. Energy cycle for the polarographic reduction of a solvated metal ion M^{z+}

Where reduction is reversible, E can be considered equal to the polarographic half-wave potential $E_{1/2}$: The exact relationship between $E_{1/2}$ and E is given by: $E_{1/2} = E - \frac{RT}{zF} \sqrt{\frac{D_{ox}}{D_{red}}}$ with diffusion coefficients D extrapolated to zero ionic strength (*32*). It is necessary to make allowance for differences in diffusion coefficients only if exact evaluations are required.

$$E \approx E_{1/2} \tag{5}$$

For the reduction of a given cation in different solvents, the terms ΣI, ΔG_{am} and ΔG_{subl} remain constant, hence a linear relation is expected between ΔG_{red} or $E_{1/2}$ and ΔG_{sv}. Consequently, a similar relationship should exist between $E_{1/2}$ and DN, provided that the entropic contributions are constant and that DN is indeed a fair measure of the solvating power of EPD solvents towards cations. With increasing donicity of the solvent, half-wave potentials are shifted towards more negative values because more energy is required to break the solvate bonds in the course of the reduction process (*24, 33—35*). Figs 3 and 4 show (*36—38*) that for the ions Co^{2+}, Ni^{2+}, Cd^{2+}, Mn^{2+} and Zn^{2+} half-wave potentials are approximately proportional to solvent donicities. The same applies to a number of redox reactions such as the reduction of Eu^{3+} to Eu^{2+}, Sm^{3+} to Sm^{2+} and Yb^{3+} to Yb^{2+} (Fig. 5) (*36*).

All these ions are known to form more or less covalent solvate bonds. For Tl^{+}, Na^{+}, K^{+}, Rb^{+} (Fig. 6) (*36, 37, 39*) the slopes of the $E_{1/2}$-DN curves are much smaller than those in figs 3—5 because solvation enthalpies and energies are smaller for monovalent cations than for divalent or higher-valent cations.

The relationship between $E_{1/2}$ and DN is less apparent for alkali metal ions than for the ions in Figs 3—5; this may be explained by the following considerations;

Apart from steric effects no exactly linear relationship can be expected between $E_{1/2}$ and DN because $E_{1/2}$ is a free-energy quantity, but DN

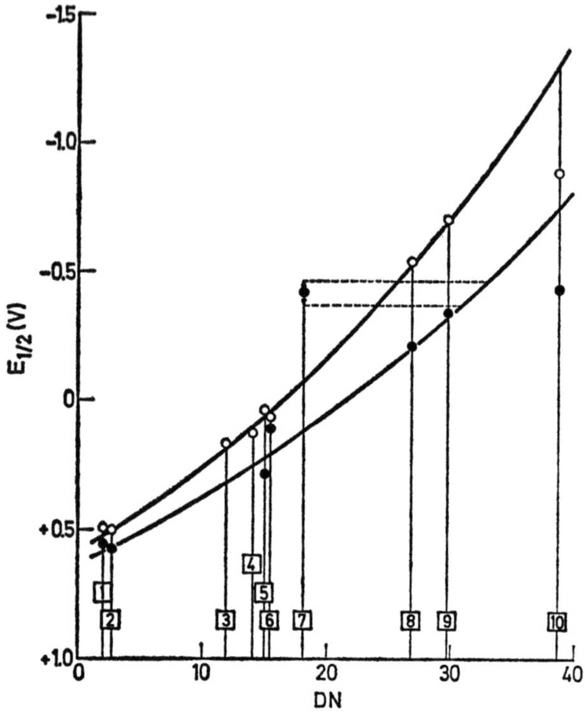

Fig. 3. Relationship between half-wave potential $E_{1/2}$ and solvent donicity DN for the polarographic reduction $Co^{2+} \rightarrow Co^0$ (—○—), $Ni^{2+} \rightarrow Ni^0$ (—●—). 1 BF, 2 NM, 3 BN, 4 AN, 5 PDC, 6 ES, 7 H_2O, 8 DMF, 9 DMSO, 10 HMPA

is an enthalpy quantity. If differences in solvation enthalpies for different solvents are large, differences in entropy terms $T\Delta S$ can be neglected and a nearly linear relationship between $E_{1/2}$ and DN may be observed, provided that ΔH_{sv} is proportional to DN. This is more likely to be the case for divalent or higher-valent cations than for alkali metal ions since solvation enthalpies increase strongly with increasing positive oxidation state of the cations.

As pointed out in the preceeding section, solvent donicities can probably be used as a measure for the solvating power only where cations with a marked tendency to covalent bond formation are considered. Recent investigations (see Sect. 5) and quantum-mechanical calculations (see Sect. 6) support the view that there are certain covalent contributions in solvate bonding, even between alkali metal ions and solvent molecules, although the electrostatic contributions are probably dominant. Therefore a *general* relationship between $E_{1/2}$ and DN would not be expected. This argument is supported by the fact that the $E_{1/2}$-DN

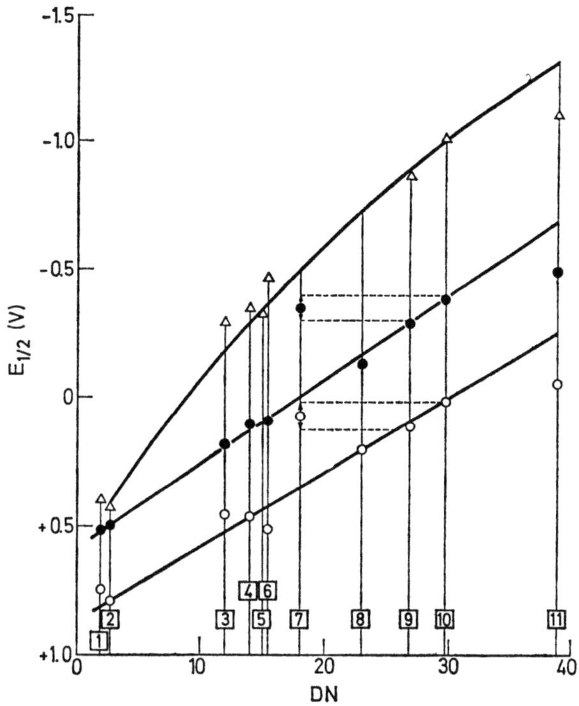

Fig. 4. Relationship between half-wave potential $E_{1/2}$ and solvent donicity DN for the polarographie reduction $Zn^{2+} \rightarrow Zn^0$ (—●—), $Mn^{2+} \rightarrow Mn^0$ (—Δ—), $Cd^{2+} \rightarrow Cd^0$ (—O—). 1 BF, 2 NM, 3 BN, 4 AN, 5 PDC, 6 ES, 7 H_2O, 8 TMP, 9 DMF, 10 DMSO, 11 HMPA

curve for Tl^+ (where solvate bonds are considerably more covalent) is a smooth curve (Fig. 6), although solvation enthalpies are in the same range as for alkali metal ions. Additional support for this conclusion comes from recent studies on the reduction of Ag^+ in various aprotic solvents (*39*). The following elementary electrostatic (and consequently over-simplified) model for the solvation of alkali metal ions may account for these results.

In Fig. 7, S denotes a polar molecule whose charge distribution in the polarizing field of the cation M^+ may be approximated as follows: Z_1, Z_2, Z_3, Z_4 denote the centres of positive and negative charges, respectively. The potential electrostatic energy between M^+ and S is determined by the arithmetic sum of the interactions between Z_1, Z_2 (attraction) and Z_1, $Z_3 + Z_1$, Z_4 (repulsion). If distance b is large compared with distance a, the potential electrostatic energy between M^+ and S may as a first approximation be described by interaction of Z_1,

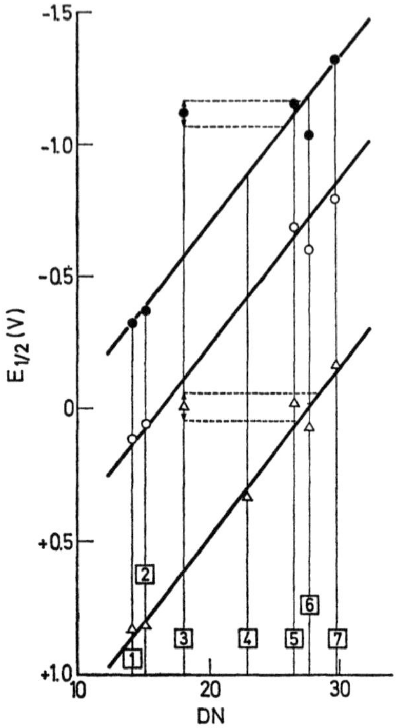

Fig. 5. Relationship between half-wave potential $E_{1/2}$ and solvent donicity DN for redox reactions: $Sm^{3+} \rightarrow Sm^{2+}$ (—●—), $Yb^{3+} \rightarrow Yb^{2+}$ (—○—), $Eu^{3+} \rightarrow Eu^{2+}$ (—Δ—). 1 AN, 2 PDC, 3 H_2O, 4 TMP, 5 DMF, 6 DMA, 7 DMSO

Z_2 only. This approximation may indeed be permissible for large organic solvent molecules (e.g. THF) where the charges Z_3 and Z_4 will be dispersed over a large number of atoms. The second assumption is that (at least for EPD molecules having the same EPD atom, e.g. an oyxgen atom) increasing donicity (i.e. increasing availability of the electron pair) is parallelled by an increasing accumulation of negative partial charge at the EPD atom. This model predicts an increase in solvating power with increasing donicity, provided that the solvents considered are sufficiently different in EPD strengths, but probably no *general* relationship (as observed for transition metal ions) can be expected because both assumptions are to be regarded as rough approximations only. In fact, polarographic reduction of alkali metal ions is definitely easier in solvents of medium donicity, such as BN, PDC, AN and ES, than in strong EPD solvents such as TMP, DMSO, DMF and HMPA, whereas half-wave potentials are very little differentiated within both groups of solvents.

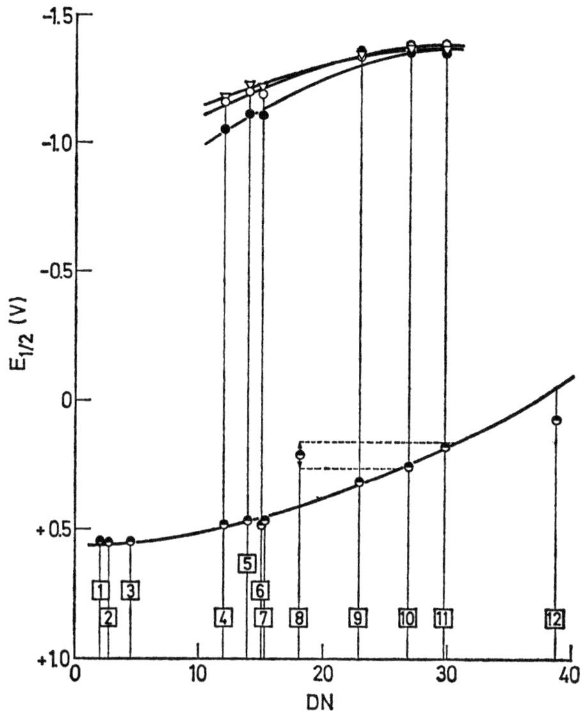

Fig. 6. Relationship between half-wave potential $E_{1/2}$ and solvent donicity DN for the polarographic reduction $Tl^+ \rightarrow Tl^\circ$ (—◓—), $Na^+ \rightarrow Na^\circ$ (—●—), $K^+ \rightarrow K^\circ$ (—○—), $Rb^+ \rightarrow Rb^\circ$ (—∇—). 1 BF, 2 NM, 3 NB, 4 BN, 5 AN, 6 PDC, 7 ES, 8 H_2O, 9 TMP, 10 DMF, 11 DMSO, 12 HMPA

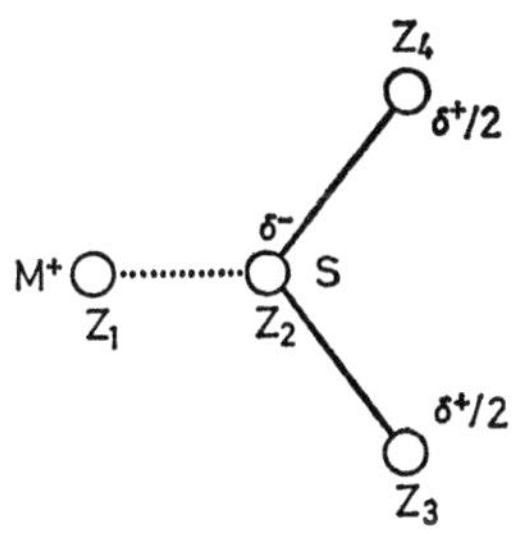

Fig. 7. Electrostatic model for the interpretation of alkali metal ion — solvent dipole interactions b: distances Z_1-Z_4 and Z_1-Z_3; a: distance Z_1-Z_2

Additional support for this model comes from UV, IR and NMR measurements of ion-pair equilibria (see Sect. 5).

It is interesting to compare half-wave potentials in nonaqueous solvents with half-wave potentials in water. Despite the limited accuracy of these values[1]), is is obvious that half-wave potentials in water are much more negative than predicted on the basis of Figs 3—6 using the donicity of water DN $\approx$ 18 as derived from spectrophotometric measurements in 1,2-dichloroethane (*40, 41*). This "anomalous" behaviour may be explained in the following way:

1. Water is a highly structured liquid, therefore the solvation entropies of cations are less negative than in aprotic solvents (*43*). Desolvation of cations in the course of a polarographic reduction is therefore entropically less favourable than in aprotic solvents and this will result in a shift towards more negative potential values ("entropic effect"). Unfortunately, very little is known about solvation entropies of cations (and anions) in aprotic solvents. *Criss* (*101*) reports a value of -5.4 e.u. for the partial molal entropy of the K^+ — ion from which a value $\Delta\Delta S_{sv} \approx 25$ e.u. is calculated for the difference in solvation entropy of K^+ in water and DMF. If a similar value is accepted for Tl^+ and if it is assumed that solvation entropies for different aprotic solvents do not differ very much, a shift in $E_{1/2}$ of -0.32 volt would be predicted. The shift actually observed (Fig. 6) is smaller, namely ~ -0.2 volt. Larger shifts are observed for transition metal ions (Figs. 3—5) for which differences in solvation entropies will probably be larger than the value calculated for Tl^+. It may be concluded from this result that the shifts in $E_{1/2}$ observed for water are essentially entropic in nature.

2. Consider a hydrated metal ion (for instance $Co(H_2O)_6{}^{2+}$) which is dissolved in a weakly coordinating medium such as 1,2-dichloroethane on the one hand, and in pure water on the other hand. In the latter case, water molecules of the first hydration shell will interact via hydrogen bonds with surrounding water molecules and this will tend to increase the EPD properties of the water molecules immediately bonded to the cation ("enthalpic effect"). This means that water in the pure liquid state will probably behave as a considerably stronger EPD as compared with its EPD properties exhibited in weakly coordinating media such as 1,2-dichloroethane[2]). If it is *tentatively* assumed that the negative shifts

---

[1]) In order to eliminate the influence of liquid-liquid junction potentials, half-wave potentials in figs. 3—6 are referred to bisbiphenyl-chromium(I)-iodide (*42*) (BBCr(I)). Since this compound is insoluble in water, $E_{1/2}$ values for water cannot be directly determined. Estimates are based on the observation that $E_{1/2}$ values for BBCr(I) versus the saturated calomel electrode (SCE) vary to only a small extent. A value of $E_{1/2}$ for BBCr(I) versus SCE of 0.68 ± 0.05 volt is proposed.

[2]) The low donicity value of water DN $\sim$ 18 as obtained from spectro-photometric measurements in 1,2-dichloroethane is supported by calorimetric measurements from which a donicity value of DN $= 19.1$ was obtained for methanol (*44*). The somewhat higher donicity of methanol can be ascribed to the +I-effect of the methyl group.

observed for water are essentially due to the "enthalpic effect", the "donicity values" of water as a measure of its EPD properties in the pure liquid state can be obtained by the simple graphic procedure shown in Figs 3,5 and 6. A mean value "DN" $=29\pm2$ is obtained from independent measurements on Tl^+, Cd^{2+}, Ni^{2+}, Sm^{3+} and Yb^{3+}. It is interesting that this value is close to the "donicity" of water in the liquid state of ≈ 33, as predicted from *Popov*'s δNa^{23}– DN relationship (see Sect. 5). Since chemical shifts for Na^{23} are independent of differences in solvation entropies, one might conclude from this result that shifts in $E_{1/2}$ are rather a consequence of the "enthalpic effect".

Summarizing the above considerations, it appears most likely that the negative shifts observed for water are due to both an entropic and an enthalpic effect. Indeed, both effects must be interrelated because the postulated enthalpic effect is conceivable only for highly structured solvents. At present it is not possible to evaluate the individual contributions of both effects because too little is known about the solvation entropies of cations in different solvents. A similar behaviour as for water is to be expected for other highly associated liquids such as formic acid, acetic acid, methanol, formic amide etc. We therefore propose to introduce (at present only in a qualitative way) the term *"bulk donicity"* DN^B as a measure of the EPD properties of associated donor solvents in the pure liquid state. The "bulk donicity" is to be distinguished from the corresponding donicity value DN, defined in section 2, as a measure of the EPD properties exhibited by the donor in dilute solutions of weakly coordinating solvents.

In contrast to water, $E_{1/2}$ in the solvent HMPA is found at more positive potential values (*45*) than would be expected from a consideration of its DN $=38.8$. This is definitely due to steric hindrance. Only four HMPA molecules instead of six are coordinated to Co^{2+}, Zn^{2+} and Ni^{2+} (*25*). Thus, only four instead of six solvate bonds have to be broken in the reduction process. Broad-line NMR measurements indicate that steric hindrance prevails even in the tetrasolvate of Co^{2+} (*27*).

Deviations from the $E_{1/2}$-DN curves may occur with EPD molecules which can coordinate in different ways. For instance, nitrobenzene coordinates towards $SbCl_5$ via an oxygen atom of the nitro group, while towards a soft EPA cation coordination via the aromatic ring appears possible. In the latter case $E_{1/2}$ values would be determined not by the donicity (*97*) (measured towards $SbCl_5$) but by the π-donor properties of nitrobenzene. Studies of the reduction of Ag^+ at the rotating platinum electrode in various EPD solvents indicate that nitrobenzene behaves as a stronger EPD than expected on basis of its donicity (*39*). In fact, Ag^+ forms complexes with various aromatic compounds (*46*). Specific EPD—EPA interactions were observed in the reduction of Cu^+, Ag^+

and — to a lesser extent of Hg^{2+} — in nitrile solvents (*39*). In particular, reduction of Cu^+ to Cu^0 in the solvents acetonitrile and benzonitrile is more difficult than would be expected according to the donicity values (*39*). The strong coordinating properties of nitrile solvents towards Cu^+ and Ag^+ have already been recognized by *Coetzee et al.* (*47*). Deviations for Ag^+ are somewhat less pronounced than for Cu^+ and quite small for Hg^{2+} (*39*). For all other solvents (except the system Ag^+-nitrobenzene) there is fairly good agreement between $E_{1/2}$ and DN (*39*).

Finally, it must be noted that standard electrode potential and half-wave potential are nearly identical only for reversible reduction processes (*32*). This is true for the alkali metal ions, Tl^+ and in most cases for Zn^{2+} and Cd^{2+}. Similar trends are observed for both reversible and irreversible reductions (Figs. 3–6) and this may justify the conclusion that partial irreversibility results in most cases in a shift of curves along the $E_{1/2}$ axis without any significant change in the general pattern.

4. Complex Formation of Co^{2+} with Halide and Pseudohalide Ions

The results of systematic investigations are available on the formation of tetrahalogeno and pseudohalogeno complexes of Co^{2+} in different EPD solvents (*24*). No ΔG values are available for the reactions

$$Co^{2+}_{sv} + 4X^- \longrightarrow [CoX_4]^{2-}$$

therefore the molar ratios $v = c_{X^-}/c_{Co^{2+}}$ required for nearly quantitative formation of $[CoX_4]^{2-}$ were estimated. Data[3]) are given in Table 2 (*14, 33–35, 43*). A qualitative relationship exists between solvent donicity and the extent of complex formation, except with water and DMA as solvents. Complex formation is nearly quantitative for all ligands in the weak EPD solvent nitromethane, higher ligand concentrations are required in solvents of medium donicity (AN, PDC and acetone) and still higher concentrations in strong EPD solvents such as DMSO or HMPA. The EPD properties of halide and pseudohalide ions were estimated from Table 2 and found to increase in the sequence $I^- < Br^- < Cl^- < NCS^- < N_3^-$ in agreement with EPD strengths measured towards $VO(acac)_2$ (*48*).

[3]) Estimated values

Table 2. *Molar ratios* $v = c_{X^-}/c_{Co^{2+}}$ *required to achieve nearly complete conversion of* Co^{2+}_{sv} *into* $[CoX_4]^{2-}$ *at* $c_{Co^{2+}} \approx 10^{-3}$ *mol/l in various EPD solvents*

Donor (DN)	X^-	I^-	Br^-	Cl^-	NCS^-	N_3^-
NM	(2.7)	5	4	4	4	4
AN	(14.1)	—	—	16	8	8
PDC	(15.1)	8	5	5	4	4
ES	(15.2)	—	—	15	8	8
Acetone	(17.0)	—	—	8	6	4
H_2O	(18.0)	>3000	>3000	>3000	>3000	>3000
DMA	(27.8)	—	—	16	13	4
DMSO	(29.8)	—	—	200	200	20
HMPA	(38.8)	—	—	460	160	27

Interpretation of data given in Table 2 is more difficult than in the case of polarographic reduction of metal cations, since complex formation is also influenced by anion solvation. Different solvents can differ considerably in their solvating ability towards anions, and this is especially true for protic and aprotic solvents (*10*, *43*, *49*, *50*).

There are therefore two main reasons for the large deviations observed for water: (a) entropic effects and enthalpic effects by hydrogen bonding interactions between water molecules of neighbouring hydration shells, as discussed in the preceding section, and (b) the strong solvating power of water towards halide and pseudohalide ions.

Free energies of hydration of halide and pseudohalide ions are 4—15 kcal/mol more negative than the corresponding values for aprotic solvents (*43*, *49*, *50*). As a consequence these ions will behave as much weaker ligands in water than in aprotic solvents. Differences in solvating power towards anions also account for the different molar ratios v observed in DMA and DMSO. Although both solvents have nearly the same donicity, complex formation is less quantitative in DMSO because anions are more strongly solvated in this solvent. Furthermore, coordination towards Co^{2+} is probably sterically more favourable for DMSO than for DMA. Determination of polarographic half-wave potentials should offer a possibility of detecting such steric effects because a shift to more positive $E_{1/2}$ values would be expected in case of steric hindrance[4]). Unfortunately no half-wave potentials are available at present for reduction of Co^{2+} in DMA. Molar ratios v for the nearly quantitative formation of $[CoX_4]^{2-}$ in HMPA are only slightly higher than for DMSO, and this appears at first sight rather unexpected. In fact, HMPA

4) This applies only if the reduction is reversible or if the degree of irreversibility is comparable in the solvents under consideration.

(DN = 38.8) is a stronger EPD than DMSO (DN = 29.8) but its ability to solvate anions is much smaller (*43*) compared to other aprotic solvents, and coordination of HMPA to Co^{2+} (as well as to several other metal cations) is definitely sterically hindered (see preceding section). The role of solvent donicity in complex formation is supported by polarographic studies of the reduction of Eu^{3+} to Eu^{2+} in different solvents using supporting electrolytes with anions of different EPD properties (Fig. 8)

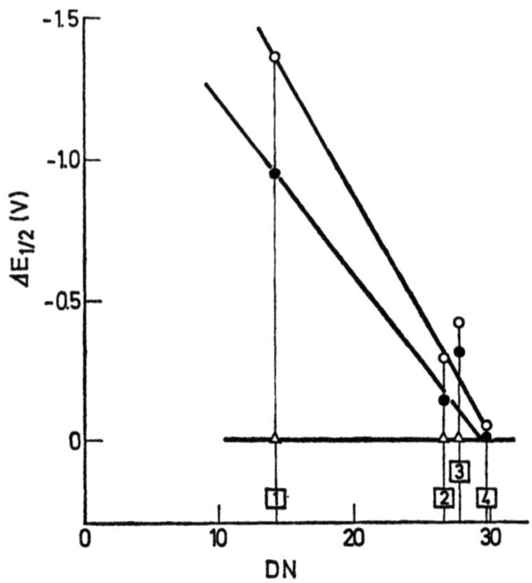

Fig. 8. Differences $\Delta E_{1/2} = E_{1/2}(Et_4N^+X^-) - E_{1/2}(Et_4N^+ClO_4^-)$ for the polarographic reduction of $Eu^{3+} \rightarrow Eu^{2+}$ using $Et_4N^+ClO_4^-$ and salts $Et_4N^+X^-$ as supporting electrolytes $X^- = J^-$ (—Δ—), Cl^- (—●—), N_3^- (—O—) 1 AN, 2 DMF, 3 DMA, 4 DMSO

(*36, 51*). Differences $\Delta E_{1/2}$ between $E_{1/2}$ values using $[Et_4N]^+[ClO_4]^-$ as supporting electrolyte and $[Et_4N]^+X^-$ ($X^- = I^-$, Cl^- and N_3^-) increase with decreasing donicity of the solvent, with the exception of DMA. Complex formation is easier in DMA than in DMF although DMA has a slightly higher donicity than DMF and this is in agreement with recent studies on solvation of Cl^- and N_3^- which are more strongly solvated in DMF than in DMA (*43*).

5. Ion-Pair Equilibria

a) General

The existence of different types of ion pairs, namely so called "intimate" or "contact ion pairs" and "solvent-separated ion pairs" was originally postulated in order to explain the stereochemical course of solvolysis and electrophilic substitution reactions (*52—55*). Since then the existence of both types of ion pairs has been widely accepted and found to be in agreement with results obtained by various physico-chemical methods, such as UV spectroscopy, conductivity measurements and electron spin resonance (*56—62*). In "contact ion pairs", cations and anions are considered to be in immediate contact with each other, while in solvent-separated ion pairs, ions are separated by solvent molecules.

b) UV Spectroscopy

Hogen-Esch and *Smid* have recently investigated the formation of contact ion pairs and solvent-separated ion pairs of alkali metal salts of various carbanions and radical ions in a number of solvents (*56, 57*). The results obtained for fluorenyl salts are listed in Table 3. The donicities of the EPD solvents 2-methyl-THF and DME[5]) are unknown but probably do not differ very much from that of THF. Toluene, according to preliminary measurements behaves as a very weak EPD towards $SbCl_5$ (*63*). In strong EPD solvents such as pyridine and DMSO, only solvent-separated ion pairs and free ions are present, while the formation of contact ion pairs is favoured in weak EPD solvents (toluene) and

Table 3. *Amount of solvent-separated ion pairs (+ free ions) of alkali metal salts of 9-fluorene (Fl) in various EPD solvents at 25°*

	% Solvent-separated ion pairs (+ free ions)				
Solvent	DN	ε	Li^+Fl^-	Na^+Fl^-	Cs^+Fl^-
dioxane	≈15	2.2	0	0	—
toluene	≈0.1	2.4	0	—	—
2-Me-THF	$DN \approx DN_{THF}$	6.3	25	0	—
THF	20.0	7.6	75	5	0
DME	$DN \approx DN_{THF}$	7.2	100	95	0
DMSO	29.8	45.0	100	100	—
pyridine	33.1	12.3	100	100	—

[5]) DME = dimethoxyethane.

solvents of medium donor strength such as dioxane, THF or 2-methyl THF.

The exceptional behaviour of DME is due to an entropy effect, since DME can act as a bidentate chelating agent. In general entropy effects should be important in ion pair equilibria since enthalpies of reaction are usually numerically small (*56, 57, 62*).

The extent of ion pair formation is apparently dependent on the nature of the cation. In dioxane, 2-methyl-THF and THF, Na-fluorenyl is considered as being present as contact ion pairs, whereas in pyridine or DMSO only solvent-separated pairs (together with free ions) are observed. In the case of Li-fluorenyl, solvent-separated pairs are formed in 2-methyl-THF and especially in THF but not in dioxane and toluene. In the solvents pyridine and DMSO formation of solvent-separated ion pairs (together with free ions) is again quantitative[6]).

The influence of counterions on

$$\text{contact ion pair} \rightleftharpoons \text{solvent-separated ion pair}$$

equilibria was discussed by *Szwarc* on basis of a simple electrostatic model (*62*). According to this approach the formation of contact ion pairs is favoured by small counterions whereas bulky counterions promote formation of solvent-separated ion pairs. Support for this model was obtained from studies on ion pair equilibria in THF solutions of sodium salts Na^+X^- with different anions X^- (*65, 66*). Obviously this model cannot be generally applied because the opposite behaviour is observed for alkali metal salts of 9-fluorene (*56*). Formation of solvent-separated ion pairs increases in the sequence $Cs^+ < K^+ < Na^+ < Li^+$. Although in THF the Stokes radius of Li^+ is larger than for Cs^+ (*62*), application of the electrostatic model to *solvated* ions is of little use, because different energies are required when different cations are (partially) desolvated on transfer from solvent-separated ion pairs to contact ion pairs. For the system studied by *Szwarc et al.*, electrostatic considerations are permissible because anions are solvated only poorly in THF.

c) IR Investigations

IR techniques have recently been used in studies of cation solvation. Bands which must be attributed to vibrations of cations within a solvate shell or a cage containing both counterion and solvent molecules were

---

[6]) In pyridine, ionized species will be present mainly as solvent-separated ion pairs because pyridine is a strong EPD medium, but a weak dissociating medium. DMSO is both a strong EPD and a strong dissociating medium and therefore ionized species will be present as free ions to a much greater extent (*64, 49, 50, 97*).

first detected by *Edgell* and coworkers for solutions of various alkali metal salts in THF (*67*). Corresponding bands for Li, Na and K salts are observed at about 400, 200 and 150 cm^{-1}, respectively. In solvents of medium EPD strength, such as THF (*67, 68*) and acetone (*69*), band frequencies are dependent on the nature of the anion; in contrast, in strong EPD solvents, such as DMSO (*70, 71*) 1-methyl-2-pyrrolidone (*72, 73*) and pyridine (*74*), the frequencies are nearly independent of the anion. This was ascribed to the formation of solvent-separated ion pairs and free ions in the strong EPD solvents and the presence of contact ion pairs in THF and acetone. *Popov* and coworkers demonstrated that variations of band frequency on changing the anion, as observed in THF and acetone, cannot be ascribed to the rather low dielectric constants of these solvents, but are more likely due to their moderate solvating power (*69, 71*). For example, nearly constant frequencies are observed for Li^+ and Na^+ salts independent from the anion in mixtures of benzene and DMSO with dielectric constants ranging from ≈ 7 to 46 (*71*). Valuable information is further provided by studies of mixtures of acetone and nitromethane (*69*). In these mixtures the intensity of the Li^+ vibrational band observed at 425 cm^{-1} is a linear function of the mole ratio acetone/$LiClO_4$ provided that the mole ratio is $\geqq 4$. It was concluded that under these conditions Li^+ is coordinated to 4 molecules of acetone. Preferential solvation of Li^+ by acetone is readily understood by comparing the donicities of the two solvents: $DN_{acetone} = 17.0$, $DN_{NM} = 2.7$, whereas elementary electrostatic considerations are not helpful: NM has a higher dipole moment ($\mu = 3.57$ Debye) than acetone ($\mu = 2.88$ Debye) (*75*). Raman and NMR results were thought to be due to formation of contact ion pairs at mole ratios acetone/$LiClO_4 < 4$ with $[ClO_4]^-$ ions entering into the coordination sphere of Li^+. If this interpretation is correct, the position of the Li^+ band in pure nitromethane solution should be influenced by successive addition of $[ClO_4]^-$ ions; replacement of NM by $[ClO_4]^-$ appears reasonable in view of the poor donor properties of NM.

The influence of various anions on the position of the Li^+ and Na^+ solvation bands was also investigated in acetone (*69*). With perchlorates, tetraphenylborates, thiocyanates, nitrates and iodides, band frequencies are independent of the nature of the anion. For bromides and chlorides a shift to lower frequencies is observed. Adding $[Bu_4N]^+[ClO_4]^-$ or $[Bu_4N]^+ I^-$ to solutions of $LiClO_4$ or LiI again does not change the bands but a shift to lower frequencies occurs on addition of $[Bu_4N]^+Br^-$ or $[Bu_4N]^+Cl^-$. The behaviour of these anions (with the exception of NCS^-) is consistent with both the expected magnitude of electrostatic interactions and the known tendency for covalent bond formation. If, however, bonds with considerable covalent character were formed, NCS^- should behave as a stronger donor than Cl^- and Br^- (*48*). In fact, it does not which favours essentially electrostatic bonding between alkali metal cations and halide or pseudohalide ions. In electrostatic interactions NCS^- should behave as a rather weak ligand due to strong charge dispersion in the molecule.

d) NMR Investigations

Selective solvation of various cations was studied in recent years in mixtures of water with solvents such as dioxane, THF, acetonitrile, propanediol-1,2-carbonate and pyridine (*76—82*). Except for pyridine-water mixtures, cations are preferentially solvated by water.

Relatively little is known about competitive solvation in mixtures of nonaqueous solvents. Complex formation between Na^+ and THF in solutions of $Na^+[AlBu_4]^-$ in hexane at molar ratios 1 : 1 (solvated contact ion pair) and 1 : 4 (solvent-separated ion pair) was reported by *Schaschel* and *Day* with proton NMR as well as IR and conductivity measurements (*83, 84*). Preferential solvation of alkali metal ions by DMSO in 1-pentanole and by acetone in nitromethane was observed by *Popov et al.* (*69, 85*).

A very remarkable relationship exists between donicity and chemical shift of the Na^{23} nucleus (*86, 87*) in solutions of sodium perchlorate or sodium tetraphenylborate in various EPD solvents. A nearly straight line (Fig. 9) is observed for 9 different EPD solvents, but water behaves

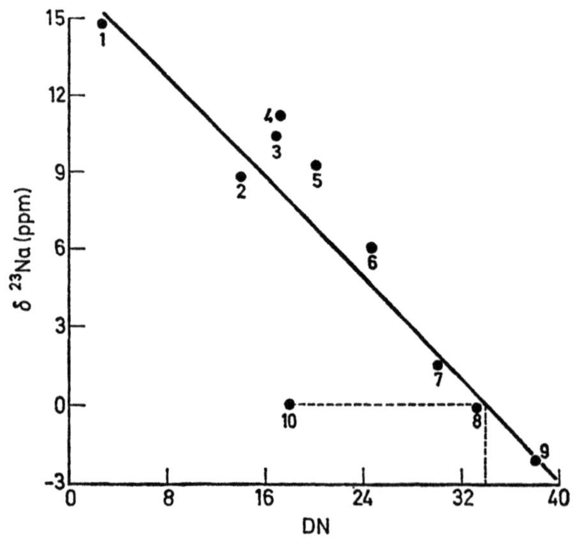

Fig. 9. Relationship between solvent donicity DN and chemical shift of the ^{23}Na nucleus in solutions of $NaClO_4$ or $Na[BPh_4]$ in various EPD solvents; reference: saturated aqueous NaCl solution

as a much stronger EPD than would be expected on the basis of its donicity (see Sect. 3).

This relationship has led to the suggestion that the chemical shift of the Na^{23} nucleus is a measure of the solvating power of an EPD solvent towards the Na^+ ion (*86*). This is not quite confirmed, because the chemical shift of the Na^{23} nucleus is mainly determined by the paramagnetic shielding term (*88*). Thus, it can safely be concluded only that covalent interactions — although possibly small — make some contribution to the solvation of Na^+ by EPD solvents and that the extent of covalency in the solvate bonds increases with increasing solvent donicity. This view is supported by preliminary measurements on Li^+ where no analogous relationship could be found (*86*). In fact, paramagnetic and diamagnetic shielding terms are much closer for Li^7 than for Na^{23}. Studies of the influence of anions on Na^{23} chemical shifts in various EPD solvents revealed (*86*, *87*), that chemical shifts for the perchlorate and tetraphenylborate in solvents of different donicity are independent of concentration, while for the iodide and to a lesser extent for the thiocyanate they do depend on concentration. These observations are in agreement with those discussed above which have been interpreted in terms of equilibria between contact ion pairs and solvent-separated ion pairs.

6. Quantum-Mechanical Considerations

Recently LCAO—MO calculations have been carried out for hydrated alkali metal ions, Be^{2+} and some halide ions (*89—94*). Although these calculations are rather qualitative in nature, two important results are obtained (*94*):

1. Hydration of both cations and anions involves a charge transfer between the central ion and the solvent molecules which is characteristic of the formation of partially covalent bonds,

2. Hydrogen-bond interactions between the water molecules of neighbouring hydration shells of cations and anions are much stronger than hydrogen-bond interactions in pure water.

Both effects are predicted in a qualitative way by the "principle of chemical functions" recently developed by one of the authors (*95—97*) (see next section).

According to LCAO—MO calculations, the most stable configuration of water molecules within the hydration sphere is a planar arrangement of individual water molecules and the cation; this is the arrangement expected for purely electrostatic interactions. On the other hand, for halide ions X^-, configurations with linear X—H—O bonds were obtained, which is contrary to elementary electrostatic considerations (*94*).

The amount of charge transfer obtained for alkali metal ions from LCAO—MO calculations is probably somewhat too high. No Raman lines attributable to ionsolvent vibrations could be detected in solutions of various alkali metal salts in THF (*68*), acetone (*69*) and 4-methylpyridine (*98*) even after prolonged periods of photographic exposure. It may therefore be concluded that covalent contributions, though present, are not dominant. In contrast, strong Raman lines were found for hydrated transition metal ions for which much stronger covalent interactions are to be expected (*99*, *100*).

7. Functional Approach to Chemical Interactions

It has been shown in the previous sections that the phenomenological approach to cation-solvent interactions is more widely applicable than any of the other approaches. It is, however, also apparent that in several cases other contributions have to be taken into consideration in addition to the application of the donicity concept.

According to the functional approach (*95*—*97*), molecular properties cannot satisfactorily account for the interaction between two reacting particles, which is decisively influenced by changes which occur in course of the reaction in the electron clouds of the reacting particles. Two main types of interaction are distinguished, namely

1. electron donor (ED)-electron acceptor (EA) interactions, characterized by complete electron transfer from ED to EA and
2. electron-pair donor (EPD)-electron-pair acceptor (EPA) interactions, characterized by formation of a coordinate link.

Each pair of functions, namely ED and EA functions as well as EPD and EPA functions respectively, are considered to be correlated functions.

According to classical definitions, a distinction is made between redox reactions and acid-base-type reactions in the broadest sense.

A redox reaction in solution does not change the *actual net charges* at the reacting particles by integral numbers, and this suggests that covalent bonding interactions between solute and solvent are involved. Likewise the formation of a coordinate bond by an acid-base-type reaction such as solvation involves a certain amount of charge transfer (see preceding section).

It may therefore be concluded that any reaction in solution may formally be considered to be the result of both ED—EA interactions and of EPD—EPA interactions, each occurring to a different extent; the sequence of these functions is regulated by the following functional principle (*96, 97*):

Any chemical interaction is started when one of the reacting entities exhibits one of the four functions (ED, EA, EPD, EPA), the reactant exhibiting the reverse *correlated* function. This brings about changes in the electron population, for which each of the reacting entities may try to compensate. Compensation is — according to the functional principle — only possible by means of the reverse *non-correlated* function.

For example, lithium metal has a strong tendency to react as ED, consequently the Li^+ ion tries to compensate for the loss of the electron by attracting water molecules (or other solvent molecules) which will function as EPD while the Li^+ ion acts as EPA. In this way the positive charge at the lithium ion is diminished on coordination due to partial transfer of charge from the EPD molecules to the cation.

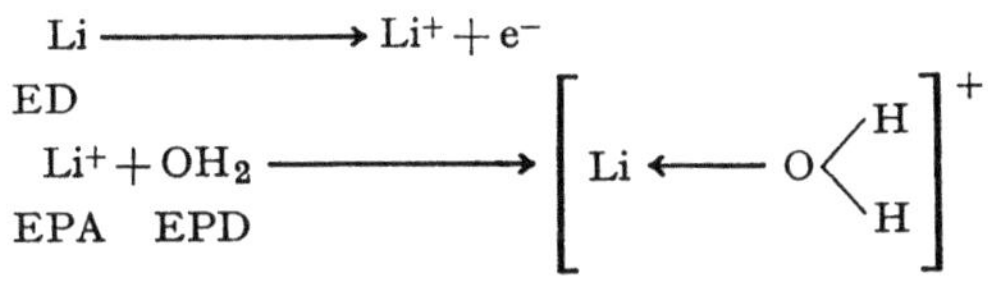

On the other hand, the oxygen atom of the coordinated water molecule which acts as EPD experiences a certain decrease in its electron population. Compensation — although only to a small extent in this case — is effected by the stronger attraction of the electrons of the O—H bonds. The oxygen atom thus functions as EA towards the hydrogen atom, which in due course will function as ED:

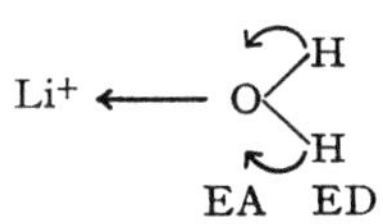

Thus the O—H bonds of the coordinated water molecules are more polarized and therefore the hydrogen atoms are more acidic than in uncoordinated water molecules. Of course, the amount of induced polarization is highly influenced by the EPA properties of the cation (Fe^{3+} is a stronger acid than Fe^{2+}).

The functional approach to chemical interactions provides a general pattern for the electron dynamics of all types of chemical interaction, as has been discussed in detail elsewhere (*97*). In this connection it may be desirable to mention briefly how a covalent bond is ionized by the nucleophilic attack of an EPD molecule (*50*):

The formation of the coordinate link between an EPD solvent molecule and an EPA atom M of a polarized covalent bond

$$\overset{\delta+}{\text{M}}-\overset{\delta-}{\text{X}}$$

leads to an increase in electron population at $\overset{\delta+}{\text{M}}$:

$$\text{EPD} \rightarrow \underbrace{\overset{\delta+}{\text{M}}-\overset{\delta-}{\text{X}}}_{\text{EPA}}$$

This causes an electron shift from $\overset{\delta+}{\text{M}}$ to $\overset{\delta-}{\text{X}}$

$$\text{EPD}-\underset{\text{ED}}{\text{M}}\overset{\curvearrowright}{-}\underset{\text{EA}}{\text{X}}$$

which increases the polarity of the M—X bond in the adduct. Increasing donicity of the EPD and increasing polarizability of the bond favour this process.

Complete transfer of an electron from $\overset{\delta+}{\text{M}}$ to $\overset{\delta-}{\text{X}}$ leads to heterolytic fission of the M—X bond and formation of ions, the cation being stabilized by coordination (solvation):

$$\text{EPD.M}-\text{X} \rightleftharpoons [\text{EPD.M}]^+ . \text{X}^-$$

In a medium of low dielectric constant the ions are present mainly as (solvent-separated) ion pairs, which are separated in a medium of high dielectric constant.

This leads to a reconsideration of the concept "contact" or "intimate" ion pair. Although these terms have been in use for about 15 years, the term "unionized species" appears more appropriate (*97*). For instance, the bond in a diatomic alkali metal halide as is known to exist in the gas phase is certainly not purely ionic (*102*) and its formation from gaseous ions can be described by the functional approach (*97*).

$$\text{A}^+ + \text{B}^- = \underset{\text{EPA}}{\overset{\delta+}{\text{A}}} \leftarrow \underset{\text{EPD}}{\overset{\delta-}{\text{B}}}$$

In a solvent of moderate solvating power such as dioxane, THF etc. these molecules are dissolved without rupture of bonds and thus without ionization. Solvation will certainly cause further polarization of the bonds.

However, as long as both atoms are in intimate contact with each other, some degree of covalency will be retained. This species — still unionized — is identical with what has been considered a "contact ion pair".

We therefore prefer the term "unionized species" instead of "contact" or "intimate" ion pair (*97*) and suggest that the term "ion pair" be restricted to solvent-separated ion pairs in the sense of *N. Bjerrum*, where cation and anion are no longer in immediate contact with each other. The degree of covalency of bonds in "unionized species" will of course depend on the EPA properties of the cation and the EPD power of the anion (*97*). In case of bonds between weak EPA ions such as alkali metal ions and weak EPD anions such as $[ClO_4]^-$ or $[BPh_4]^-$, the situation may approach the limiting case of an ideal contact ion pair.

8. Conclusion

The functional approach provides a unified pattern for the description of cation-EPD-solvent interactions in that it takes into account the respective covalent contributions in the solvate bonds. This approach is qualitative by its very nature, but a phenomenological property, namely the donicity of the electron-pair donor (the solvent molecule) can be used to obtain an astonishingly large number of semi-quantitative data. Though far from ideal, this approach is the best available at present to reconcile hitherto unrelated facts and goes some way to make up for the lack of means to calculate thermodynamic data for cation-solvent interactions, particularly for interactions with strong covalent contributions.

The applicability of donicities to cation-solvent interactions is most convincingly demonstrated by the polarographic reduction of various metal ions in solvents of different donicity. The observed variation of half-wave potentials with solvent donicity can be explained neither in terms of the Born equation nor by simple microscopic electrostatic models in view of the magnitude of the dipole moments of solvent molecules. The concept also provides the basis for an interpretation of complex formation reactions and the behaviour of electrolytes (ion pair equilibria) in a large number of EPD solvents.

The new concept restricts the term "ion pair" to "solvent-separated ion pairs" in its original sense, so-called "contact ion pairs" being regarded as unionized species with polarized covalent bonds.

The authors are grateful to Professor A. I. *Popov*, Michigan State University and to Professor M. C. *Day*, Louisiana State University for stimulating discussions. The support of this work by the "Fonds zur Förderung der wissenschaftlichen Forschung in Österreich" is gratefully acknowledged.

9. References

1. *Rosseinsky, D. R.:* Chem. Rev. *65,* 467 (1965).
2. *Desnoyers, J. E., Jolicoeur, C.:* In: Modern Aspects of Electrochemistry (ed. *J. O'M. Bockris,* and *B. E. Conway*), vol. **5**, chapter 1. New York: Plenum Press 1969.
3. *Pauling, L.:* The Nature of the Chemical Bond, 3rd ed. Ithaca-New York: Cornell University Press 1960.
4. *Goldschmidt, V. M.:* Skrifter det Norske Videnskaps. Akad. Oslo I. Mat. Nat. Kl. (1926), p. 21.

5a. *Gourary, B. S., Adrian, F. J.:* Solid State Phys. *10,* 127 (1960).

5b. *Grunwald, E., Baughman, G., Kohnstamm, G.:* J. Am. Chem. Soc. *82,* 5801 (1960).

6. *Arnett, E. M., McKelvey, D. R.:* J. Am. Chem. Soc. *88,* 2598 (1966).
7. *Parker, A. J., Alexander, R.:* J. Am. Chem. Soc. *90,* 3313 (1968).
8. *Haberfield, P., Clayman, L., Cooper, J. S.:* J. Am. Chem. Soc. *91,* 787 (1969).
9. *Parker, A. J.:* In: Advances in Physical Organic Chemistry (ed. *V. Gold*). London-New York: Academic Press 1967.
10. *Alexander, R., Ko, E. C. F., Parker, A. J., Broxton, T. J.:* J. Am. Chem. Soc. *90,* 5049 (1968).
11. *Fuchs, R., Bear, J. L., Rodewald, K. F.:* J. Am. Chem. Soc. *91,* 5797 (1969).
12. *Born, M.:* Z. Physik *1,* 45 (1920).
13. *Bjerrum, N., Larsson, E.:* Z. Physik. Chem. (Leipzig) *127,* 358 (1927).
14. *Latimer, W. M., Pitzer, K. S., Slansky, C. M.:* J. Chem. Phys. *7,* 108 (1939).
15. *Strehlow, H.:* Z. Elektrochem. *56,* 119 (1952).
16. *Passoth, G.:* Z. Physik. Chem. (Leipzig) *203,* 275 (1954).
17. *Laidler, K. J., Pegis, C.:* Proc. Roy. Soc. (London) *A 241,* 80 (1957).
18. *Noyes, R. M :* J. Am. Chem. Soc. *84,* 513 (1962).
19. *Conway, B. E., Bockris, J. O'M.:* Modern aspects of electro-chemistry, chapter 2 (1954).
20. *Buckingham, A. D.:* Discussions Faraday Soc. *24,* 151 (1957).
21. — Quart. Rev. *13,* 183 (1959).
22. *Morf, W. E., Simon, W.:* Helv. Chim. Acta *54,* 794 (1971).
23. *Gutmann, V., Wychera, E.:* Inorg. Nucl. Chem. Letters *2,* 257 (1966).
24. — Coordination Chemistry in Non-Aqueous Solutions. Wien: Springer 1968.
25. *Donoghue, J. T., Drago, R. S.:* Inorg. Chem. *1,* 866 (1962).
26. — — Inorg. Chem. *2,* 1158 (1963).
27. *Gutmann, V., Weisz, A., Kerber, W.:* Monatsh. Chem. *100,* 2096 (1969).
28. *Drago, R. S., Donoghue, J. T., Herlocker, D. W.:* Inorg. Chem. *4,* 836 (1965).
29. *Pearson, R. G.:* J. Am. Chem. Soc. *85,* 3533 (1963).
30. *Tamres, M., Searles, S.:* J. Phys. Chem. *66,* 1099 (1962).
31. *Drago, R. S., Wayland, B. B.:* J. Am. Chem. Soc. *87,* 3571 (1965).
32. *Heyrovsky, J., Kuta, J.:* Grundlagen der Polarographie. Berlin: Akademie Verlag 1965.
33. *Gutmann, V.:* Topics in Current Chemistry *27,* 59 (1972).
34. — Rec. Chem. Progr. *30,* 169 (1969).
35. — Chimia *23,* 285 (1969).
36. — Allgem. Prakt. Chem. *21,* 116 (1970).
37. — *Schmid, R.:* Monatsh. Chem. *100,* 2113 (1969).
38. — *Cechak, H.:* Monatsh. Chem. in press.
39. — *Duschek, O.:* in preparation.
40. — *Steininger, A., Wychera, E.:* Monatsh. Chem. *97,* 460 (1966).

41. — *Mayer, U.:* Monatsh. Chem. *98*, 294 (1967).
42. *Vlček, A. A.:* Progr. Inorg. Chem. *5*, 211 (1963).
43. *Mayer, U., Gutmann, V.:* Monatsh. Chem. *101*, 912 (1970).
44. *Olofsson, G.:* Acta Chem. Scand. *22*, 1352 (1968).
45. *Cechak, H., Gutmann, V.:* in preparation.
46. *Beverwijk, C. D. M., van der Kerk, G. J. M., Leusink, A. J., Noltes, J. G.:* Organometall. Chem. Rev. *5*, 215 (1970).
47. *Coetzee, J. F., McGuire, D. K., Hedrick, J. L.:* J. Phys. Chem. *67*, 1814 (1963).
48. *Gutmann, V., Mayer, U.:* Monatsh. Chem. *99*, 1383 (1968).
49. — Angew. Chem. *82*, 858 (1970); Angew. Chem. Intern. Ed. Engl. *9*, 843 (1970).
50. — Chem. Brit. *7*, 102 (1971).
51. — *Peychal-Heiling, G.:* Monatsh. Chem. *100*, 1423 (1969).
52. *Grunwald, E.:* Anal. Chem. *26*, 1696 (1954).
53. *Winstein, S., Clippinger, E., Fainberg, A. H., Robinson, G. C.:* J. Am. Chem. Soc. *76*, 2597 (1954).
54. — *Robinson, G. C.:* J. Am. Chem. Soc. *80*, 169 (1958).
55. *Cram, D. J., Mateos, J. L., Hauck, F., Langmann, A., Kopecky, K. R., Nielsen, W. D., Allinger, J.:* J. Am. Chem. Soc. *81*, 5774 (1959).
56. *Hogen-Esch, T. E., Smid, J.:* J. Am. Chem. Soc. *88*, 307 (1966).
57. — — J. Am. Chem. Soc. *88*, 318 (1966).
58. *Chang, P., Slates, R. V., Szwarc, M.:* J. Phys. Chem. *70*, 3180 (1966).
59. *Hirota, N.:* J. Phys. Chem. *71*, 127 (1967).
60. *Adam, F. C., Weissman, S. I.:* J. Am. Chem. Soc. *80*, 1518 (1958).
61. *Atherton, N. M., Weissman, S. I.:* J. Am. Chem. Soc. *83*, 1330 (1961).
62. *Szwarc, M.:* Accounts Chem. Res. *2*, 87 (1969).
63. *Gutmann, V., Weihs, P., Mayer, U.:* unpublished.
64. — *Mayer, U.:* Monatsh. Chem. *100*, 2048 (1969).
65. *Roberts, R. C., Szwarc, M.:* J. Am. Chem. Soc. *87*, 5542 (1965).
66. *Cserhegyi, A., Jaguar-Grodzinski, J., Szwarc, M.:* J. Am. Chem. Soc. *91*, 1892 (1969).
67. *Edgell, W. F., Watts, A. T., Lyford, J., Risen, W.:* J. Am. Chem. Soc. *88*, 1815 (1966).
68. — *Lyford, J., Wright, R., Risen, W., Watts, A.:* J. Am. Chem. Soc. *92*, 2240 (1970).
69. *Wong, M. K., McKinney, W. J., Popov, A. I.:* J. Phys. Chem. *75*, 56 (1971).
70. *Maxey, B. W., Popov, A. I.:* J. Am. Chem. Soc. *89*, 2230 (1967).
71. — — J. Am. Chem. Soc. *91*, 20 (1969).
72. *Wuepper, J. L., Popov, A. I.:* J. Am. Chem. Soc. *91*, 4352 (1969).
73. — — J. Am. Chem. Soc. *92*, 1493 (1970).
74. *McKinney, W. J., Popov, A. I.:* J. Phys. Chem. *74*, 535 (1970).
75. Handbook of Chemistry and Physics, 48th ed. Cleveland, Ohio: The Chemical Rubber Co., 1968.
76. *Fratiello, A., Douglass, D. C.:* J. Chem. Phys. *39*, 2017 (1963).
77. — *Miller, D.:* J. Chem. Phys. *42*, 796 (1965).
78. *Maciel, G. A., Hancock, J. K., Lafferty, L. F., Mueller, P. A., Musker, W. K.:* Inorg. Chem. *5*, 554 (1966).
79. *Cogley, D. R., Butler, J. N., Grunwald, E.:* J. Phys. Chem. *75*, 1477 (1971).
80. *Fratiello, A., Christie, E. G.:* Trans. Faraday Soc. *61*, 306 (1964).
81. — *Douglass, D. C.:* J. Mol. Spectry. *11*, 465 (1963).
82. *Hinton, J. F., Amis, E. S.:* Chem. Rev. *67*, 367 (1967).
83. *Schaschel, E., Day, M. C.:* J. Am. Chem. Soc. *90*, 503 (1968).
84. *Hammonds, C. N., Day, M. C.:* J. Phys. Chem. *73*, 1151 (1969).

85. *Maxey, B. W., Popov, A. I.:* J. Am. Chem. Soc. *90*, 4470 (1968).
86. *Erlich, R. H., Roach, E., Popov, A. I.:* J. Am. Chem. Soc. *92*, 4989 (1970).
87. — *Popov, A. I.:* J. Am. Chem. Soc. *93* 5620 (1971).
88. *Bloor, E. G., Kidd, R. G.:* Cdn. J. Chem. *46*, 3425 (1968).
89. *Hoyland, J. R., Kier, L. B.:* Theoret. Chim. Acta *15*, 1 (1969).
90. *Diercksen, G. H. F., Kraemer, W. P.:* Chem. Phys. Letters *5*, 570 (1970).
91. *Burton, R. E., Daly, J.:* Trans. Faraday Soc. *66*, 1281 (1970).
92. *Lischka, H., Plesser, Th., Schuster, P.:* Chem. Phys. Letters *6*, 263 (1970).
93. *Schuster, P., Preuß, H. W.:* Chem. Phys. Letters *10* (1971), in press.
94. *Russegger, P., Lischka, H., Schuster, P.:* Theoret. Chim. Acta (1971), in press.
95. *Gutmann, V.:* Allgem. Prakt. Chem. *21*, 289 (1970).
96. — Monatsh. Chem. *102*, 1 (1971).
97. — Chemische Funktionslehre. Wien: Springer 1971.
98. *Popov, A. I.:* private communication.
99. *Hester, R., Plane, R., Walraten, G.:* J. Chem. Phys. *38*, 249 (1963).
100. — — Inorg. Chem. *3*, 768 (1964).
101. *Criss, C. M., Luksha, E.:* J. Phys. Chem. *72*, 2966 (1968).
102. *Herzberg, G.:* Molecular Spectra and Molecular Structur, vol. I. Spectra of Diatomic Molecules. Princeton, New Yersey-London: D. Van Nostrand Company, Inc., 1950, reprinted 1967.

Received January 12, 1972

Acid Salts of Carboxylic Acids, Crystals with some "Very Short" Hydrogen Bonds

J. Clare Speakman

Chemistry Department, The University, Glasgow, Scotland

"We have... broideries of intricate design".

(James Elroy Flecker, The Golden Road to Samarkand)

"...'tis ever our old customers carbon and hydrogen
pirouetting with oxygen in their morris antics:
the chemist booketh all of them as C, H, O..."

(Robert Bridges, The Testament of Beauty)

Table of Contents

I. Introduction 142
II. Some Phase-Rule Considerations 143
III. The Nature of Acid Salts of Monobasic Acids — an Historical Note... 145
IV. The X-Ray Study of Acid Salts 146
V. The Dimensions of the Carboxyl Group and of the Carboxylate Ion... 147
VI. Some Acid Salts of Type *B* 148
VII. Acid Salts of Type *A* 151
VIII. Summary of the Hydrogen Bonding in Type *A* Acid Salts 158
IX. Di-*p*-chlorophenyl Hydrogen Phosphate 159
X. Pseudo-Type *A* Acid Salts 160
XI. Basic Salts of Monoacidic Bases 161
XII. Acid Salts of Mixed Type and of More Complex Composition 163
XIII. Acid Salts of Dicarboxylic Acids 165
XIV. Some Acid Salts of Type A_2 166
XV. Some Acid Salts of Type B_2 175
XVI. Some Acid Salts of Oxalic Acid 176
XVII. Anomalous Acid Salts of Dicarboxylic Acids 178
XVIII. Summary of Results for the Acid Salts of Symmetrical Dicarboxylic Acids 181
XIX. The $H_5O_2^+$ Ion 182
XX. A General, Elementary Consideration of "Very Short" O···H···O Bonds 184
References 197

I. Introduction

In elementary chemical theory, a monobasic acid (HX) forms only one series of salts with a univalent cation (M^+): the neutral salts, MX. A dibasic acid (H_2Y) may form two series: the acid salts, MHY, as well as the neutral salts, M_2Y. Historically this was the basis for the distinction between monobasic and dibasic (or polybasic) acids. As *Ida Freund* (*1*) wrote in 1904, the "basicity is the number of stages in which the replaceable hydrogen of an acid can be substituted by metals". The phrase, "acid salt of a monobasic acid", is a contradiction in terms. Such salts should not exist.

However, from time to time crystalline substances whose compositions corresponded to MHX_2 were discovered, recorded in the literature, and often forgotten. Potassium hydrogen bifluoride, KHF_2, was known to Berzelius and has been of great interest to chemists for a century. It may stand here as representative of a large number of acid salts of monobasic inorganic acids. Similarly anomalous salts of dibasic inorganic acids also occur, the mineral *Trona* being an example. The early history of this compound is outlined by *Mellor* (*2*); its composition was ultimately established as $Na_3HY_2 \cdot 2H_2O$, where $H_2Y = H_2CO_3$, so that it became known as "sodium sesquicarbonate".

Anomalous acid salts are very common with monocarboxylic acids. Sodium hydrogen di-formate, $NaH(HCO_2)_2$, was first described by *Bineau* (*3*) in 1847; ammonium hydrogen bis-glycollate, $NH_4H(HOCH_2 \cdot CO_2)_2$, by *Heintz* (*4*) in 1861; and potassium hydrogen di-benzoate, $KH(C_6H_5CO_2)_2$, by *Gerhardt* (*5*) in 1853. The late *Neville Smith* (*6*) surveyed the literature and noted at least a hundred examples, many of which had been discovered accidentally. But most monocarboxy-acids yield acid salts. In some cases an acid salt is more easily made than the neutral salt; it may crystallize preferentially when one is trying to prepare the neutral salt.

This review is concerned essentially with acid salts of carboxylic acids. It deals first with some acid salts of monocarboxy-acids whose structures have been studied by crystallographic methods; it then goes on to some salts of dicarboxy-acids. For the physical chemist the main interest of these structures arises because many of them involve hydrogen bonds of a particularly important kind. This must justify a digression on some basic salts of monoacidic bases, which have similar hydrogen bonding in exactly analogous anti-structures. The review ends with a tentative discussion of these compounds in relation to the general theory of the hydrogen bond.

Some questions have been asked about the proper naming of acid salts. The modern rule is to use the prefix "di"- (for MHX_2), except where it would be otherwise misleading; then "bis-" is to be preferred. For example, whilst the acid salt formulated above was correctly named potassium hydrogen di-benzoate, the wholly analogous salt $KH(C_6H_5 \cdot CH_2 \cdot CO_2)_2$ is potassium hydrogen bis-phenylacetate, since "diphenylacetate" might imply a salt of the well-known diphenylacetic acid, $(C_6H_5)_2CH \cdot CO_2H$. Again we have written "bis-glycollate" because a diglycollic acid exists, with formula, $CH_2(OH) \cdot CO_2CH_2 \cdot CO_2H$.

II. Some Phase-Rule Considerations

Conditions for the thermodynamic stability of an acid salt, MHX_2, are usefully systematized in the isothermal phase-diagram for the ternary system H_2O—HX—MX. When there is no stable acid salt, the triangular diagram takes on the form shown in Fig. 1 (a). The point A represents the solubility of the acid in water, and the curve AB the effect on this solubility of adding the salt MX to the solution; C represents the solubility of this salt in water and CB the effect of added acid on this solubility.

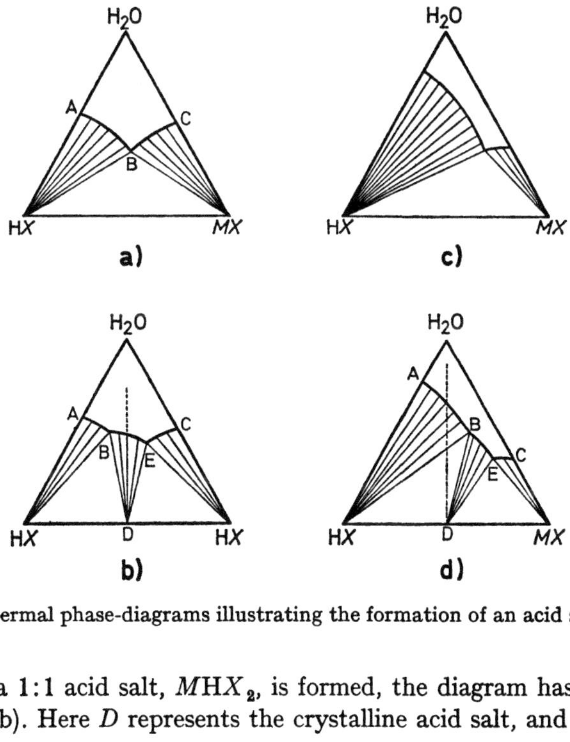

Fig. 1. Isothermal phase-diagrams illustrating the formation of an acid salt

When a 1:1 acid salt, MHX_2, is formed, the diagram has the form shown in (b). Here D represents the crystalline acid salt, and the curve

BE shows how its solubility varies as we change the proportions of free acid and salt dissolved in the water. When the diagram has the form shown in (b), where a straight line drawn from *D* to the apex of the triangle, representing pure water, cuts the curve *BE*, then the acid salt is stable in contact with a solution containing the appropriate concentration of dissolved material of composition $HX:MX = 1:1$. The acid salt is stable *vis à vis* its own solution. Evaporation of such a solution may lead to the precipitation of MHX_2.

But this is not necessarily so. Carboxy-acids are often much less soluble in water than are their salts. In such a case the ternary diagram will have the lop-sided appearance represented in Fig. 1(c); when a solid acid salt exists, the lop-sidedness may lead to the effect shown in (d). The curve *BE* now lies so far over to the right that it is no longer intersected by the direct line from the apex to *D*. The acid salt is said to be "decomposed by water"; in contact with aqueous solution, it is stable only when the solution is relatively richer in the neutral salt than corresponds to the ratio $HX : MX = 1:1$. If, over-optimistically, one tries to recrystallize a genuine specimen of the acid salt from water, crystals of the free acid will be produced. An example, amongst many, is potassium hydrogen di-benzoate (*7*).

The difficulty of crystallizing such acid salts can often be circumvented by using a mixed solvent, such as aqueous alcohol or aqueous acetone. Such a medium will increase the low solubility of HX and decrease the high solubility of MX; with these solubilities roughly equalized, the pseudo-ternary phase-diagram may become of the type betokened by Fig. 1(b).

The phase-diagrams for a number of ternary systems involving mandelic acid, $C_6H_5 \cdot CH(OH) \cdot CO_2H$, were studied by *Ross* (*8*), who thereby was able to establish the compositions of a variety of interesting acid salts from racemic and optically-active forms of this acid. Strictly speaking, an acid salt MHX_2 exists only in the solid state, for in solution it breaks down into free acid and the ionized salt M^+X^-.

Incidentally, an aqueous solution of MHX_2 is an *optimum buffer solution*, with a pH-value very close to the pK_A of the acid HX. It offers the advantage that the buffer may be prepared from a single solid compound, rather than by the usual method of mixing precisely equivalent amounts of two solutes (*9*). There is the disadvantage that the effective molar solubility of the acid salt is limited by the solubility of the free acid, which is often low for any but the simplest carboxylic acids.

When a higher concentration of the acid salt can be attained, the view of its existing in solution as HX and M^+X^- is too simple. Complex ions of the general formula $((HX)_mX_n)^{n-}$ will be formed, and in increasing proportion as the concentration increases. The tendency can run riot in favourable circumstances. Potassium hydrogen bis-phenylacetate, for example, in contact with a relatively large volume of water, achieves a solubility of only about 0.1 M, which is the solubility of phenylacetic acid in a dilute solution of its potassium salt. But a solution more than ten times as concentrated can be made by warming the solid acid salt with a small volume of water. It remains stable on cooling; but the free acid is — paradoxically — precipitated by addition of water. Some process of solubilization is evidently in operation in the concentrated solutions. We shall shortly be describing double anions of the type HX_2^-, which exist in some crystalline acid salts; something similar to these must be capable of existing in solution, with a tendency to polymerize at higher concentrations.

III. The Nature of Acid Salts of Monobasic Acids — an Historical Note

The phrase "acid salt of a monobasic acid" is formally a contradiction of terms. But when a useful theory is being developed, it pays to ignore a certain number of conflicting facts. As the value of the notion of the basicity of an acid became clear, the occurrence of anomalous acid salts was generally ignored. Some chemists would not know of their existence, and few could have known of their commonness. Others faced the difficulty by arguing that a monocarboxy-acid like acetic is really dimeric, and with some justification. Such acids are largely dimerized in non-polar solvents and also in the vapour phase immediately above their boiling points. *Bennett* (*10*) in 1916 made careful measurements of the surface tension of pure acetic acid over a wide range of temperature up to 150 °C. Applying a modified version of the Eötvös equation to his results, he concluded that the true formula was $C_4H_8O_4$, with "sodium binacetate", $NaH(C_4H_6O_4)$ as its legitimate acid salt. Indeed, when crystal-structure analysis came to be applied to crystalline monocarboxy-acids, almost all were found to consist of doubly hydrogen-bonded dimeric molecules (1), though acetic acid is an exception.

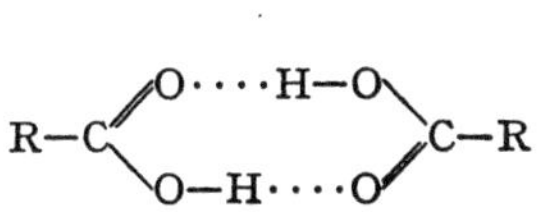 (1) 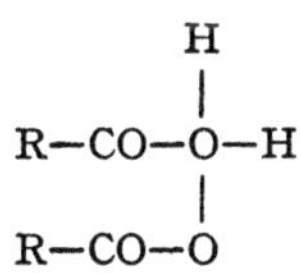 (2) 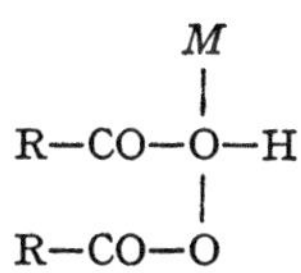 (3)

Until the development of X-ray methods, the nature of crystalline solids was necessarily a matter for speculation. The concept that a crystal of composition MHX_2 was simply an alternative form of the salt

MX with "acid of crystallization", analogous to the "water of crystallization" in (say) $NaCl . 2 H_2O$, was plausible (and sometimes correct); but as a working hypothesis it was neither very interesting nor very fruitful. Before the importance of hydrogen bonding had been recognized, association was often attributed to the ability of oxygen to increase its valency to four. On this basis, a carboxylic acid dimer was formulated as (2), and the acid salt as (3).

IV. The X-Ray Study of Acid Salts

A. A Note on Units, on the Diagrams, and on Other Matters

In 1942 *J. M. Robertson* began to establish a powerful school of X-ray crystal-structure analysis at Glasgow. When I joined his Department a few years later, it seemed appropriate to apply his methods, and the equipment and know-how of his school, to gain some firm knowledge of the structures of crystalline acid salts of carboxylic acids.

Throughout this review we follow the custom of crystallographers in expressing interatomic distances in Ångstrom units: 1 Å = 100 pm. For energy quantities we use the kilocalorie: 1 kcal = 4.184 kJ. Estimated standard deviations are given in parentheses, whereby 1.542 (3) Å means 1.542 ± 0.003 Å.

The structural diagrams have been drawn in a uniform style. Carbon atoms are represented by cross-hatched circles; hydrogens by very small, oxygens by moderate-sized, and metallic cations by large open circles; nitrogens by single-hatched circles. When necessary, symmetry elements in the diagrams are represented by the conventional symbols of the crystallographer's "International Tables", except that a centre of symmetry is marked by an asterisk.

For symmetry groups Hermann-Mauguin symbols are used. The following Schoenflies equivalents may be listed:

$\bar{1}$	C_i	m	C_s	3	C_3
2	C_2	mm	C_{2v}	$\bar{3}$	S_6
222	D_2	$2/m$	C_{2h}		

B. Two Fundamental Types of Acid-Salt Structures

Acid salts are ideally suited for X-ray study. They generally crystallize well. They necessarily contain a metal atom, which may be of atomic number high enough for the heavy-atom method of phase-determination; failing this, the metal can often be replaced isomorphously, whereupon the more powerful method of isomorphous replacement becomes available. When this work began it was also important that many of these acid salts crystallize with one of the unit-cell axes short, so that the structure could profitably be studied in projection, the only approach that was then computationally practicable.

The concept, mentioned above, of an acid salt as $M^+X^-.HX$, does indeed turn out to be valid in a minority of the cases studied. The two X-units of the formula are crystallographically non-equivalent and characterizable as anion and neutral molecule. Such structures we have classified as Type *B* (*11*).

In a majority of the acid salts studied a more interesting situation obtains. The two X-units of the formula are equivalent. Such structures are classified as Type *A* .

The crystallographic argument at its simplest can be illustrated by an example in the monoclinic space group $P\,2_1/c(C^5_{2h})$. This space group is of order four; when an atom is placed in a general position x, y, z in the unit cell, three others are generated by the symmetry elements at positions

$$-x,\ -y,\ -z;$$
$$x,\ -y,\ \tfrac{1}{2}+z;$$

and

$$-x,\ y,\ \tfrac{1}{2}-z.$$

If there are only two molecules of formula MN_2 in the unit cell, each molecule must be centrosymmetric. Atom M must lie at a centre of inversion, which could be taken as the origin (0, 0, 0); whilst an N-atom at x, y, z must be matched by a second atom, symbolized by $\bar{N}$, at $-x$, $-y$, $-z$. The cell must also contain an identical MN_2 molecule at another centre of inversion. A symmetrical $NM\bar{N}$ molecule, with M in a special position, is a formal consequence of the crystal symmetry. Such a situation, involving two molecules of MHX_2 in the cell, implies a symmetrical $[XHX]^-$ ion, with the two X-moieties linked by an O··H··O bond across a symmetry element of the crystal. These are the Type *A* acid salts, the oxygen-analogues of the well known bifluorides such as K^+ $[FHF]^-$.

V. The Dimensions of the Carboxyl Group and of the Carboxylate Ion

This subject was surveyed by *Hahn* (*12*) and more recently, by *Nahringbauer* (*13*). *Marsh* and *Donohue* (*14*) have dealt authoritatively with the special case of amino-acids.

In elementary resonance theory, the carboxylate anion is represented as a hybrid of the two forms (4) and (5). In the unperturbed anion, both forms contribute equally, so that the C—O distances and C—C—O angles

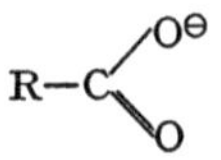 (4) 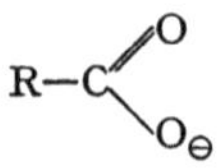 (5) 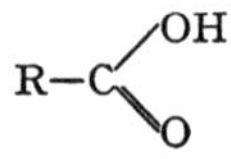 (6)

should be equal. On the other hand, in the neutral molecule (6) these distances and angles should differ.

Pauling's standard covalent radii give C=O = 1.21 Å, C—O = 1.43 Å, but a C—O distance as long as 1.43 Å is never realized in a carboxyl group. Actual distances show considerable variability, with 1.23 and 1.31 Å as average values in neutral carboxyl. As for the angles, classical stereochemistry would predict that R—C—OH should be tetrahedral (~110°), and R—C=O about $\frac{1}{2}(360°-110°) = 125°$. Average angles observed are 115° and 122°. In carboxylate anions, the average C—O distance is 1.26 A, the R—C—O angle ~117½°. In amino-acids the angles tend to differ a little, but there are special environmental influences at work here.

It has been observed (*15*) that the *sum* of C—O distances, over a wide range of carboxyl groups, whether ionized or not, does not differ much from 2.52 Å.

VI. Some Acid Salts of Type *B*

The most carefully studied structure is rubidium hydrogen bisglycollate (*16*), $RbHX_2$, where $HX = HO \cdot CH_2 \cdot CO_2H$ (Fig. 2).

With four $RbHX_2$ units in a cell belonging to the monoclinic space group $P2_1/c$ (of order four), the two glycollate residues of the formula are not crystallographically equivalent. Differentiation between anion (X^-) and acid molecule (HX) can be made from three types of evidence of increasing validity. First, the anion makes more O···Rb contacts than does the neutral molecule. Secondly, the C—O distances and C—C—O angles unambiguously distinguish the ionized and non-ionized carboxyl groups. Thirdly, the hydroxylic H-atoms were directly located in a "difference" synthesis. A more recent study (*17*) of the isomorphous K salt has led to similar conclusions.

A much earlier example (*18*) was found in a series of isomorphous hydrated acid salts of salicylic acid: $MHX_2 \cdot H_2O$, where M = K, Rb

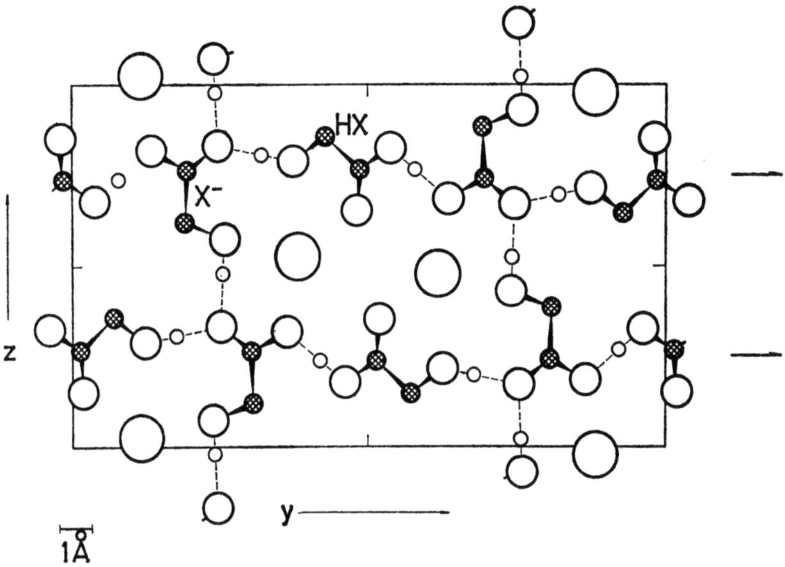

Fig. 2. Crystal structure of rubidium hydrogen bis-glycollate, *a*-axis projection

Table 1. *Some dimensions in rubidium hydrogen bis-glycollate*

	X		H*X*
C—O	1.241(14) Å	C=O	1.212(14) Å
C—O	1.264(14)	C—OH	1.308(14)
O—C—O	123.5(1.0)°	O=C—OH	123(1)°
C—C—O	120(1)	C—C=O	125(1)
C—C—O	116(1)	C—C—OH	112(1)

and NH_4, and H*X* is represented by formula (7). The structure of the ammonium salt, as found by two-dimensional methods, is shown in Fig. 3.

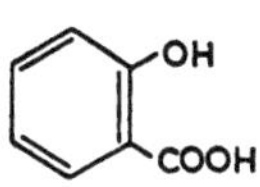

(7)

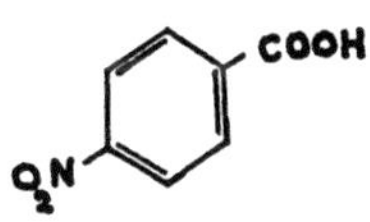

(7a)

The salicylate residues are crystallographically non-equivalent, but the analysis was not accurate enough to identify their respective states of ionization. However, the allocation of hydrogen atoms shown in the diagram seems reasonable.

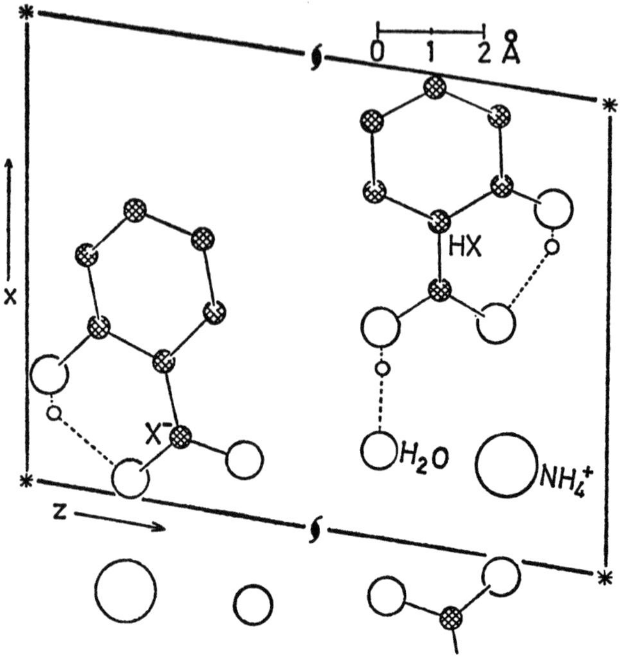

Fig. 3. Crystal structure of ammonium hydrogen disalicylate hydrate, *b*-axis projection

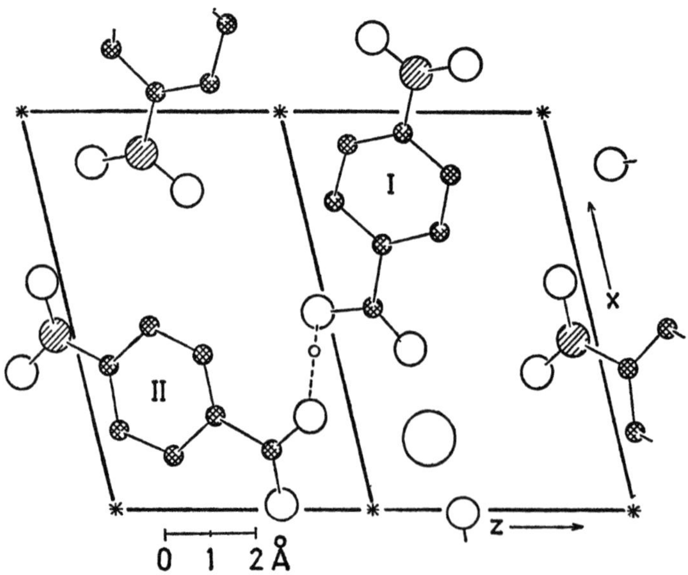

Fig. 4. Crystal structure of potassium hydrogen di-*p*-nitrobenzoate, *b*-axis projection

Also of Type *B* is potassium hydrogen di-*p*-nitrobenzoate, KHX_2, where $HX = p\text{-}NO_2 \cdot C_6H_4 \cdot CO_2H$, (7a). Though still two-dimensional, this analysis (*11*) was rather more precise. With two KHX_2 units in a triclinic cell of space group $P\bar{1}$ (order two), the two nitrobenzoate residues of the formula are non-equivalent. The structure (Fig. 4) is more complex because the K^+-ion makes a contact with one of the oxygen atoms of a nitro-group, as well as with those of the carboxyls. The allocation of hydrogen atoms is indirect; it is based on the dimensions of the two carboxyl groups: in I, C—O = 1.32 and 1.22 Å, C—C—O = 113° and 123°; in II, C—O = 1.26 and 1.24 Å, C—C—O = 120° and 118°.

VII. Acid Salts of Type *A*

A. Potassium Hydrogen Bis-phenylacetate

This was the first compound of its type to be studied structurally (*19*). As it has attracted some attention and illustrates many of the problems, it will be discussed in some detail. Other Type *A* crystals will be treated more cursorily.

The formula is KHX_2, where $HX = C_6H_5 \cdot CH_2 \cdot CO_2H$ (8). (Isomorphous NH_4 and Rb salts may be prepared.) It crystallizes in the

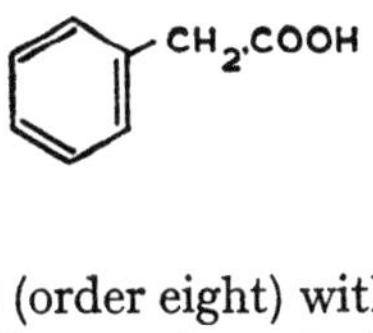

(8)

monoclinic space group $I2/a$ (order eight) with four KHX_2 units per cell. The K^+-ion and the acidic H atom have to be in special positions, and the two phenylacetate residues are equivalent. As shown in Fig. 5, K^+ lies on a twofold axis and the acidic H atom is effectively at a crystallographic centre of symmetry, where it links the two acidic radicals into an $[XHX]^-$ ion *via* a short hydrogen bond. In the original two-dimensional analysis (*20*) the O···H···O distance was estimated to be 2.55 Å. Though accuracy better than ±0.10 Å was explicitly disclaimed, various authors have subsequently cited this bond-length without regard to its uncertainty.

On the basis of a later study (*21*), the general features of the structure were confirmed, but the O···H···O distance was amended to 2.443 (4) Å. Other relevant dimensions are listed in Table 2.

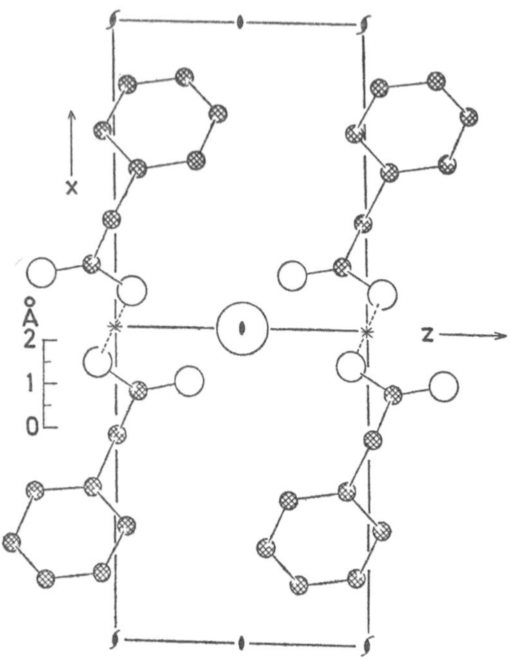

Fig. 5. Crystal structure of potassium hydrogen bis-phenylacetate, *b*-axis projection

Table 2. *Some dimensions in potassium hydrogen bis-phenylacetate*

		O—C—OH	124.7(3) °
C—O	1.223(3) Å	C—C—O	120.6(3)
C—OH	1.291(3)	C—C—OH	114.8(2)
O···H···O	2.443(4)	C—O····O	113.9(2)

Bacon and *Curry* studied this acid salt by neutron diffraction, first at room temperature (*22*), and then at 120°K (*23*). Only *x*- and *z*-coordinates were determined, but, as the *b*-axis is short, interatomic distances are insensitive to the *y*-coordinates, which were simply taken from the early X-ray investigation. Fig. 6 reproduces the "difference" synthesis (*23*), from which the neutron-scattering density of the K, O, and C atoms have been subtracted, so that the (negative) peaks due to the protons show up much more clearly. The most important feature of this map, for our purposes, is that the peak representing the acidic hydrogen atom is roughly circular.

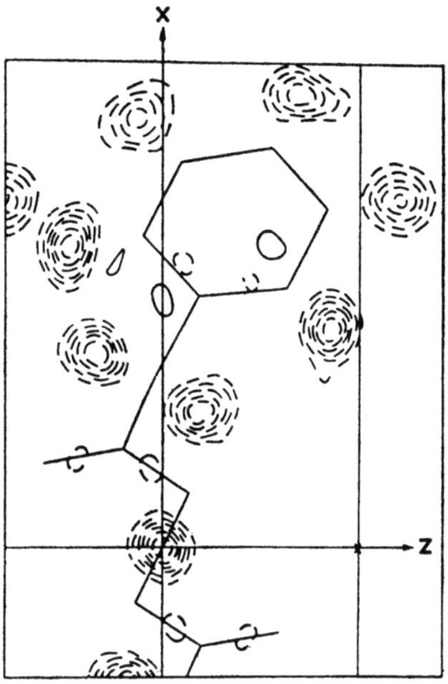

Fig. 6. "Difference" neutron-scattering density corresponding to Fig. 5. (From a photograph kindly supplied by Prof. G. E. Bacon. Negative density shown by broken contours)

The infrared spectrum of this (solid) acid salt (*24, 25, 26*) will be discussed in section *XX* B. Here we simply state that it is "anomalous" without explanation of the term.

B. Sodium Hydrogen Di-acetate

Amongst organic compounds, cubic crystals are rare. Some examples are known, e.g. hexamethylenetetramine, where the molecule is more or less spherical, or where it achieves effective sphericity by "free rotation" in the crystal; but, amongst the salts of carboxylic acids, sodium hydrogen di-acetate, $NaHX_2$, where HX = acetic acid, is unique. The cubic cell, with edge about 16 Å, contains 24 $NaHX_2$ molecules. The space group, $Ia3$, is of order 48. In 1922 *Wyckoff* (*27*) recorded his surprise that "such a simple organic substance... has a very complicated (crystal) structure." Though acid acetates of the other alkali cations, and of NH_4^+, are known, none of them is isomorphous with $NaHX_2$. Replacement of the Na^+ ion causes a complete change of crystal structure. KHX_2, for instance, crystallizes in the monoclinic system[1]).

---

[1]) And has a Type *B* structure (personal communication from Dr. *M. Currie* (*113*)).

These circumstances made the solution of the structure tedious. Once solved, it proved to have the elegance arising from its very high symmetry, and it was duly refined with three-dimensional X-ray data (*28*).

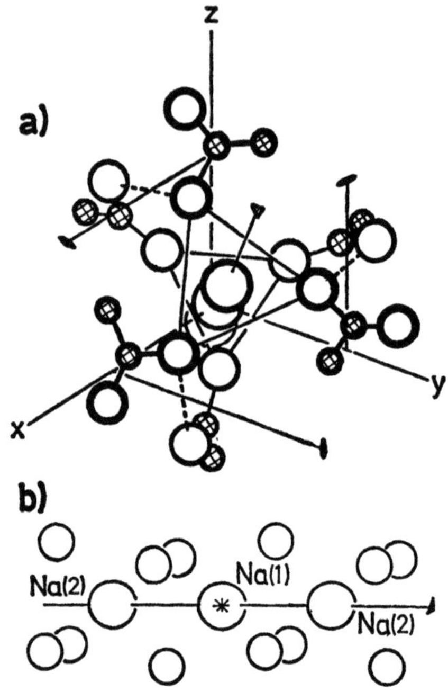

Fig. 7. (a) View of the crystal structure of sodium hydrogen di-acetate, near the origin of the cubic cell.
(b) The "triad" of sodium ions, with its oxygen environment

Table 3. *Some dimensions in sodium hydrogen di-acetate*

		O—C—OH	121.7(7)°
C—O	1.243(10) Å	C—C—O	122.3(7)
C—OH	1.295(10)	C—C—OH	116.1(7)
O···H···O	2.444(10)	C—O····O	110.8(5)

The atoms of the acetate residue are in the general 48-fold positions. The acidic H atoms lie on twofold axes in special positions of multiplicity 24. The 24 Na^+ ions are of two kinds: Na(1) is in an 8-fold position of symmetry 3, whilst Na(2) is in a 16-fold position of symmetry 3. Fig. 7

concentrates attention on one corner of the cubic cell. Na(1) is situated at the corner itself, through which runs a cube-diagonal (threefold axis). On this diagonal lies a pair of Na(2) ions, 3.30 Å on either side of Na(1), to give a "triad" of sodium ions (Fig. 7(b)).

The space group has twofold axes of symmetry, one-quarter of the way along each of the three principal axes, as shown in Fig. 7(a). Pairs of acetate moieties, related by a twofold axis, are so arranged as to give each Na^+ an immediate environment of six O-atoms, and hence to cushion the three positive charges on each triad. Each acetate pair is linked by a very short hydrogen bond across a dyad axis. Other dimensions are given in Table 3.

The infrared spectrum of sodium hydrogen di-acetate is notably "anomalous" (see section XX B and Fig. 19).

C. Potassium Hydrogen Di-aspirinate

The acid potassium salt of aspirin (H*X* = acetyl-salicyclic acid (9)) was thought to be of possible pharmacological value as a means of administering aspirin in a form of reduced acidity. The crystal structure was determined by *Manojlović* and *Speakman* (*29*), that of the isomorphous Rb salt by *Grimvall* and *Wengelin* (*30*). Subsequently, *Hamilton* and coworkers (*31*) followed with a neutron-diffraction study.

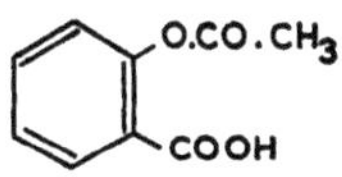

(9)

The cell (space group *C*2/*c*, of order 8) contains four KHX_2 units, and Type *A* symmetry is required. In general lay-out the structure resembles that of potassium hydrogen bis-phenylacetate. Some details are listed in Table 4. The infrared spectrum is "anomalous".

Table 4. *Some dimensions in potassium hydrogen di-aspirinate (X = X-ray, N = neutron diffraction results)*

	X	N		X	N
			O—C—OH	124.1(3)°	124.4(3)°
C—O	1.221(4) Å	1.217(4) Å	C—C—O	121.4(3)	121.1(3)
C—OH	1.284(5)	1.284(3)	C—C—OH	114.5(3)	114.4(3)
O···H···O	2.455(5)	2.448(4)	C—O····O	111.2(4)	111.3(3)

D. Some Acid Salts of Trifluoroacetic Acid

The compounds MHX_2, where $M=$ K, Rb and Cs, and $HX = CF_3 \cdot CO_2H$ all crystallize in the same monoclinic space group $I2/a$ (or its equivalent) with four molecules per cell. Hence they are all structurally of Type *A*. The K and Rb salts are isomorphous but the Cs salt has different cell dimensions and a different crystal structure (*32*).

In the Cs salt, as in most of the acid salts of aromatic acids so far studied, the arrangement corresponds to that in a jam sandwich. It is based on a layer of cations and acidic hydrogen atoms (collectively, the jam) between two layers of acid residues (the bread). The oxygen atoms of the carboxyl groups (the butter) face the cations and hydrogens, to which they are firmly attached by electrovalency or hydrogen bonding. The layering in KHX_2 is more confused. This is connected with the presence of a cation···F contact, which is not made in the Cs salt. The K^+···F distance, 2.962(4) Å, is within the range of the K^+···O contacts 2.84, 2.85 and 2.97 Å. (To be sure, the ionic radius of fluorine is slightly smaller than that of oxygen; but the two symmetry-related F atoms are clearly just as much part of the environment of the K^+ ion as are the other three pairs of O atoms.) The layers of trifluoroacetate residues have electronegative atoms on both sides; the bread has been buttered on both sides.

In both structure-patterns, the hydrogen bonds lie across centres of symmetry. Relevant interatomic distances and angles are shown in Table 5. All the infrared spectra are "anomalous".

KHX_2 in both normal and deuteriated forms has been studied by neutron diffraction (*111*). Some results of this work will be discussed in Sect. *XX* G.

Table 5. *Some dimensions in potassium hydrogen di-trifluoroacetate (X = X-ray, N = neutron diffraction results), and in caesium hydrogen di-trifluoroacetate*

	X	N		X	N
KHX_2			O—C—OH	128.4(4)°	129.3(2)°
C—O	1.212(6) Å	1.214(2) Å	C—C—O	120.0(4)	119.5(2)
C—OH	1.268(6)	1.266(3)	C—C—OH	111.6(4)	111.2(2)
O···H···O	2.435(7)	2.437(4)	C—O····O	115.6(4)	114.6(2)
$CsHX_2$			O—C—OH	127(2)°	
C—O	1.20(3) Å		C—C—O	119(2)	
C—OH	1.32(3)		C—C—OH	114(2)	
O···H···O	2.38(3)		C—O····O	114.5(2.0)	

E. Potassium Hydrogen Di-*p*-Hydroxybenzoate Monohydrate

This acid salt, $KHX_2 \cdot H_2O$, where HX = (10), has a normal Type *A* structure so far as the carboxyl-function of the acid is concerned. Interest attaches to the phenolic group, which makes contacts with another, symmetry-related, group, as well as with the water molecule (whose

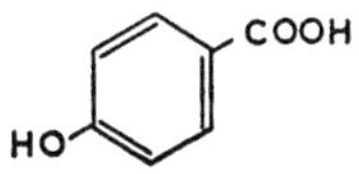

(10)

presence was revealed only by the X-ray analysis (*33*)). The crystal structure is shown in Fig. 8, and relevant interatomic distances and angles, based on a later refinement (*34*), in Table 6. The infrared spectrum shows " anomalous" features.

The second hydrogen bond, also across a centre of symmetry, between phenolic groups is much longer than the one between carboxyl groups, and its C—O···O angle is much less favourable for strong bonding. The oxygen atom of the water molecule lies on a twofold axis and makes a pair of contacts with phenolic oxygens at 2.639 Å; this certainly ranks as

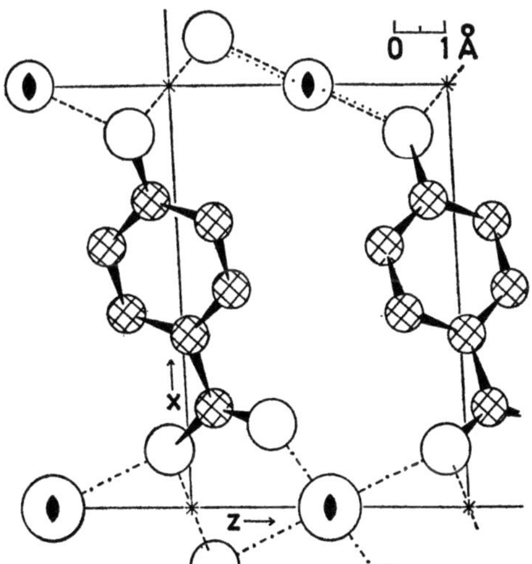

Fig. 8. Crystal structure of potassium hydrogen di-*p*-hydroxybenzoate hydrate, *b*-axis projection. (The oxygen atoms near the top, sited on twofold axes, belong to water molecules)

Table 6. *Some dimensions in potassium hydrogen di-p-hydroxybenzoate hydrate*

		O—C—OH	123.7(4)°
C—O	1.232(5) Å	C—C—O	120.7(4)
C—OH	1.284(6)	C—C—OH	115.6(4)
O···H···O (carboxylic)	2.458(6)	C—O····O	111.4(4)
O···H···O (phenolic)	2.692(7)	C—O····O	122.8(5)

a hydrogen bond. A second pair of contacts with other water molecules, O···O being 3.202 Å, represents possible, though questionable, hydrogen bonding. The problem is that the number of available protons is not equal to the number of presumptive hydrogen bonds; it has been suggested (*34*) that there is some disorder of the hydrogen atoms in this region. The electron-density "difference" synthesis shows a peak at the centre of the inter-carboxyl bond, but gives no clear picture of the positions of the phenolic and water hydrogen atoms. The O atoms of the water molecule have vibrational parameters generally larger than those of the other O atoms. We hope to study this material by neutron diffraction. Specific heat measurements might also be worth making.

VIII. Summary of the Hydrogen Bonding in Type *A* Acid Salts

The main interest of these structures is in their "very short", and apparently symmetrical, O···H···O bonds (summarized in Table 7), which include most of the shortest O···H···O bonds so far measured, apart from those in chelated systems (see Sects. XIV *A* and XX *E*).

These bonds occur in chemically different environments, and there is no reason to suppose they have exactly the same length. None the less, the average length, weighted according to the standard deviation of each, is strikingly low — 2.447(3) Å — as also is the weighted standard deviation of the average. The agreement between the C—O···O angles, too, is close, the average being 112(2)°. This is near the tetrahedral angle, but perhaps significantly larger, at least in the sense that no individual angle is less than 109½°. Following *Donohue* (*35*), such angles may be regarded as mandatory in very strong O···H···O bonds.

Table 7. *Summary of the inter-carboxyl hydrogen bonds in some Type A acid salts of monocarboxylic acids, MHX_2*

HX	*M*	space group	Molecules per cell	Symmetry of bond		O···H···O	C—O···O	Ref.
phenylacetic	K	*I2/a*	4	$\bar{1}$		2.443(4) Å	113.8°	*(21)*
benzoic	K	*C2/c*	4	$\bar{1}$		2.51(4)		*(107)*
p-HO-benzoic	K	*P2/c*	2	$\bar{1}$		2.458(6)	111.4	*(34)*
acetic	Na	*Ia3*	24	2		2.444(10)	110.8	*(28)*
cinnamic	NH_4	*C2/c*	4	$\bar{1}$		2.51(3)	110	*(108)*
p-Cl-benzoic	K	*C2/c*	4	$\bar{1}$		2.457(13)	110.2	*(109)*
m-Cl-benzoic	K	*C2/c*	4	$\bar{1}$		2.437(7)	110.2	*(105)*
trifluoro-acetic	Cs	*A2/a*	4	$\bar{1}$		2.38(3)	114.5	*(32)*
	K	*C2/c*	4	$\bar{1}$	(X)	2.435(7)	115.6	*(32)*
					(N)	2.437(4)	114.6	*(105)*
aspirin	K	*C2/c*	4	$\bar{1}$		2.455(5)	111.2	*(29)*
	Rb					2.48(2)	114	*(30)*
o-NO_2-benzoic	Rb	*P*$\bar{1}$	1	$\bar{1}$		2.43(6)		*(11)*
anisic	K	*Pbcn*	4	2		2.476(18)	111	*(110)*

[Weighted mean of nine most precise values 2.445(2) Å]

IX. Di-*p*-chlorophenyl Hydrogen Phosphate

Though this is not an acid salt, and does not even involve any carboxyl groups, its structure is analogous to that of a Type *A* acid salt, as was predicted by *Hadži* and *Novak* *(36)* from its infrared spectrum, and confirmed by X-ray work *(37)*.

(11)

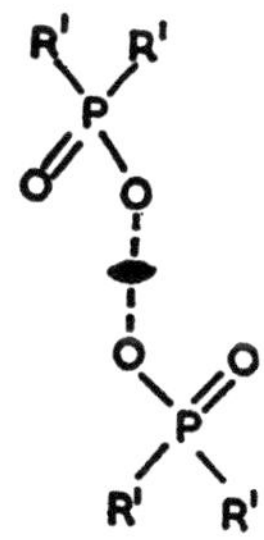

(12)

The corresponding diphenyl hydrogen phosphate *(38)* crystallizes with infinite chains of O···H···O linked molecules, spiralling round a screw-axis as suggested in (11), $R = C_6H_5$. In contrast the chloro-derivative

(12), $R' = p\text{-}Cl \cdot C_6H_4$, has discrete dimeric units in which a pair of phosphate residues is joined by an O···H···O bond across a crystallographic twofold axis. (This is exactly the situation in sodium hydrogen di-acetate.) The hydrogen bond is very short, 2.40(2) Å. The P—O···O angle, however, is 129°.

X. Pseudo-Type *A* Acid Salts

In Type *A* acid salts there is — by definition — strict crystallographic symmetry of the hydrogen bond. The $[XHX]^-$ ion is bisected by a symmetry element of the crystal. This is usually a centre of inversion, in which case the proton has — formally at any rate — to be exactly at the midpoint of the O···H···O bond. Sometimes it is a twofold axis, which allows more latitude in placing the proton, which must lie on the line bisecting the bond, but not necessarily at its mid-point, though it is unlikely to be far away. In Sect. *XXD* we suggest that the symmetry of the crystallographic environment may have some influence in shortening the bond. Against this we must set the fact that some crystals with very similar structures do not use crystal symmetry in forming their hydrogen bonds.

The best authenticated example (*39*) is potassium hydrogen diformate, KHX_2, where $HX = H \cdot CO_2H$. This crystallizes with eight KHX_2 units per cell in the orthorhombic space group *Pbca* (of order 8), and no molecular symmetry is required. There is a well-defined, discrete dimeric anion $[XHX]^-$; the two formate residues — denoted *L* and *M* in Table 8 — are non-equivalent, but their dimensions hardly differ. Despite the lack of formal symmetry, the hydrogen bond is "very short", close to the mean of the bonds summarized in Table 7.

Table 8. *Some dimensions in potassium hydrogen di-formate*

	L		*M*
C—O	1.220(6) Å		1.240(6) Å
C—OH	1.257(8)		1.251(7)
O—C—OH	125.3(6) °		124.1(5) °
C—O····O	113.6(4)		113.4(4)
O···H···O		2.447(6)	

Structures of this sort may be classified as pseudo-Type *A*. At present the known examples are few.

XI. Basic Salts of Monoacidic Bases

These compounds are antithetic to the acid salts we have been describing, and therefore merit mention, though they are not covered by the title of the review. Just as a carboxylic acid H*X* may form a salt MHX_2 in addition to the conventional neutral salt *MX*, so a monoacidic base *B* may – with a strong acid such as HCl – form not only a neutral salt *B*HCl, but also a basic salt B_2HCl. Such a basic salt is sometimes called a "hemi-hydrochloride": *i.e.* *B*. ½HCl. A long known example is aniline hemi-hydrochloride, $(C_6H_5 \cdot NH_2)_2$HCl, but its crystal structure is not known. A brief report (*40*) on the structure of acetamide hemi-hydrochloride ($B = CH_3 \cdot CO \cdot NH_2$) showed the crystals to have two B_2HCl molecules in a cell of the space group $P2_1/c$ (order 4). This implies that the acidic hydrogen atom, as well as the chloride ion, is in a special position, and that the two acetamide residues are equivalent. The structure, in fact, embodies a centrosymmetric dimeric cation of the kind suggested by (13).

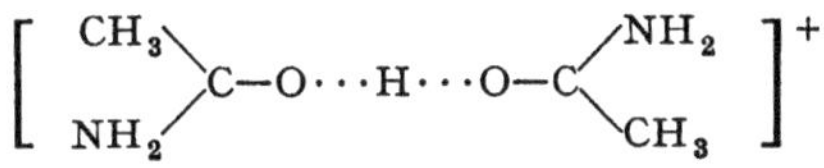

(13)

A brief report of neutron-diffraction work (*41*) at —150° gives the O···H···O distance as 2.418(11) Å. The infrared spectrum is "anomalous" (*26*).

Hadži (*42*) reported a similar spectrum from the basic salt B_2. HBr, where *B* = α-picoline-N-oxide, and predicted a similar structure. This was confirmed by X-ray work (*43*). The crystal contains the dimeric ion (14), with a hydrogen bond (O···H···O ≈ 2.40 Å) lying across a centre of symmetry.

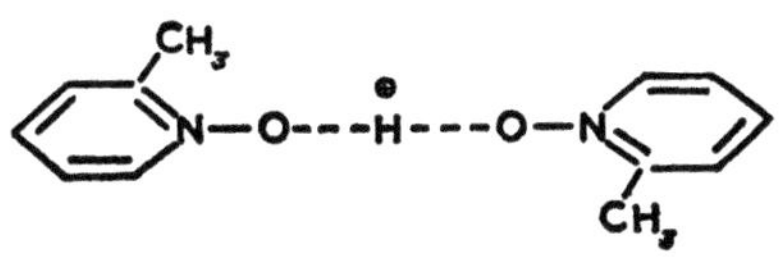

(14)

(Further work on this salt is being done.)

Cook (*44*) has prepared a number of crystalline basic salts of various organic bases with acids such as $HAsF_6$. From their infrared spectra he predicted that many of them should have Type *A* structures of the sort described in this section. So far only two structures have been successfully elucidated. One is of an arsenate ester (*45*), the other of a basic salt B_2. $HAsF_6$, with *B* = 1-methyl-2-quinolone (15), which does indeed embody the dimeric cation $[BHB]^+$, but merely with virtual — not strict crystallographic — symmetry (*46*). Nevertheless, the hydrogen bond is "very short" with O···H···O = 2.439(12) Å. The spectrum (Fig. 22(b)) resembles those we have been describing as "anomalous".

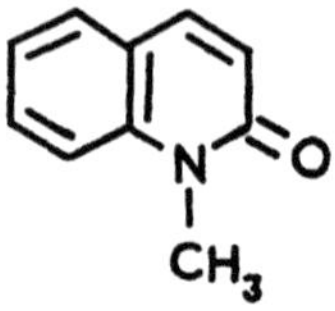

(15)

An especially interesting basic salt of pseudo-Type *A* is $B_2 . HClO_4$, B = mesitaldehyde, $(CH_3)_3C_6H_2(CHO)$, which crystallizes, with two formula-units per cell, in the space group $P2_1/m$ (order four) (*47*). Special symmetry is required, and a Type *A* structure containing centrosymmetric $[BHB]^+$ cations might reasonably have been expected. However, the crystal symmetry is used in a different way: the cations lie in mirror planes[2]), one of which is shown in Fig. 9. The mesitaldehyde molecules linked by a given proton are thus non-equivalent, and crystallographic symmetry-centres operate between the dimeric units rather than within them. The O···H···O bond connecting the base-residues is not required to be symmetrical, but it is short: O···H···O = 2.459(10) Å. The infrared spectrum has not been published.

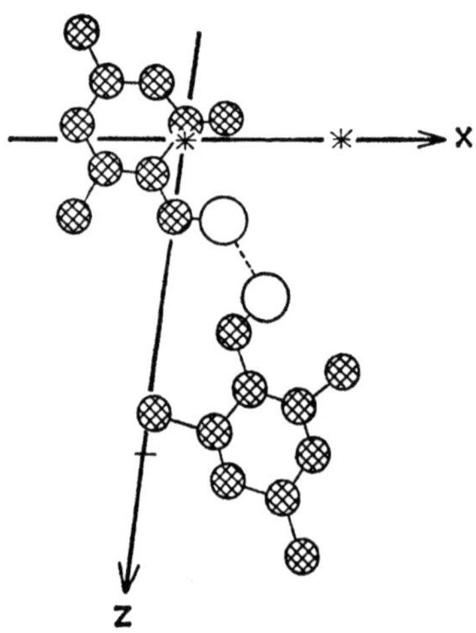

Fig. 9. Crystal structure of mesitaldehyde hemi-perchlorate, *b*-axis projection. (This view shows only the atoms lying in one of the mirror-planes of the crystal)

[2]) This is one of the rare exceptions to a rule: aromatic molecules, which would be planar in the gaseous state, never crystallize with their rings on a mirror-plane.

XII. Acid Salts of Mixed Type and of More Complex Composition

The straightforward acid salt of a monobasic acid has the formula MHX_2. Not surprisingly, more complex compositions have been discovered. We may cite as examples KH_2X_3, HX = acetic acid, and $NH_4H_3X_4$, HX = *dl*-mandelic acid (*6*, *8*). From another angle of classification, structures have been found which cannot be cleanly allocated to Type *A* or *B*. Two examples will be discussed, but the field is at present largely unexplored, and many more complicated systems could certainly be prepared. Just as in the crystal chemistry of the silicates, permutation of a few simple *motifs* can lead to a plethora of structural schemes.

A. Ammonium Hydrogen Di-acetate

Berthelot recognized that one form of "ammonium acetate" was acidic, and he realized that it was an acid salt like "sodium biacetate". The preparation and identity of this substance were clarified in 1902 by *Reik* (*48*), who showed its composition to be NH_4HX_2 (HX = acetic acid). The crystal structure has now been carefully determined by *Nahringbauer* (*49*), with results that may delight the collector of topochemical curiosities.

This compound crystallizes in the triclinic space group $P\bar{1}$ of order two. The number of NH_4HX_2 units per cell turns out to be three, which is unusual, except for trigonal or hexagonal crystals. Structure analysis reveals that the compound is better represented as

$$NH_4^+XHX^- . 2\,(NH_4^+X^- . HX).$$

The NH_4^+ ion and the dimeric XHX^- of the first set are sited on crystallographic centres of symmetry. (A centre of symmetry is, of course, impossible for the tetrahedral NH_4^+ ion, which must be disordered in this case.) In the second set, the NH_4^+, X^- and HX are in general positions, duplicated by operation of crystallographic centres of inversion. Thus the crystal contains both Type *A* and Type *B* structure: NH_4HX_2 and $2\,(NH_4X . HX)$ respectively. The structural plan, as it affects the acetate residues, is represented schematically by Fig. 10.

This identification is supported by the bond-lengths and angles listed in Table 9. The sums of the C—O distances in each carboxyl are nearly constant; their differences differ in a systematic manner:

$$\Delta(HX) > \Delta(XHX^-) > \Delta(X^-).$$

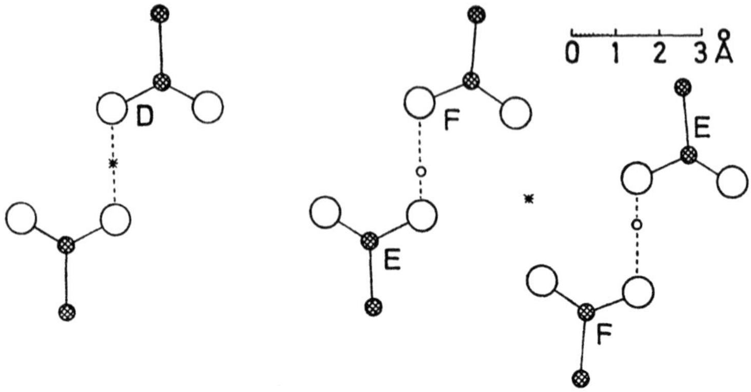

Fig. 10. Schematic view of the anion-structure in ammonium hydrogen di-acetate. (D is the acetate residue constituting half of the centrosymmetric XHX^- ion; E denotes HX, and F X^-)

Table 9. *Some dimensions in ammonium hydrogen di-acetate* (*see Fig. 10*)

	Residue D	Residue E		Residue F
C—O	1.221(4) Å	1.209(4) Å		1.252(4) Å
C—OH	1.279(4)	1.325(4)		1.278(4)
	(Δ = 0.058)	(Δ = 0.116)		(Δ = 0.026)
O—C—OH	124.3(3) °	122.5(3) °		123.1(3) °
C—C—O	120.3(3)	125.2(3)		120.3(3)
C—C—OH	115.5(3)	112.4(3)		116.6(3)
O···H···O	2.474(6) Å		2.517(6) Å	
C—O····O	114.9(3) °	114.7(3) °		110.7(3) °

The hydrogen bond lying across a centre of symmetry is 0.04 Å shorter than the bond between HX and X^-.

B. Rubidium Di-hydrogen Tri-monochloroacetate

RbH_2X_3 ($HX = ClH_2 \cdot CO_2H$) crystallized out when we tried to prepare the acid salt $RbHX_2$, which we hoped might be isomorphous with triclinic KHX_2. There are four RbH_2X_3 units in a cell of space group $P2_1/c$, so that no special symmetry is required. However, the projected structure (Fig. 11) comprises two crystallographically independent, centrosymmetric XHX^- anions (Type A) and one discrete HX molecule, which appears to be hydrogen bonded to one of the dimeric anions (*50*). At the time of writing, this structure is not sufficiently refined to justify reporting bond-lengths.

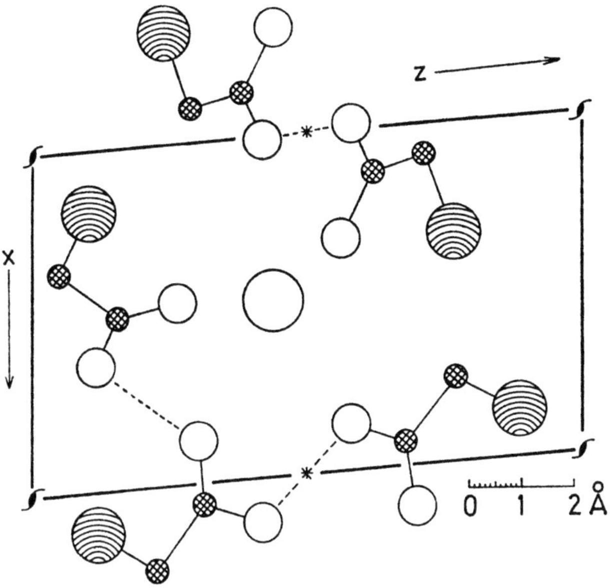

Fig. 11. Crystal structure of rubidium di-hydrogen tri-monochloro-acetate, *b*-axis projection. (Circles with curved hatching represent chlorine atoms)

XIII. Acid Salts of Dicarboxylic Acids

Unlike those of monobasic acids, acid salts of dibasic acids do not surprise the unsophisticated by their mere existence. A dicarboxylic acid, H_2Y, should yield two series of salts, MHY and M_2Y, and we should expect the primary anion in the former to have one carboxyl group ionized, the other not. This situation is found only in a few cases. Examples are ammonium hydrogen phthalate (*51*) and ammonium hydrogen *d*-tartrate (*52*). A crystal structure of this sort, with differentiated $-CO_2H$ and $-CO_2^-$ groups, resembles the Type *B* acid salts of monobasic acids (Sect. VI). When a dicarboxylic acid is involved, the differentiated groups belong to the same molecule. We classify such acid salts as Type B_2.

Amongst the acid salts of the simpler dicarboxylic acids, at any rate, we find much more frequently structures corresponding to Type *A*. For instance, in potassium hydrogen malonate the two carboxyl groups of the mono-ionized anion are crystallographically equivalent. Such structures we classify as Type A_2.

XIV. Some Acid Salts of Type A_2

A. Potassium Hydrogen Maleate

This salt is by no means an ordinary example of Type A_2, but it ranks first for description because it has been intensively studied over the past twenty years. Maleic acid, $H_2Y = C_4H_4O_4$ (16), forms the expected acid salt KHY, which crystallizes in the orthorhombic space group *Pbcm* of order eight. As there are only four KHY units per cell, molecular symmetry is implied. For this reason, and because of its "anomalous" spectrum, *Cardwell, Dunitz* and *Orgel* (*53*) suspected that the hydrogen-maleate anion had the symmetrical structure represented by (18), rather than the obvious (17). The acidic hydrogen atom was thought to lie on a

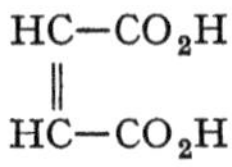
(16)
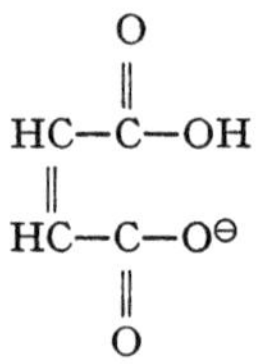
(17)
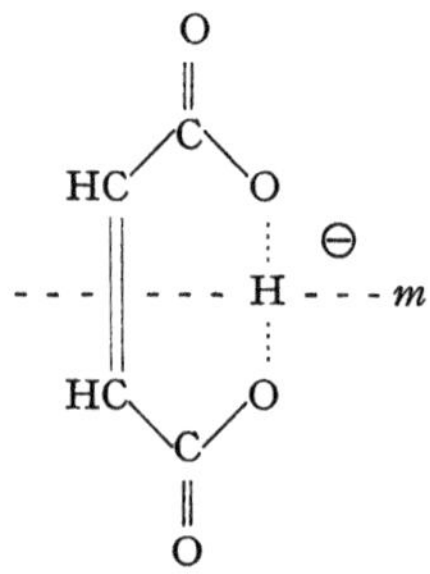
(18)

crystallographic mirror-plane (m), near to — though not necessarily at — the O···O mid-point. This looked like a symmetrical hydrogen bond at a time (1953) when such a concept had great novelty.

This led *Darlow* and *Cochran* to undertake, about 1956, the then arduous task of a three-dimensional structure analysis with X-rays, though their final results (*54*) were not published until 1961. The proposed structure was confirmed, and the length of the hydrogen bond was found to be 2.437(4) Å.

Meantime *Peterson* and *Levy* (*55*) had independently made a two-dimensional neutron-diffraction study. This gave only a projection of the structure, but located the proton more precisely, and supported the idea that it was indeed at the centre of the bond.

A technical difficulty afflicts the attempt to determine precisely the length of a bond in a crystallographically symmetrical position, such as this O···H···O bond, as well as all the hydrogen bonds in Type A structures. In general, when the positions of two atoms are known, each with a standard deviation of σ, the derived interatomic distance has a standard deviation of $\sqrt{2}\sigma$. But when the atoms are symmetry-related, the errors in their positions are correlated, so that the overall standard deviation of the distance rises to 2σ. When we want to measure some interatomic distance to within (say) 0.005 Å or less, this is a tiresome restriction[3]). To evade it, *Ellison* and *Levy* later (*56*) carried out a thorough neutron-diffraction analysis of the acid potassium salt of chloromaleic acid (19). The acid salt, $KH(C_4ClHO_4)$, crystallized in the

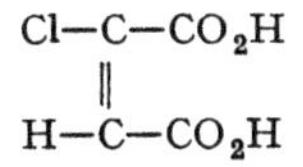

(19)

space group *Pbcn* (also of order 8), but with eight molecules per cell, so that no special symmetry is required in the acid-anion. The crystal structure is quite different from that of the unsubstituted acid maleate, but the geometry of the acid-anion is much the same. The hydrogen bond is as short as any so far measured: O···H···O = 2.403(3) Å.

Table 10. *Some dimensions in the hydrogen-anions of potassium hydrogen maleate (by X-ray diffraction), and of potassium hydrogen chloromaleate (by neutron diffraction)*

	Maleate	Chloromaleate		
		=C(Cl)CO_2		=CH · CO_2
C=C	1.348(4) Å		1.344(2) Å	
C—C	1.498(3)	1.516(2) Å		1.505(2) Å
C—O	1.235(2)	1.226(2)		1.240(2)
C—OH	1.284(3)	1.277(3)		1.280(3)
O··H··O	2.437(4)		2.403(3) Å	
		[O··H··O	175.4(4)°]	
O—C—OH	122.7(3) °	123.3(2) °		122.7(2) °
C—C—O	117.0(3)	118.6(3)		117.0(3)
C—C—OH	120.3(3)	118.0(1)		120.3(2)
C—O··O	109.3(2)			

Table 10 compares relevant X-ray results for the maleate with the neutron results for the chloromaleate anion. Because of the absence of symmetry in the latter, nearly twice as many dimensions need to be declared; those in the last column refer to the carboxyl group more distant from the chlorine substituent.

[3]) There are subtler, and perhaps more important, reasons for suspicious misgivings about crystallographic work on hydrogen bonds across symmetry elements. Unless the observational data are free from error (which they are not), electron-density calculations almost always produce peaks, or hollows, at special positions. "Rubbish" tends to collect there. Moreover, the space-group symmetry has to be built into the calculation. Therefore accurate assessment of the density genuinely due to a hydrogen atom at such a site becomes particularly uncertain. However, these hydrogen bonds across symmetry elements *do* occur. They are "there" before us. Somebody might as well study them by crystallographic methods.

The hydrogen-maleate anion has exact *m*-symmetry. As its atoms are nearly coplanar, there is a close approach to *mm*-symmetry. The first thing to notice is that the dimensions of the carboxyl groups are anomalous. As was pointed out in Section V, the almost universal rule for an non-ionized, or partly ionized, carboxyl group is that the longer C—O bond is associated with the smaller C—C—O angle. Here the rule is broken: the lengths of the C—O bonds differ in the expected way, but the difference between the angles is —3° instead of the normal +6°. The 9° anomaly is experimentally highly significant.

Darlow (*57*) has considered the strain in this anion in terms of a simple model. Given normal bond-lengths and angles, and a coplanar $O_2CHC:CHCO_2$ system, the O···H···O distance would be only 1.47 Å, which would impose an intolerable strain. This might be relieved were the ring of atoms to become buckled, but this does not occur because of the double bond, and perhaps also because the seven-membered ring is quasi-conjugated. The only other way of increasing the separation between the oxygen atoms is to change the bond-angles. (To change the lengths of covalent bonds would be much harder.) The consequent strains, defined (*57*) as the differences between the actual and ideal angles, are: +9° for C=C—C, —4° for C—C—O, and +6° for C—C—OH.

Thus the O···H···O bond is constrained to be "very short" by the external "pincer movement" of the chelation, rather than be any intrinsic attraction between the oxygen atoms. To be sure, this statement may rest on an improper distinction between attractive and repulsive forces. A molecule adopts the shape which embodies a balance of all forces at every atom. But it is surely significant that most of the shortest O···H···O bonds have been found in chelated structures such as this.

As Table 10 shows, the introduction of the chlorine atom into this rather rigid system does not make much difference in the dimensions. However, the C—C—O angles alter by about 2°, in such a sense as to reduce the anomaly in the carboxyl groups: the strains +6° and —4° in the maleate are reduced to +4° and —2° in the chloromaleate. The angles are now "the right way round", though only just.

Although the proton is no longer constrained by symmetry to be equidistant from the two oxygen atoms, the O···H distances of 1.206 and 1.199(5) Å are not significantly different. The most interesting feature is the overall shortening of the hydrogen bond by 0.037 Å, which may be connected with the electron-withdrawing influence of the chlorine substituent. Chloromaleic acid is even stronger, in its first dissociation, than maleic. And in a general way strong acidity favours strong hydrogen bonding (Sect. XX). *Ellison* and *Levy* (*56*) express the opinion that the hydrogen bond is "effectively centered".

In summary, the *cis*-configuration and rigidity of the hydrogen-maleate anion force the carboxyl groups into close proximity. The negative charge on one carboxyl attracts the H atom of the other. A "very short" and probably symmetrical, hydrogen bond results. Although phthalic acid has a similarly rigid system, the symmetrical hydrogen-phthalate anion has still to be found.

B. Potassium Hydrogen Malonate

The simplest series of dicarboxy-acids has the general formula, $HO_2C\cdot(CH_2)_n\cdot CO_2H$. The first six members ($n=0\cdots5$) are oxalic, malonic, succinic, glutaric, adipic and pimelic acids. All form acid salts, which have been — or are being — examined by diffraction methods. The first member of an homologous series is usually anomalous; the second member sets the trend. We shall therefore leave the acid oxalates till later and begin by considering an acid malonate.

Potassium hydrogen malonate, KH*Y*, where $H_2Y=HO_2C\cdot CH_2\cdot CO_2H$, forms beautiful crystals, whose morphology was described by *Haushofer* in 1882. There are four KH*Y* units in a cell belonging to the space group *C*2/*m* (order eight). Special symmetry is required; in fact the K^+ ion lies on a mirror-plane, whilst the malonate residue has its CH_2 group on a dyad axis. The acidic H atom is at a centre of symmetry. The anion structure consists of infinite hydrogen-bonded chains of the form:

$$\underset{*}{\cdots H\cdots}O_2C\cdot \underset{\uparrow}{HCH}\cdot CO_2\underset{*}{\cdots H\cdots}O_2C\cdot \underset{\uparrow}{HCH}\cdot CO_2\underset{*}{\cdots H\cdots}\;,$$

where the centres are marked by asterisks and the axes by arrows. Potassium hydrogen malonate clearly belongs to Type A_2.

This structure was studied independently by *Parthasarathy* (*58*) and by a group at Glasgow (*59*). The disagreement between the structures originally reported has now been resolved (*60*). A neutron-diffraction analysis, yielding more precise locations for the hydrogen atoms, has also been published (*61*). The structure is shown in Fig. 12, relevant dimensions in Table 11.

Table 11. *Some dimensions in potassium hydrogen malonate (X = X-ray, N = neutron diffraction results)*

	X	N		X	N
			O—C—OH	123.1(3)	122.5(3) °
C—O	1.223(4) Å	1.217(3) Å	C—C—O	124.4(3)	125.0(2)
C—OH	1.293(4)	1.293(4)	C—C—OH	112.5(3)	112.5(2)
O···H···O	2.459(5)	2.468(6)	C—O····O	111.7(2)	111.6(2)

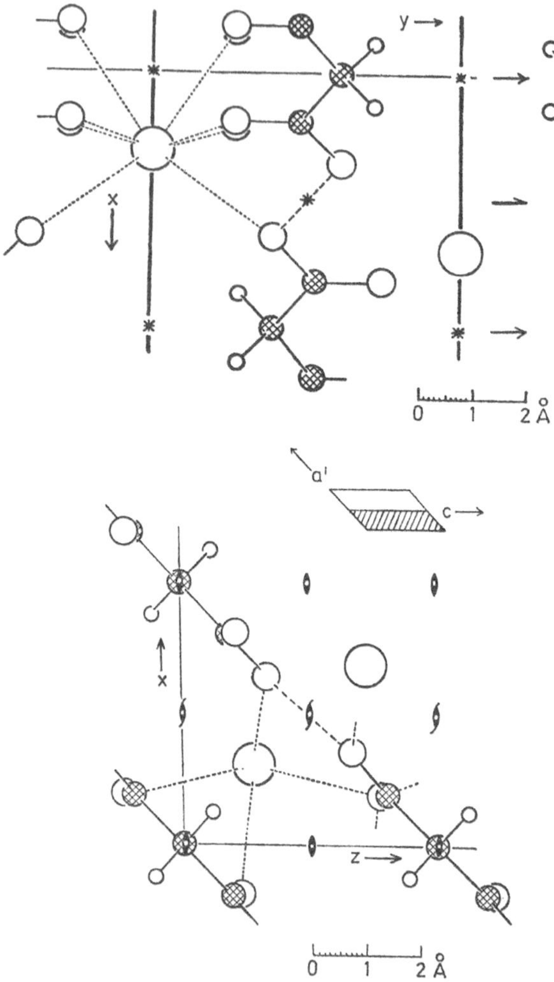

(a)

(b)

Fig. 12. Crystal structure of potassium hydrogen malonate: (a) *c*-axis projection; (b) *b*-axis projection, with inset showing relationship between habit and structure

The two C—CO_2 units of the anion, related by the twofold axis, have an angle of less than 10° between their respective mean planes, so that the entire anion is roughly planar. As the hydrogen bonds are almost coplanar with the carboxyls, the whole malonate chain constitutes a nearly flat ribbon (parallel to the prominently developed (201) face of the crystal, as indicated in Fig. 12(b)).

The infrared spectrum (Fig. 20(a)) of potassium hydrogen malonate is "anomalous".

C. Some Acid Salts of Succinic Acid

The morphology of the acid potassium salts of succinic acid ($H_2Y = HO_2C \cdot CH_2 \cdot CH_2 \cdot CO_2H$) was studied by *Marshall* and *Cameron* (*62*), who commented on the tendency, which we too have noticed, for some of these salts to form crystals with curved faces and edges. They also studied the phase-diagram for the system H_2Y—K_2Y—H_2O, which is a good example of the effect represented in Fig. 1(d). A 1:1 aqueous solution of KOH and H_2Y yields, on cooling or evaporation, crystals of composition KH_3Y_2. To prepare KHY one must use a solution containing a higher proportion of alkali.

Potassium hydrogen succinate crystallizes with four KHY_2 molecules in a cell belonging to the space group $I2/a$ (equivalent to $C2/c$). This at once suggests a Type A_2 structure, as was borne out by a detailed X-ray and neutron analysis by *McAdam* (*63*). The hydrogen-succinate ions, lying on symmetry-centres, are linked into infinite chains by hydrogen bonds lying across twofold axes:

$$\cdots\underset{\uparrow}{H}\cdots O_2C\cdot CH_2\cdot\underset{*}{}CH_2\cdot CO_2\cdots\underset{\uparrow}{H}\cdots O_2C\cdot CH_2\cdot\underset{*}{}CH_2\cdot CO_2\cdots\underset{\uparrow}{H}\cdots .$$

This is an inverse use of the symmetry elements used in the malonate chains. Fig. 13 gives two views of the structure, and Table 12 relevant dimensions. Hydrogen atoms excepted, the whole anion is virtually planar, in contrast to the succinic acid molecule in the crystal of the β-form (*64*), which is twisted by 11° about the central C—C bond.

The monohydrated acid Cs salt (*63a*), $CsHY \cdot H_2O$, crystallizes with two formula-units in a cell of space group $P2_1/m$ (order 4) and has a Type A_2 structure. The accuracy of the X-ray analysis is limited by the presence of Cs. This atom and the water molecule lie on mirror-planes; infinite chains of hydrogen-succinate anions — themselves centrosymmetric — are linked by O···H···O bonds across centres of symmetry. The water molecule makes a pair of equivalent hydrogen bonds with carbonyl oxygen atoms: O—H···O = 2.73(3) Å.

Table 12. *Some dimensions in potassium hydrogen succinate* (*X = X-ray, N = neutron diffraction results*

	X	N		X	N
			O—C—OH	123.5(2)°	123.3(5)°
C—O	1.225(4) Å	1.223(7) Å	C—C—O	122.9(2)	122.9(4)
C—OH	1.301(2)	1.299(7)	C—C—OH	113.6(2)	113.8(4)
O···H···O	2.446(4)	2.444(10)	C—O····O	114.3(2)	115.2(4)

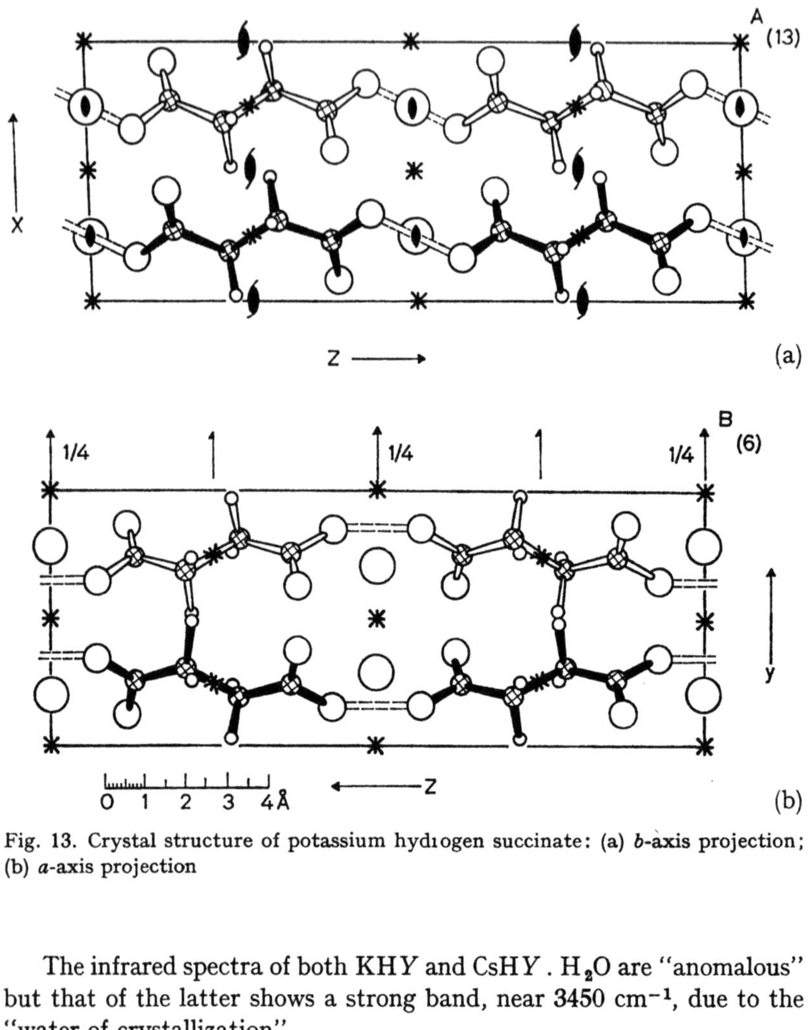

Fig. 13. Crystal structure of potassium hydrogen succinate: (a) *b*-axis projection; (b) *a*-axis projection

The infrared spectra of both KH*Y* and CsH*Y* . H_2O are "anomalous" but that of the latter shows a strong band, near 3450 cm^{-1}, due to the "water of crystallization".

D. Some Acid Salts of Glutaric Acid

Acid salts of this acid ($H_2Y = HO_2C \cdot CH_2 \cdot CH_2 \cdot CH_2 \cdot CO_2H$) have been studied by *Macdonald* (*65*). The structure of RbH*Y* — the first to be elucidated — is remarkable (*66*). Its general lay-out is similar to those of the acid malonate and acid succinates, but it has been marshalled into a much trimmer array. The space group is the orthorhombic *Cmma* (of order 16). As the cell contains only four RbH*Y* units, severe symmetry restrictions are imposed. The Rb^+ ions lie on points of 222-symmetry, the hydrogen-glutarate residues have *mm*-symmetry, and the hydrogen bonds, which link these residues into infinite chains, lie across points of 2/*m*-symmetry. Only nine parameters are needed to position the ten heavier atoms of $RbC_5H_7O_4$. Such highly symmetrical arrangements are common amongst simple inorganic crystals, but are rare amongst organic compounds. The crystal structure (Fig. 14(a)) is of Type A_2. The dimensions (Table 13) are of limited accuracy because of the presence of Rb.

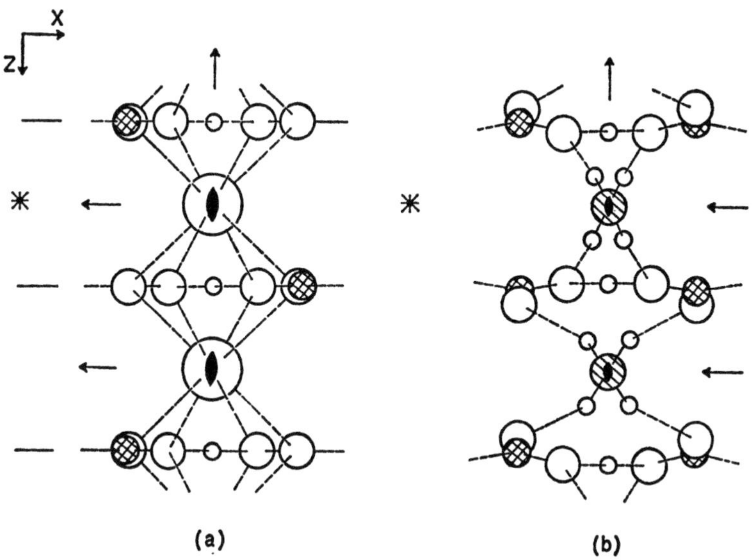

Fig. 14. Crystal structures of acid glutarates in the region of their cations and of the hydrogen bond: (a) rubidium (or potassium) salt; (b) ammonium salt

X-Ray data have been collected for KH*Y*, which is isomorphous with RbH*Y* but less affected by absorption errors. Refinement has led to a more precise analysis (*105*), some interim results of which have been included in Table 13.

Table 13. *Some dimensions in acid glutarates*, *M*H*Y*

M	Rb	NH_4	K
C—O	1.194(15) Å	1.223(6) Å	1.225(2) Å
C—OH	1.284(13)	1.279(5)	1.299(3)
O···H···O	2.40(2)	2.460(7)	2.445(3)
O—C—O	124.5(1.0) °	123.3(3) °	122.4(2) °
C—C—O	122.5(1.0)	122.1(3)	123.3(2)
C—C—OH	113.0(1.0)	114.5(4)	114.3(2)
C—O····O	112.5(1.0)	113.5(3)	110.2(3)

Table 14. *Cell parameters of acid glutarates, MHY, and of potassium hydrogen pimelate (KHPim)*

M		Rb	NH_4	K	KHPim
a	Å	18.55	18.59	18.476	22.46
b	Å	7.57	7.56	7.282	7.45
c	Å	5.29	5.27	5.133	5.04
Z		4	4	4	4
Space group		*Cmma*	*Pmaa*	*Cmma*	*Cmma*

The ammonium salt, NH_4HY, crystallizes in the same system as Rb*HY*, with almost identical cell dimensions (Table 14), but it gives some weak reflexions that systematically infringe the condition for *C*-centring and show that this crystal must belong to a different space group, *Pmaa*.

The debasement of symmetry when Rb^+ is replaced by NH_4^+ is attributable to the formation of N—H···O bonds (Fig. 14(b)). In Rb*HY* each Rb^+ makes contact with eight oxygen atoms, in two sets of four: one set comprises the oxygens of four equivalent carbonyl groups, with $Rb^+ \cdots O = 3.078(7)$ Å; the other, oxygens of four equivalent hydroxyl groups, with $Rb^+ \cdots O = 2.946(6)$ Å. Two cation-sites, equivalent in Rb*HY*, become non-equivalent in NH_4HY, but each ammonium ion makes four N—H···O contacts. The carbonyl oxygens move 0.32 Å towards one site, and the hydroxylic oxygens move 0.19 Å in the opposite direction towards the other cation-site. This twisting of the carboxyl group is communicated in a lesser degree to the methylene groups. The glutarate residue becomes non-planar. This structure accounts for the extra X-ray reflexions from NH_4HY and has been confirmed by a neutron-diffraction analysis (*66*). Dimensions are included in Table 13.

All these acid glutarates have "anomalous" spectra.

E. Acid Salts of Adipic and Pimelic Acids

Several acid salts of adipic acid ($H_2Y = HO_2C \cdot (CH_2)_4 \cdot CO_2H$) have been prepared, but the crystals are unsuitable for X-ray diffraction. However, from the next homologue, pimelic acid ($H_2Y = HO_2C \cdot (CH_2)_5 \cdot CO_2H$), K*HY* can be obtained as suitable crystals. Crystallographic data are included in Table 14. Two cell dimensions are nearly the same as in the glutarates, but one is increased to accommodate the extra methylene groups. This, together with similarities between corresponding X-ray photographs, suggests that the acid pimelate is isostructural with the acid potassium and rubidium glutarates. Further work is planned.

XV. Some Acid Salts of Type B_2

The crystalline acid salts of dicarboxylic acids so far described have been of Type A_2. Their hydrogen-anions are not well represented by the conventional formula (20), but rather by the over-elaborate (21). Type B_2 structures corresponding to (20) are not numerous amongst the acid

(20)

(21)

salts so far studied. One well authenticated example is ammonium hydrogen *d*-tartrate (*52*), where ionized and non-ionized carboxyl groups have their dimensions differentiated with a clarity worthy of a textbook (Table 15).

Table 15. *Some dimensions in the ionized and non-ionized carboxyl groups of ammonium hydrogen tartrate*

(Ionized)				(Non-ionized)			
		O—C—O	124.8°			O=C—OH	124.5°
C—O	1.261(4) Å	C—C—O	118.2	C=O	1.220(4) Å	C—C=O	123.1
C—O	1.263(4)	C—C—O	117.1	C—OH	1.311(4)	C—C—OH	112.5

Another example is sodium bicarbonate (*66a*), NaH*Y*, if we admit carbonic acid ($H_2Y = H_2CO_3$) as prototype of a dicarboxylic acid. The hydrogen-carbonate units are linked by hydrogen bonds into infinite chains (Fig. 15(a)), much like the chains in Type A_2 acid salts. If the hydrogen atoms are ignored, the bicarbonate chains might appear to be centrosymmetric. But each unit is actually related to its neighbours by a

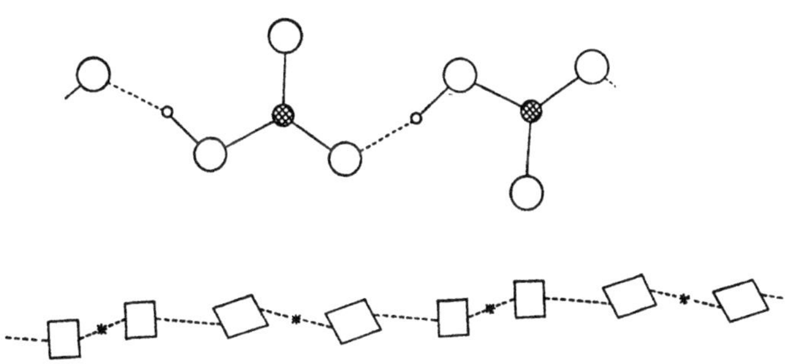
(a)
(b)

Fig. 15. (a) HCO_3 chains in sodium bicarbonate; (b) schematic representation of infinite chains of hydrogen-anions in potassium hydrogen mesotartrate

glide-plane of the crystal. There is no symmetry element at the centre of the hydrogen bond, and the participating O atoms are not equivalent. The H atom is about 1.1 Å from one O and 1.5 Å from the other (*67*). The C—O distances unambiguously support the implied attachment: C—O = 1.275 Å, C—OH = 1.345 Å, and C=O (carbonyl) = 1.246 Å. The O—H···O distance is 2.611 Å, the C—O(H)···O angle 112°.

The infrared spectrum of sodium bicarbonate (Fig. 20(c)) resembles those of Type *A* crystals.

A recent analysis of sodium hydrogen fumarate (*67a*) indicates a structure of Type B_2.

XVI. Some Acid Salts of Oxalic Acid

A. Hydrazinium Hydrogen Oxalate

Oxalic acid ($H_2Y = HO_2C \cdot CO_2H$) and its salts have been the subjects of extensive, and sometimes intensive, crystal-structure analysis for about 40 years (*68–71a*). To the middle of 1970, the most detailed study of an acid oxalate was that of hydrazinium hydrogen oxalate ($H_2N \cdot NH_3^+$ HY^-) by both X-ray (*72*) and neutron (*73*) diffraction. With two formula units in a cell of space group $P2_1/m$, this has a Type A_2 structure: centrosymmetric HY^- ions are linked into infinite chains by "very short" hydrogen bonds across centres of symmetry. Dimensions are in Table 16. This symmetry implies that the hydrogen bonds are "*trans*" with respect to the oxalate residue and that the carboxyl groups are parallel. Analysis shows them to be virtually coplanar.

Table 16. *Some dimensions in hydrazinium hydrogen oxalate* (*X* = *X-ray*, *N* = *neutron-diffraction results*

	X	N		X	N
			O—C—OH	125.7(2)°	126.1(3)°
C—O	1.230(3) Å	1.236(4) Å	C—C—O	120.4(3)	119.8(3)
C—OH	1.279(3)	1.280(4)	C—C—OH	113.9(3)	114.1(4)
O···H···O	2.450(4)	2.448(6)	C—O····O	114.5(2)	114.5(3)

B. Potassium Hydrogen Oxalate

The acid potassium salt, contrarily, crystallizes with four KH*Y* units per cell in $P2_1/c$, no special molecular symmetry being implied. The infinite chains of HY^- units are topologically like those in the hydrazinium salt; but the hydrogen bonding is *"cis"* at each oxalate residue, the residues are significantly non-planar, and the hydrogen bonds do not lie across any symmetry element of the crystal (Fig. 16). From her dimensions for the carboxyl groups, *Pedersen* (*74*) deduced that *C* is ionized, *D* not. This has now been confirmed by a more definitive analysis (*74a*), some details from which are shown in Table 17. The overall length of the hydrogen bond being 2.52 Å, the acidic hydrogen would not be centrally placed (see Sect. XX); and this X-ray study duly finds the hydrogen in a (qualitatively) appropriate position. This is clearly a Type B_2 structure, very like that of the hydrated (acid) sodium salt so far as the oxalate chains are concerned (see Sect. XVI *C*). The infrared spectrum of KH*Y* is "anomalous" (Fig. 20(d) (*74*)).

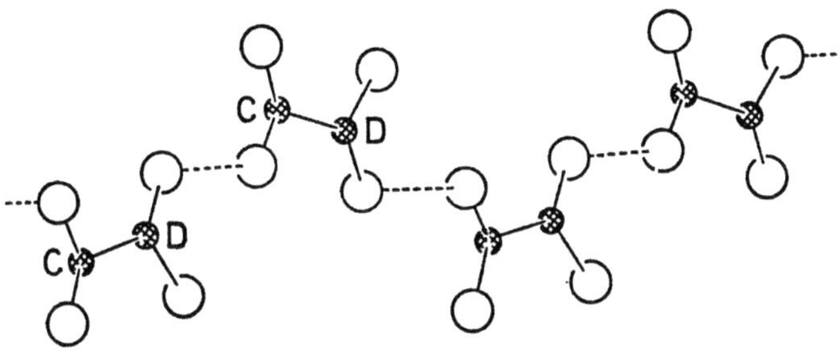

Fig. 16. Hydrogen-anion chains in potassium hydrogen oxalate

Table 17. *Some dimensions in potassium hydrogen oxalate (By X-rays: standard deviations less than 0.004 Å and 0.4°)*

Ionized carboxyl (*C*)				Non-ionized carboxyl (*D*)			
		O—C—O	127.7 °			O—C—O	124.7 °
C—O	1.233 Å	C—C—O	116.8	C—O	1.211 Å	C—C—O	120.5
C—O(H)	1.256	C—C—O(H)	115.5	C—OH	1.308	C—C—OH	114.7
		O—H···O		2.525 Å			

C. Sodium Hydrogen Oxalate Monohydrate

Two recent reports (*74b, 74c*) of this structure show it to be an ideal example of a Type B_2 structure. At the ionized end of the hydrogen-oxalate residue C—O = 1.242(2) and 1.251(2) Å, and C—C—O = 115.2(2) and 117.4(2)°; at the other end, C—OH = 1.313(2) and C=O = 1.207(2) Å, and C—C—O = 113.1(2) and 121.5(2)°. There are infinite hydrogen-bonded chains with O—H···O = 2.571(2) Å.

The corresponding lithium salt, LiH*Y* . H_2O, has a very different, and apparently unsymmetrical, structure (*74d*).

XVII. Anomalous Acid Salts of Dicarboxylic Acids

A. General

The acid salts of dicarboxy-acids so far described have compositions in the ratio of MHY. But just as monocarboxy-acids sometimes form salts of anomalous composition, such as MHX_2, so also do dicarboxylic acids. We have already mentioned the "superacid" succinate, KH_3Y_2. A variety of such compositions is known, and examples are certain to multiply as search is turned in that direction.

A series of superacid oxalates has been known for over a century — the tetroxalates (or "quadroxalates") of composition $MHY.H_2Y.2\,H_2O$, where H_2Y = oxalic acid. The atom M may be K, Rb, Tl or NH_4, and the members of the series are isomorphous despite the considerable range of cation-size.

Sodium sesquicarbonate, $Na_3HY_2.2\,H_2O$, where H_2Y = carbonic acid, was mentioned in the introduction. Its structure (*75*) is partially of Type A_2. Salts of the cations *cis*- and *trans*-$[Coen_2Cl_2]^+$ with maleic, glutaric, malonic, thiodiacetic and other acids have been prepared, with variously superacid compositions (*76*).

A recent case of some complexity is $Na_2H_4Y_3$, where H_2Y = cyclobutane-*trans*-1,3-dicarboxylic acid. The crystal structure (*77*) corresponds to the formulation, $2\,Na^+\,Y^{2-}.\,2\,H_2Y$. The double anion is centrosymmetric and can be identified because its C—O distances (1.25, 1.28 Å) and C—C—O angles (120°, 117°) are nearly equal. The centrosymmetry of the anion requires the four-membered ring to be planar. In contrast, the cyclobutane ring of the non-ionized molecule is buckled. The two H_2Y molecules — related by a crystallographic centre of symmetry — are identifiable by having different C—O distances and C—C—O angles: 1.32, 1.21 Å and 114°, 123° in one carboxyl group; 1.32, 1.22 Å and 112°, 124° in the other.

The structure of KH_3Y_2, where H_2Y = glutaric acid, contains infinite chains of unsymmetrical HY^- anions linked by short — but not symmetrical — O—H···O bonds. These chains are cross-linked through H_2Y molecules, bound at each end by longer hydrogen bonds between their hydroxylic oxygen atoms and the non-hydroxylic oxygen atoms of the chains (*105*).

B. The Crystal Structures of the Tetroxalates

As the tetroxalates crystallize beautifully, provide a lavish series of isomorphs, and have been well known for many years, it is surprising that their crystal structure was not determined until 1964 when *Haas*

(*78*) untertook work on potassium tetroxalate by accident. He writes, "The molecular weight of the unit-cell contents···was only slightly greater than that of the supposed formula: $K_3Al(C_2O_4)_3$. 3 H_2O, but when the positions of all the non-hydrogen atoms were established by X-ray diffraction methods, the unit cell was found to contain two formula weights of $K(HC_2O_4)$. $(H_2C_2O_4)$. 2 H_2O."

Haas' analysis was an accurate one, but we prefer to describe the tetroxalate structure in terms of the isomorphous ammonium compound, NH_4HY. H_2Y. 2 H_2O, for which a detailed neutron-diffraction analysis has been carried out (*79*).

There are two NH_4HY . H_2Y . 2 H_2O units per cell of the space group $P\bar{1}$. No special molecular symmetry is implied, but it turns out the two H_2Y molecules lie on non-equivalent centres of symmetry. Hence the asymmetric unit of the crystal can be expressed as $\frac{1}{2}(H_2Y)^{I}$. $\frac{1}{2}(H_2Y)^{II}$. MHY . 2 H_2O. The identification of the two H_2Y molecules is confirmed by the differences between appropriate bond-lengths and angles (Table 18). The third oxalate residue must correspond to HY^-. Its B_2 character is supported by its dimensions and confirmed by the location of all eleven protons, in positions indicated in Fig. 17.

Table 18. *Some dimensions in ammonium tetroxalate (neutron diffraction results). Some dimensions in potassium tetroxalate (By X-rays) are added for comparison*

	$(H_2Y)^{I}$	$(H_2Y)^{II}$	(HY^-)	
C—C	1.549(5) Å	1.544(5) Å	1.549(3) Å	
C—O	1.211(4)	1.206(4)	1.212(4) Å	1.230(4) Å
C—OH	1.288(4)	1.296(4)	1.291(4)	1.247(4)
O—C—O	126.3(3) °	126.5(3) °	124.8(3) °	127.5(3) °
C—C—O	120.6(2)	121.6(2)	121.8(2)	117.5(2)
C—C—OH	113.0(2)	112.0(2)	113.3(2)	115.0(2)
	K^+····O	NH_4^+···O		
	2.870(4) Å	2.950(4) * Å		
	2.870(4)	2.938(3) *		
	2.879(4)	2.937(4)		
	2.893(4)	3.069(4)		
	2.915(4)	2.974(4) *		
	2.931(4)	2.927(4) *		
	2.924(4)	3.166(4)		
	3.044(4)	3.247(4)		
O—H····O	O—H	C—O····O	O—H····O	
2.472(4) Å	1.069(7) Å	113.4(2)	178.2(6) °	
2.500(4)	1.102(7)	114.7(2)	177.1(7)	
2.520(5)	1.040(7)	113.1(2)	173.3(7)	
2.721(4)	0.966(7)		169.5(6)	
2.745(4)	0.973(6)		167.4(6)	
2.833(5)	0.949(8)		164.8(7)	
2.899(5)	0.945(8)		158.2(7)	

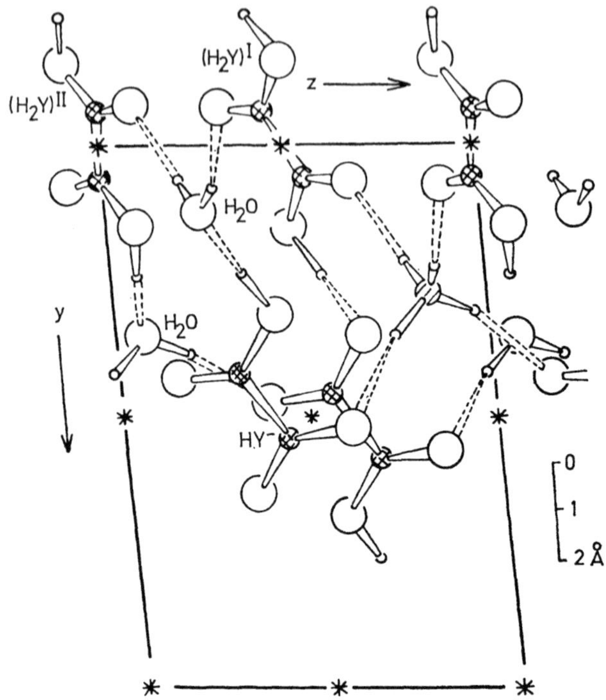

Fig. 17. Crystal structure of ammonium tetroxalate, *a*-axis projection

Location of these protons provides information in another direction. K^+ or Rb^+ ions in crystals of this sort make contacts with 6, or 7, or sometimes more, oxygen atoms, at distances ranging from 2.7 Å upwards for K^+, and from 2.8 Å for Rb^+. In an isomorphous ammonium salt, NH_4^+ will tend to form N—H···O bonds. There are then only four protons available for a larger number of possible contacts. Two obvious possibilities exist: either the protons are localized in four of the N···O contacts, which thereby attain special status; or there is some disordering, so that some, or all, of the four protons are shared between two N···O contacts. A situation analogous to the latter obtains in the higher-temperature forms of the ammonium halides. The former possibility is realized in ammonium tetroxalate.

The four N—H···O contacts (asterisked in Table 18) fit the tetrahedral symmetry of the ammonium ion very well; the four N—H···O angles are all within the range 160—165°, and the six H—N—H angles within the range 106—113°. There is one N···O contact, without a proton, which is not significantly longer than the longest N—H···O distance. When K^+ is replaced by NH_4^+, there happens to be a particular orientation of the latter which enables it to make four tetrahedrallydirected hydrogen bonds; along with this, there is a general, minor repositioning of the environment.

The neutron-diffraction results also establish the positions of the protons in seven different O—H···O bonds, and hence help to define the Nakamoto-Margoshes-Rundle curve. (See Sect. XX *A*.) These bonds all have their proton nearly colinear with their oxygen atoms, and the short ones have a carboxyl group as proton-donor.

C. Potassium Hydrogen Mesotartrate

A low-temperature X-ray study of potassium hydrogen mesotartrate, KH*Y*, where H_2Y = *meso*-$HO_2C \cdot CH(OH) \cdot CH(OH) \cdot CO_2H$ has recently been carried out by *Kroon* and *Kanters* (*79a, 114*). Though not compositionally anomalous, this acid salt has a structure that is an abnormal variant of Type A_2.

It is easy to imagine a Type A_2 crystal structure for KH*Y*: centrosymmetric hydrogen-mesotartrate anions linked into chains by hydrogen bonds across centres of symmetry. In fact, neither mesotartaric acid nor its anion adopts a centrosymmetric conformation in any of the crystals so far studied.

The structure of potassium hydrogen mesotartrate is rather complex. The space group is $P\bar{1}$ (order two), with four KH*Y* molecules per cell, and hence two KH*Y* per asymmetric unit. The structure is a highly contorted modification of an ideal A_2 arrangement. The anion-chain is complicated because it contains two sorts of hydrogen-anions with different conformations, and because weaker intermolecular hydrogen bonds are formed by the alcoholic hydroxyl groups. In Fig. 15(b) the crystallographically distinct anions are betokened by squares and rhombs. There are three kinds of inter-carboxyl hydrogen bonds: that between a square and a rhomb is necessarily non-symmetric (O···O = 2.543(1) A); the other two (O···O = 2.452(2) and 2.483(2) A) lie across centres of symmetry.

XVIII. Summary of Results for the Acid Salts of Symmetrical Dicarboxylic Acids

The free acids of the series $HO_2C \cdot (CH_2)_n \cdot CO_2H$, where $n = 0, 1, 2, 3, 4$ and 5 have known crystal structures, from which one can draw the following generalizations: (a) The molecule is planar, or nearly so, as far as its carbon and oxygen atoms are concerned; and (b) the molecule is symmetrical about its mid-point, this being a crystallographic inversion centre when n is even, and a twofold axis when n is odd. Malonic acid is an exception (*80*): one carboxyl group is twisted through nearly 90° about the C—C bond, as is necessary to prevent two oxygen atoms — one from each carboxyl — getting in one another's way.

The simple acid salts, MHY, conform to these rules, and with rather greater regularity in that malonic acid is no longer exceptional; its hydrogen-anion is nearly planar, though the oxygen atoms just maintain a proper van der Waals separation. Hydrazinium hydrogen oxalate conforms to the rules, but the potassium salt — as we have seen — is obdurately unorthodox. Acid salts of adipic acid have not yet been studied.

In all these acid salts, the hydrogen-anions are linked to give structures of Type A_2 by O···H···O bonds across crystallographic symmetry elements. Table 19 collects some of the more important results for comparison with those in Table 7.

Table 19. *Summary of inter-carboxyl hydrogen bonds in some Type A_2 acid salts of dicarboxylic acids, MHY*

H_2Y	M	space group	Molecules per cell	Symmetry of bond	O···H···O	C—O···O	Ref.
malonic	K	$C2/m$	4	$\bar{1}$	(X) 2.459(5) Å	111.7°	(*60*)
					(N) 2.468(6)	111.6	(*61*)
succinic	K	$I2/a$	4	2	(X) 2.446(4)	114.3	(*63*)
					(N) 2.444(10)	115.2	(*63*)
glutaric	Rb	$Cmma$	4	$2/m$	2.40(2)	112.5	(*66*)
	NH_4	$Pmaa$	4	2	2.460(7)	113.5	(*66*)
	K	$Cmma$	4	$2/m$	2.445(3)	110.2	(*105*)
oxalic	N_2H_5	$P2_1/m$	2	$\bar{1}$	(X) 2.450(4)	114.5	(*72*)
					(N) 2.448(7)	114.5	(*73*)

[Weighted mean 2.450(2) Å]

XIX. The $H_5O_2^+$ Ion

The hydrogen ion of the aqueous-solution chemist is conventionally written as "H^+", but no electrochemist of the last 50 years has taken this symbol other than in a Pickwickian sense. The ion is certainly hydrated; and the formula "H_3O^+" is often used to indicate the hydration explicitly. Even this should not be taken literally but rather as including implicitly all such other, more hydrated species as may be in equilibrium with protons: H_3O^+, $H_5O_2^+$, $H_7O_3^+$, etc. We do not yet have any firm description of liquid water in detailed molecular terms, and therefore to rationalize the cationic content of an aqueous acid in terms of these species may be artificial. However, there is direct evidence for the existence of the first three in crystals.

Solid nitric acid, for example, contains the entities H_3O^+ and NO_3^-. The species $H_5O_2^+$ is also found in crystals, and sometimes it is centred upon a point of inversion. We then have an exact Type *A motif*, with a "very short" hydrogen bond; $[H_2O\cdot\cdot H\cdot\cdot OH_2]^+$ is a special case of the general cation $[BHB]^+$ which occurs in the symmetrical basic salts of some monoacidic bases (Sect. XI).

The best example, and the longest known, occurs in the cobalt complex originally formulated as $[Co\,en_2Cl_2]Cl\,.\,HCl\,.\,2\,H_2O$ (*en* = ethylenediamine). With two formula units per cell in the space group $P2_1/c$, a Type *A* structure is indicated; and a reasonable guess would be that the hydrogen-dichloride ion, $[Cl\cdot\cdot H\cdot\cdot Cl]^-$, is involved. The actual organization (*81*) is $[Co\,en_2Cl_2]^+2\,Cl^-\,.\,H_5O_2^+$. There are two oxygen atoms, 2.66 Å apart across a centre of symmetry, which must correspond to the two molecules of "water of crystallization", bonded through the acidic hydrogen atom. This hydrogen was not located in the initial X-ray work, but the above deductions have now been confirmed by a neutron-diffraction analysis (*82*). The $O\cdots H\cdots O$ distance is revised to 2.50(3) Å, with the proton unmistakably at the centre of $H_5O_2^+$. Similar Type *A* structures occur in coordination compounds of some other metals.

However, in most crystals containing $H_5O_2^+$, the ion does not have strict symmetry. These then are, at best, pseudo-Type *A* structures. For instance, a neutron-diffraction study (*83*) of $HAuCl_4\,.\,4\,H_2O$ shows the mid-point of $H_5O_2^+$ to lie at a crystallographic centre of symmetry, although the $O\cdots O$ distance is as large as 2.57(1) Å. There appear to be two half-protons, each about 0.99 Å from an oxygen atom but well off the $O\cdots O$ line. *Williams* and *Peterson* (*83*) hint at spectroscopic reasons for suspecting a mystery clouding the true nature of this $H_5O_2^+$ ion.

The same authors have briefly reported (*84*) neutron-diffraction work on the hexahydrate of anthranilic acid. From our point of view, the important finding is a discrete $H_5O_2^+$ ion, with $O\cdots O = 2.44$ Å, but without special crystallographic symmetry; the proton is definitively located 0.05 Å from the bond-centre. The hydrates of hydrogen chloride, $HCl\,.\,2\,H_2O$ and $HCl\,.\,3\,H_2O$ (*85*) also contain $H_5O_2^+$ ions without special symmetry, though both hydrogen bonds are "very short" (2.414(7) and 2.43(1) Å).

It has recently been proposed (*86*) that $H_5O_2^+$ is "an entity produced when two oxygens interact through a *proton* (with $O\cdots H\cdots O$ linear)

...and that the entity may exist in environments other than crystalline ...and may be a form of the hydrated proton in water." Calculations on its stability are reported.

Derry and *Hamor* (*86a*) have recently published evidence for the existence of a discrete $H_7O_3^+$ ion in crystals of composition $C_{14}N_2H_{16}^{2+}$ $2\,Br^-.\,H_2O.\frac{1}{3}\,HBr$.

XX. A General, Elementary Consideration of "Very Short" O···H···O Bonds

A. Some Properties of Hydrogen Bonds

The principal interest of the acid salts described in this review is that many of them, those of Type *A*, contain some of the shortest known hydrogen bonds between oxygen atoms. A tentative discussion of these short bonds within the framework of hydrogen bond theory may be permissible. To facilitate such discussion, the background will be sketched. Detailed, and more authoritative, accounts will be found elsewhere (*87, 88, 89*).

The hydrogen bond results from a relatively weak attraction between one entity *A*H and another *B*: *A*—H···B. The atoms *A* and *B*, or the relevant atoms of *A*H and *B* if these be molecules, must be strongly electronegative. Firm hydrogen bonding is largely confined to systems where *A* and *B* are F, O, or N. Another condition for hydrogen bonding, closely related to the first, is that *A*H must be a reasonably strong acid and *B* a reasonably strong base. However, if the acidic and basic strengths become too high, the tendency towards hydrogen bonding over-reaches itself; the proton is completely transferrred from *A* to *B* and an ionic bond results; *A*—H···*B* becomes $A^\ominus$ H—$B^\oplus$.

Hydrogen bonding was recognized as a phenomenon about 1905, though the name was not applied until 1921 (*88, 89*). Rational discussion of its mechanism could hardly begin until the results of various physical studies became available from about 1926. Especially important were the structures of crystals as elucidated by X-ray diffraction, for instance, by *Robertson* (*90*). The primitive crystallographic evidence of hydrogen bonding was the location of two oxygen atoms — for it is with O···H···O bonding that we are almost exclusively concerned here — in positions too far apart for them to be covalently bonded to each other, and too close for them not to be bonded at all.

Covalently bound oxygen atoms would be less than 1.5 Å apart, but no hydrogen bonds less than 2.4 Å long are known. The upper limit for hydrogen bonding might be set at about 2.8 Å, twice the van der Waals radius of oxygen. This limit must be raised a little, however, since a hydrogen atom, covalently bonded to one of the oxygen atoms, must lie roughly along the O···O direction. We should then add to the 1.4 Å radius of one oxygen atom about 1.0 Å for the O—H distance at the other oxygen, plus a further 0.9 Å (say) for the van der Waals radius of hydrogen; the sum is about 3.3 Å. (This simple view has been questioned by *Bellamy* and *Pace* (*91*).) In practice, accepted hydrogen bonds in crystals have O···O from 2.4 up to about 3.1 Å[4]).

Before 1950 X-ray analysis could not readily detect the low electron density due to a hydrogen atom. Hydrogen bonding was merely inferred from the evidence explained above. In the more exact X-ray work now normally expected, hydrogen atoms can be located, and we then get direct evidence for their involvement in hydrogen bonding. To be sure, X-rays usually find the hydrogen atom in the wrong place... For reasons that are now understood, refinement procedures lead to a residual electron-density peak usually about 0.1 Å too close to *A* in an A—H···O bond.

In a very weak O—H···O bond, with O···O $>$3.0 Å, the O—H distance differs little from 0.95 Å, the distance in an unperturbed hydroxyl group. It is reasonable to expect that O—H will increase as the overall O···O distance diminishes; the approach of a second oxygen atom will weaken the original O—H bond. Unfortunately, it is impracticable to observe such changes directly as we progressively shorten an isolated hydrogen bond. All we can do is to compare O—H and O···O distances in a series of different bonds, which we deem to be more or less comparable. The accumulation of results from neutron-diffraction analyses enabled *Nakamoto, Margoshes* and *Rundle* (*92*) to plot O—H vs. O···O distances, and hence to sketch a curve relating these distances in an average sense. (Fig. 18).

The curve is reasonably well defined for O···O $>$2.6 Å, but at shorter distances is becomes increasingly uncertain. Were the increase of O—H with decrease of O···O to continue long enough, we should arrive at a truly *symmetrical hydrogen bond*, with the proton vibrating about a single position of minimum energy at the mid-point. A situation of this sort is supposed to obtain in the FHF^- ion (*93*), with F···F = 2.264(3) Å. If analogous symmetry can be realized in the O—H···O bond, the possibilities are represented by the straight line, of slope 0.5, drawn at the left-hand side of Fig. 18. We do not yet know for certain

---

[4]) *Intra*molecular O···O distances much less than 2.8 Å may occur in molecules where hydrogen bonding is out of the question. For instance, in 1,6-8,13-bis-oxido-[14]-annulene (*91a*) the oxygen atoms are only 2.55 Å apart. This is analogous to the situation in hexachloroethane, where there are, inevitably, Cl···Cl contacts of 2.8 and 3.1 Å — much below 3.6 Å, the sum of the van der Waals radii of the two chlorine atoms.

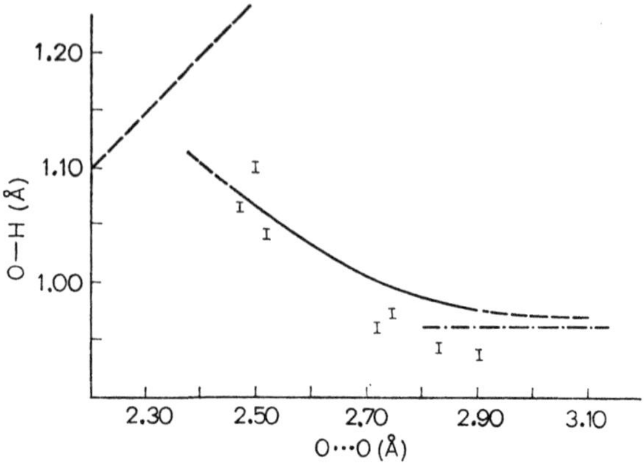

Fig. 18. Relationship between O—H and O····O distances in hydrogen bonds. (The seven points marked correspond to the O—H···O bonds described in Table 18, the lengths of the vertical lines represent the standard deviations of O—H. The curve is based on that given in reference *88*.)

whether such truly symmetrical O···H···O bonds exist, but the "very short" bonds presented in this review are plausible condidates.

The ideas sketched in the previous paragraph were clearly stated in 1952 by *Donohue* (*35*), who estimated that the O···O distance in a symmetrical hydrogen bond should be about 2.30 Å. Bonds anywhere nearly as short as this are unknown. The finding of "very short" bonds in the Type *A* crystals has therefore been a source of mitigated embarrassment to some experimentalists. These bonds certainly seem symmetrical, but they are still much too long to satisfy the Donohue criterion...

The increasing strength of hydrogen bonds as O···O diminishes is reflected in their dissociation energies. Not many numerical data are available; but, whilst the energy of a long bond may be less than 2 kcal/mole, energies exceeding — and perhaps far exceeding — 10 kcal/mole attach to "very short" bonds.

Twenty five years ago, hydrogen bonds, such as those in ice with O···O$>$2.7 Å came to be called "long", and those between 2.7 and 2.5 Å "short'. As examples of bonds shorter than 2.5 Å multiplied, there was nothing for it but to call them "very short", as we have been doing in this review. Perhaps the quotation marks may now be dropped?

B. Spectroscopic Evidence of Hydrogen Bonding

The free hydroxyl group has a stretching-frequency near 3,700 cm^{-1}, as observed in gases or in inert solvents. Striking changes are observed when hydrogen bonding occurs, whether in gases, solutions, or crystals.

The absorption peak moves to lower frequencies and becomes much broader, implying a large increase of integrated intensity. Over a range of weak (and long) O—H···O bonds there is a good correlation between the overall length and the peak frequency, which may fall as low as 2,000 cm^{-1}.

However, with short and very short bonds, the spectra become more complicated, especially for crystals of Type *A*. (For an account of these spectra, and for a theoretical interpretation of the "A, B and C bands", see *Claydon* and *Sheppard* (*94*).) The O—H stretching band disappears from the region above 2,000 cm^{-1}; instead there is general absorption over several hundred cm^{-1}, culminating around 1000 cm^{-1}. Various peaks are superimposed on this background, and there is often a "negative peak" or "window". Such spectra have been classified as Type (ii) by *Hadži* (*95*), who has termed the wide absorption the "D band". These are the "anomalous" spectra noted earlier in this review. Examples are shown in Figs. 19 and 20.

Type *A* crystals always give spectra of Type (ii), and indeed *Hadži* has successfully predicted this crystal type from spectral data. It must be emphasized, however, that Type (ii) spectra are also given by systems not of Type *A* (*95a*), and not even crystalline. But they are always associated with very short hydrogen bonds.

Figs. 19 and 20 show the infrared spectra of some acid salts and related crystalline compounds. The spectrum in Fig. 19(a) is from an acid salt of Type *B*, and it approximates to a superposition of the spectra of free acid and neutral salt. The other spectra are all of *Hadži*'s Type (ii), which is shown in its starkest form in Fig. 19(c) (sodium hydrogen diacetate) with a "window" near 950 cm^{-1}. Potassium hydrogen malonate, a Type A_2 structure, has a similar spectrum (20 (a)). The picoline-N-oxide hemi-hydrobromide (19(d)) is a Type *A* basic salt; *Cook's* basic salt (Sect. XI) is of pseudo-Type *A* and gives the spectrum (20(b)). Sodium bicarbonate (20(c)) and potassium hydrogen oxalate (20(d)) are acid salts of intermediate character (see Sect. XV and XVI *A*).

Vibrational transitions may also be studied by neutron inelastic scattering, which carries some technical advantages to compensate for its much severer experimental difficulties. This method has been applied to acid salts, notably by *Waddington* and his co-workers (*96*).

Still another source of information is nuclear quadrupole resonance spectroscopy. *Waddington* and *Smith* (*96a*) have used the ^{35}Cl resonance to obtain evidence of genuine symmetry in the ions $ClHCl^-$ and $ClDCl^-$ in a salt known (*97*) to have a Type *A* structure with respect to the HCl_2^- ion. *Hadži* and his co-workers (*97*a) have used deuteron resonance to study the structure of KDX_2, where HX = trifluoroacetic acid (see Sect. VII *D* and XX *G*). The coupling constant and asymmetry para-

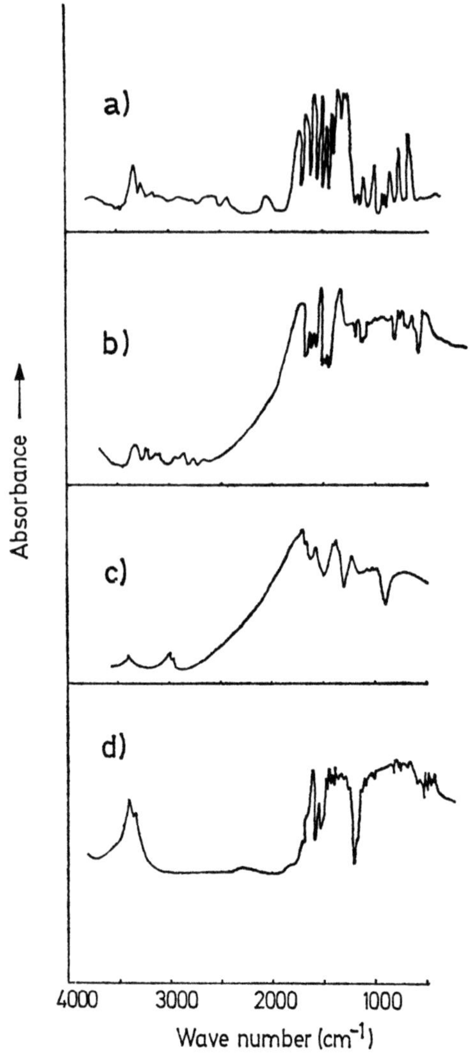

Fig. 19. Infrared spectra of some crystalline acid salts and related compounds: (a) potassium hydrogen di-*p*-nitrobenzoate (Type *B*); (b) rubidium hydrogen di-*o*-nitrobenzoate (Type *A*); (c) sodium hydrogen di-acetate; (d) α-picoline-N-oxide hemi-hydrobromide

meter are almost identical with those for potassium deuterium maleate (Sect. XIV *A*); if the evidence for genuine symmetry of the hydrogen bond in the maleate is accepted, then the bond in the acid trifluoroacetate must also be symmetrical.

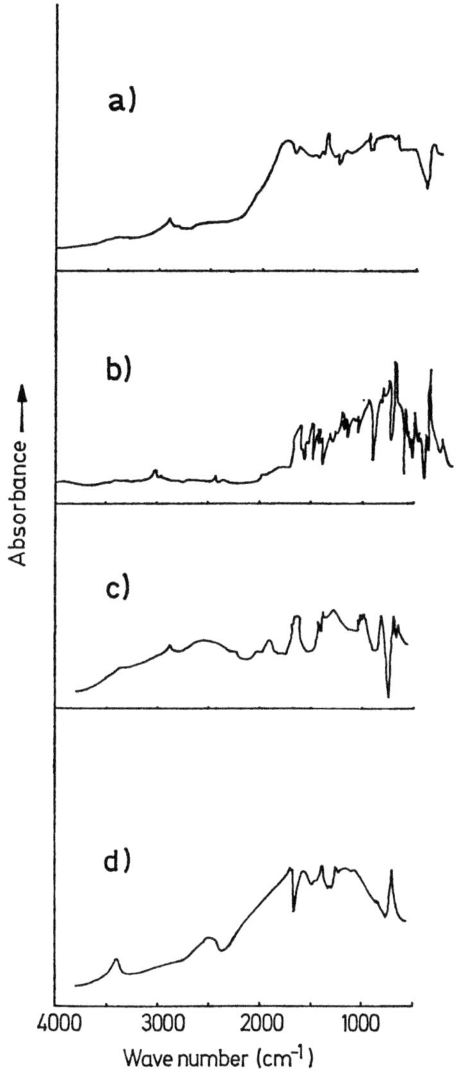

Fig. 20 Further infrared spectra: (a) potassium hydrogen malonate; (b) pseudo-Type *A* basic salt described in section XI; (c) sodium bicarbonate; (d) potassium hydrogen oxalate

C. The Theory of Hydrogen Bonding

This is dealt with in monographs (*87, 88, 89*) and *Bratoz* (*98*) has summaized the quantum-mechanical approach, which starts with an inherent difficulty: the hydrogen-bonding energy is only a small fraction of the total energy of a molecular system.

In unsophisticated terms, two "mechanisms" have been suggested. The first is a classical electrostatic one, and it can be made to seem simple. The A—H bond is always highly polar in the sense (22), and the hydrogen-acceptor atom, B, being electronegative, carries a full, or partial, negative formal charge. The electrostatic attraction between the oppositely charged atoms B and H exceeds the repulsion between the similarly

$$\overset{\delta-}{A}\text{—}\overset{\delta+}{\text{H}}\quad\overset{\delta-}{B} \tag{22}$$

charged A and B, which are further apart. There is general agreement that this is indeed a source of bonding energy in all hydrogen bonds, and the principal one in long, weak bonds.

The second mechanism involves the wave-mechanical effect of electron-delocalization, but we can present it in a crude valence-bond formalism. Atom B has a lone pair of electrons, which we show on the right-hand oxygen atom of (23). We can also write the alternative bond-diagram (24).

O—H :O (23)
O: H——O (24)

$$\left.\begin{array}{l}\text{O—H :O}\\ \text{O: H—O}\end{array}\right\} \quad (25)$$

In "resonance" terms, the system is a hybrid of (23) and (24), and the bonding is attributed to the consequent resonance energy. In most hydrogen bonds this is at most a minor contribution, since (24) will have a much higher energy than (23) because the nonbonded atoms in (24) are crushed together and the formal bond is unduly stretched.

However, in a short bond the O—H distance has increased and the O···O distance decreased, so that this objection becomes less important. In the limiting case of a symmetrical bond the two forms would be equivalent, as represented by (25). The resonance energy would be much larger, perhaps large enough to compare with the electrostatic binding energy or even to exceed it. This could be especially important in a Type A crystal structure, where the total environments of the two oxygen atoms are identical by crystal symmetry.

Moving to a different aspect, we now consider possible forms of the potential-energy diagram as the proton varies its position along the line between two oxygen atoms, supposed to be fixed. For the most general case of a long hydrogen bond, the curve has the form shown in Fig. 21 (a): two energy-minima separated by a substantial barrier; the lower minimum corresponds to the covalent attachment of the proton to one of the oxygens and not to the other. In case (b) the bond is again long with a substantial barrier, but the two minima are now equivalent in energy. This corresponds to the situation in ice; O···O being 2.76 Å, there are two possible positions for the proton about 1.0 Å from either oxygen atom. (Of course, the protons in ice are disordered; but in any particular bond, over a time-interval that is long compared with the vibration-period, the situation will be as described.)

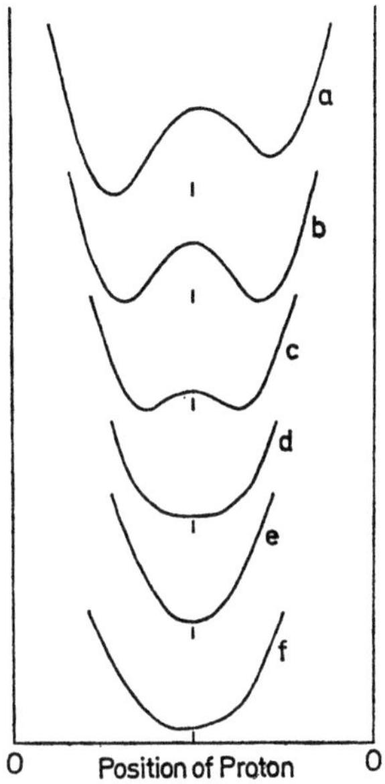

Fig. 21. Rough sketches of possible potential-energy curves for O—H···O bonds, showing how the energy may depend on the position of the proton along the O···O line

As the symmetrically environed bond becomes shorter, the two energy-minima move closer together and the barrier between them becomes lower, as suggested in (c). In the limit, the barrier has disappeared, and we have the parabolic potential-well sketched in (e), in which the proton would vibrate harmonically. Somewhere between (c) and (e) the barrier will have subsided below the zero-point level of vibrational energy. As this state of affairs is approached, the vibrational levels become split, and the description of the system becomes more complicated. The effective potential-well can perhaps be regarded as flattened, as suggested in (d); the vibration of the proton along the bond would be anharmonic.

Even when we ignore these final complications, it is virtually impossible, by diffraction measurements alone, to distinguish between situations (d) or (e), and (c) where the two minima are fairly close together, as Ibers has shown in his discussion (*89, 93*) of the FHF^- ion. Either of two models is equally competent to reproduce the observed neutron-diffraction data: the proton may be vibrating with a r.m.s. amplitude of (say) 0.25 Å about a central position; or it may be disordered, vibrating with perhaps half this amplitude about two equally probable positions 0.1 Å on either side of the centre. Only when the double-minimum situation expands to something like (b) can diffraction data distinguish it from the others. However, Ibers goes on to argue that a distinction between (c) and (d) or (e) may be reached through spectroscopic considerations in the case of the HF_2^- ion, which — he concludes — has a genuinely symmetrical bond.

D. The Relevance of Crystal Symmetry in Very Short O···H···O Bonds

If we say set aside chelated systems where compression of the bond may be operative, then the crystallographically symmetrical O···H···O bonds of the Type *A* acid and basic salts constitute a substantial set of examples, which include most of the shortest known hydrogen bonds. Can the formal crystallographic symmetry be gainsaid?

In many Type *A* acid salts the choice of space group is not unambiguous. With potassium hydrogen bis-phenylacetate, for example, the systematically absent reflexions merely indicate that the group is either *I a* or *I*2/*a*. The former includes no centre of symmetry; if it were correct, the two acid radicals would not be strictly equivalent, and the hydrogen bond between them not necessarily symmetrical. The actual structure analysis was based on the latter, more symmetrical space group. The choice was deemed to have been vindicated because it led to a satisfactory refinement. (This is accepted practice amongst crystallographers.) The structure found must then necessarily conform to the higher symmetry assumed. However, the actual structure may not be exactly symmetrical. Referring to Fig. 5, a small off-centre shift of the hydrogen atom and *very small* shifts of the oxygen atoms, in directions tending to constitute a $-CO_2H$ group at one side and $-CO_2^-$ at the other, all the other atoms remaining where they are, would make hardly any difference to the calculated structure factors. The symmetry would then be *I a*, though very close to *I*2/*a*. (The various statistical tests that may be invoked to distinguish between centrosymmetric and non-centrosymmetric space groups are useless in face of these fine distinctions. Least-squares refinement is equally indecisive.)

Some Type *A* crystals have space groups that are indicated unambiguously by the absent reflexions. Sodium hydrogen di-acetate, for instance, crystallizes in *Ia*3, and potassium hydrogen di-anisate in

Pbcn. Unfortunately such evidence is not unalienable. A well known case of its failure is azulene, which has two $C_{10}H_8$ molecules in a cell of $P2_1/a$, another unambiguous space group. This formally requires the molecule to be centrosymmetric, which is chemically impossible. This crystal has a disordered structure, with molecules having, randomly, either of two centrosymmetrically related orientations. This effect in azulene is a gross one, easily detected by careful structure analysis. Disorder on this scale can safely be ruled out the Type *A* crystal structures described above. But slight disorder, sufficient to impose pseudo-symmetry on a slightly unsymmetrical O···H···O bond, is still conceivable, even when the space group appears to be unambiguous. This, of course, is just another way of stating the difficulty discussed by *Ibers* (*89, 93*). From crystal-structure analysis alone we may conclude only that the structure is very close to being symmetrical.

We may now move this question to its more speculative converse: Does the crystal symmetry have any effect on the bond? Electron delocalization makes a contribution to the bonding energy of short hydrogen bonds. One condition for "resonance" is that the canonical forms should have nearly equal energies. Crystal symmetry of the sort obtaining in a Type *A* structure guarantees exactly equal energies. This will tend to increase the bonding energy, and to make the bond shorter. The shorter the bond, the more likely it is to be symmetrical. (See Sect. XX *C*.)

The cases listed in Tables 7 and 19 give general support to this idea. We may also draw attention to three cases where failure to make use of crystal symmetry (which would be chemically feasible) is associated with hydrogen bonds that are significantly longer than those in Type *A* structures: sodium bicarbonate (Sect. XV), potassium hydrogen oxalate (XVI *B*) and sodium hydrogen oxalate hydrate (XVI *C*).

A noteworthy example is potassium hydrogen mesotartrate (Sect. XVII *C*), both because of the high accuracy of the analysis and because symmetrical and unsymmetrical hydrogen bonds occur in the same crystal and in chemically similar situations. The unsymmetrical bond has O···O = 2.543(1) Å. The two symmetrical ones are very significantly shorter at 2.452(2) and 2.483(2) Å.

E. Further thoughts on Hydrogen-Chelates, and a Very Short O···H···O Bond that is Unsymmetrical

In Sect. XX *A* we pointed out that increasing strength of the acid, H*A*, and/or of the base, *B*, favours formation of a stronger hydrogen bond, but that, if the strength(s) became too great, the hydrogen bond would be disrupted by ionization to A^- and HB^+. In a chelated system, such as the hydrogen-maleate anion, this disaster is prevented by the stereochemistry. The ionized $-CO_2^-$ presents itself face-to-face with the non-ionized $HO_2C,-$ and stronger bonding cannot be evaded by dissociation. Thus the very short bonds in the maleate and chloromaleate systems, and some other chelates, represent special cases.

Examples of very short hydrogen bonds that are not symmetrical are of special interest. We have already given one example in Sect. XIX. Another is the cation represented in (26) whose structure (in the hydrated chloride) has been determined by X-ray (*99*) and subsequently be neutron-diffraction analysis (*100*). This cation is a hydrogen-chelate, like the hydrogen-maleate anion, and the hydrogen bond is similarly short: O···O = 2.420(5) Å. However, there is here no question of Type *A*

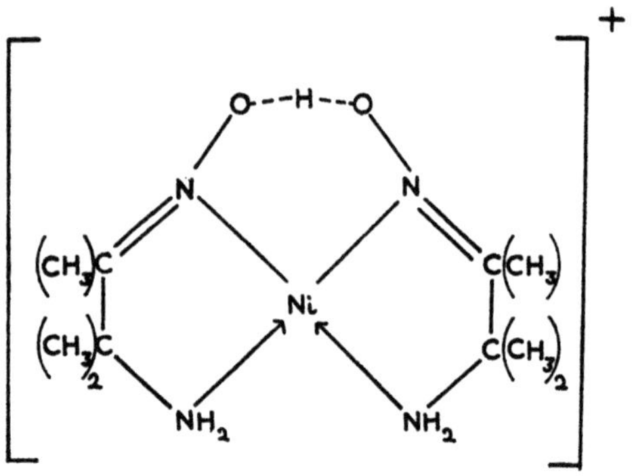

(26)

structure: there are four cations in a cell belonging to the space group $P2_1/c$. Though the ion has chemical symmetry, its environment in the crystal is not symmetrical, and, in particular, the environments of the two oxygen atoms are different. The proton has been precisely located; the O···H···O angle is 170°, and the O···H distances differ very significantly (1.242(5) and 1.187(5) Å).

The explanation offered by Hamilton and co-workers runs as follows: When the top of the energy barrier, in a close double-well of the kind represented in Fig. 23(c), lies below the level of zero-point vibration, the potential-energy curve will have a nearly flat floor. So long as the environment is symmetrical, the effective minimum is exactly at the mid-point, as in Fig. 21(d), but when the environment is unsymmetrical, the point of minimum energy may easily be displaced to one side, as suggested by the sketch in 21(f). "The analysis of the thermal motion ... leads to the conclusion that the hydrogen atom moves in a broad, single-minimum potential. That small changes in intermolecular environment can lead to significant asymmetry in the bond seems to provide further evidence for the broad, flat nature of the potential" (*100*). The position of the minimum of a very nearly flat potential is sensitive to extremely small perturbations (*100a*).

F. A Note on Corrections for Torsional Vibrations

In precise crystallographic work, bond-lengths are often corrected for the systematic shortening effects of molecular libration (*101*). In the Glasgow work covered by this review, we have ignored such corrections, and, in citing results from other laboratories, we have — wherever possible — selected the uncorrected lengths. The usual corrections are based on a model that may be inappropriate when the molecule is tied heterogeneously to its environment by hydrogen bonds and — more strongly — by electrovalency, as in acid salts. In such cases the calculated corrections are in any case small, and they are of dubious significance so far as intermolecular distances are concerned.

G. Some Effects of Deuteriation

In principle, a study of the effects of replacing hydrogen in O···H···O by deuterium ought to provide valuable information for illuminating our theory of hydrogen bonding. Much thought and effort has been put into this study. So far the results are disappointing and sometimes confusing. This topic is well covered by *Ibers* and *Hamilton* (*89*). The difficulties arise from two causes: (a) very precise work is needed to secure valid measurements of the small changes following deuteriation; (b) hydrogen-bonded systems are usually too complex for rigorous theoretical comprehension.

Deuteriation generally increases the overall O···O distance slightly, except perhaps in very long bonds. The increase is often about 0.02 Å, but there is no simple correlation between length and isotope effect. Hypochromous acid, $HCrO_2/DCrO_2$, and the corresponding cobalt compounds may be exceptional (*102, 103*). Though they have Type *A* crystal structures, their hydrogen bonds are unusually long, and the isotope effect is large: *e.g.* O···H···O = 2.49, O···D···O = 2.55 Å. The infrared spectra suggest that the potential-energy curves for the two systems are very different, that for the protonated acid being of the kind drawn in Fig. 21 (d), whilst that for the deuteriate resembles (c). Perhaps hydrogen bonds of this length constitute a special, border-line case between short and very short bonds.

Two experiments made in Glasgow may be worth mention. Potassium hydrogen diaspirinate, KHX_2, has been studied with X-rays and neutrons (Sect. VII C). Through the kindness of Dr. *Armin Wilson* we were provided with a sample of KDX_2, from which Dr. *M. Currie* collected over 2,000 diffractometer data (*104*). A satisfactory refinement led to O···D···O = 2.439(5) Å, to be compared with O···H···O = 2.455(5) by X-rays and 2.448(4) Å by neutrons. The shortening is of no better then "possible significance". Much worse, it later transpired that the deuterium content of the crystal was questionable, though it had been exposed to the atmosphere only transiently.

Potassium hydrogen di-trifluoroacetate, KHX_2, is a Type *A* acid salt (Sect. VII *D*), with a special attraction for the spectroscopist because it contains only the one hydrogen atom (Sect. XX B). Large single crystals of both KHX_2 and KDX_2 were grown in Ljubljana and studied, by neutron diffraction, in Britain (*111*). Two lines of evidence suggest that its hydrogen bond is genuinely symmetrical, with a potential-energy curve of the form sketched in Fig. 21 (d).

First, the distances, O··H··O = 2.437(4) and O··D··O = 2.437(3) Å, do not differ significantly. (In this case the deuterium-content of KDX_2 is not in doubt; it can be used as a variable parameter in the least-squares refinement based on neutron data, and it turned out to be not less than 90%.) According to a suggestion by *Rundle* (*106*) and a more detailed theoretical treatment by *Singh* and *Wood* (*112*), a zero isotope effect is to be expected in a short O··H··O bond with a single potential well.

Secondly, the results of precise neutron analysis enabled us to estimate the vibrations of the proton within its O···O environment. We found r.m.s. amplitudes of about 0.17 Å along, and about 0.07 Å normal to, the bond; and a similar difference for the deuteron in KDX_2. Such anisotropy resembles that found in the HF_2^- ion (*93*). The discussion of *Hamilton* and his co-workers (*31*) applies here also, and it leads to the deduction that we have a flattened potential well, effectively with a single minimum.

H. Some Conclusions

In all Type *A* crystals, the hydrogen bond has a symmetrical potential-energy curve. In many acid salts, with open structures, the O···H···O distance in the double-anion, $R \cdot CO_2HO_2C \cdot R$, is close to 2.44 Å. In a smaller number of chelated, but otherwise similar, structures, the O···H···O distance is somewhat shorter; and there are stereochemical reasons for this. For the latter cases, the hydrogen bond probably is genuinely symmetrical, with a single-well potential. Some, at least, of the open Type *A* structures may also contain symmetrical O···H···O bonds.

Acknowledgements. **This review should be dedicated to the large number of able research associates who have participated in the work described. — Mr. *William Cook* drew most of the diagrams.**

References

1. *Freund, Ida:* The Study of Chemical Composition, p. 519. Cambridge 1904.
2. *Mellor, J. W.:* Comprehensive Treatise on Inorganic and Physical Chemistry, Vol. II, p. 777. Longmans 1922.
3. *Bineau, M. A.:* Ann. Chim. Physik *19*, 7, 291 (1847).
4. *Heintz, W.:* Ann. Physik *114*, 451 (1861).
5. *Gerhardt, C.:* Annalen *87*, 150 (1853).
6. *Smith, N.:* Ph. D. Thesis. Sheffield 1949.
7. — *Speakman, J. C.:* Trans. Faraday Soc. *44*, 1031 (1948).
8. *Ross, J. D. N., Morrison, T. J.:* J. Chem. Soc. *1933*, 1016 (and later papers).
9. *Speakman, J. C., Smith, N.:* Nature *155*, 698 (1945).
10. *Bennett, G. M.:* Trans. Chem. Soc. *107*, 351 (1915).
11. *Shrivastava, H. N., Speakman, J. C.:* J. Chem. Soc. *1961*, 1151.
12. *Hahn, T.:* Z. Krist. *109*, 438 (1957).
13. *Nahringbauer, I.:* Hydrogen Bond Studies. University of Upsala 1970.
14. *Marsh, R. E., Donohue, J.:* Advan. Protein Chem. *22*, 236 (1967).
15. *Manojlović, Lj., Speakman, J. C.:* J. Chem. Soc. *A 1967*, 978.
16. *Golič, L., Speakman, J. C.:* J. Chem. Soc. *1965*, 2521.
17. *Mayer, R. F., Keve, E. T., Skapski, A. C.:* J. Chem. Soc. *A 1968*, 2288.
18. *Downie, T. C., Speakman, J. C.:* J. Chem. Soc. *1954*, 757.
19. *Speakman, J. C.:* Nature *162*, 695 (1948).
20. — J. Chem. Soc. *1949*, 3357.
21. *Manojlović, Lj., Speakman, J. C.:* Acta Cryst. *B 64*, 323 (1968).
22. *Bacon, G. E., Curry, N. A.:* Acta Cryst *10*, 524 (1957).
23. *Bacon, G. E., Curry, N. A.:* Acta Cryst. *13*, 717 (1960).
24. *Davies, M., Thomas, W. J. O.:* J. Chem. Soc. *1951*, 2858.
25. *Hadži, D., Novak, A.:* Nuovo Cimento *2* (x) Supplement, 715 (1955).
26. *Albert, N., Badger, R. M.:* J. Chem. Phys. *29*, 1193 (1958).
27. *Wyckoff, R. G. W.:* Am. J. Sci. *4*, 175 (1922).
28. *Speakman, J. C., Mills, H. H.:* J. Chem. Soc. *1961*, 1164.
29. *Manojlović, Lj., Speakman, J. C.:* J. Chem. Soc. *A 1967*, 971.
30. *Grimvall, S., Wengelin, R. F.:* J. Chem. Soc. *A 1967*, 968.
31. *Sequeira, A., Berkebile, C. A., Hamilton, W. C.:* J. Mol. Struct. *1*, 283 (1967—8).
32. *Golič, L., Speakman, J. C.:* J. Chem. Soc. *1965*, 2530.
33. *Skinner, J. M., Speakman, J. C.:* J. Chem. Soc. *1951*, 185.
34. *Manojlović, Lj.:* Acta Cryst. *B* 24. 326 (1968).
35. *Donohue, J.:* J. Phys. Chem. *56*, 502 (1952).
36. *Hadži, D., Novak, A.:* Proc. Chem. Soc. *1960*, 241.
37. *Calleri, M., Speakman, J. C.:* Acta Cryst. 17, 1097 (1964).
38. *Dunitz, J. D., Rollett, J. S.:* Acta Cryst. *9*, 327 (1956).
39. *Larsson, G., Nahringbauer, I.:* Acta Cryst. *B 24*, 666 (1968).
40. *Hughes, E. W., Takei, W. J.:* Struct. Reports *19*, 509 (1955).
41. *Peterson, S. W., Worsham,* J. E.: ORNL Report *1959*, 69.
42. *Hadži, D.:* J. Chem. Soc. *1962*, 5128.
43. *Mills, H. H., Speakman, J. C.;* Proc. Chem. Soc. *1963*, 216.
44. *Cook, D.:* Chem. Ind. (London) *1963*, 607.
45. *Currie, M., Speakman, J. C.:* J. Chem. Soc. *A 1969*, 1648.
46. *Calleri, M., Speakman, J. C.:* J. Chem. Soc. *A 1969*, 1644.
47. *Fisher, C. D., Jensen, L. H., Schubert, W. M.:* J. Am. Chem. Soc. *87*, 33 (1965).
48. *Reik, R.:* Sitzungsber. Wien Akad. *111*, 842 (1902).
49. *Nahringbauer, I.* · Acta Chem. Scand. *23*, 1653 (1969).

50. *O'Keane, G. P.:* B. Sc. Thesis. Glasgow 1970.
51. *Okaya, Y., Pepinsky, R.:* Acta Cryst. *10*, 324 (1957).
52. *van Bommel, A. J., Bijvoet, J. M.:* Acta Cryst. *11*, 61 (1958).
53. *Cardwell, H. M. E., Dunitz, J. D., Orgel, L. E.:* J. Chem. Soc. *1953*, 3740.
54. *Darlow, S. F., Cochran, W.:* Acta Cryst. *14*, 1250 (1961).
55. *Peterson, S. W., Levy, H. A.:* J. Chem. Phys. *29*, 948 (1958).
56. *Ellison, R. D., Levy, H. A.:* Acta Cryst. *18*, 260 (1965).
57. *Darlow, S. F.:* Acta Cryst. *14*, 1257 (1961).
58. *Parthasarathy, R.:* Science *161*, 179 (1968).
59. *Ferguson, G., Sime, J. G., Speakman, J. C., Young, R.:* Chem. Commun. *1968*, 162.
60. *Sime, J. G., Speakman, J. C., Pathasarathy, R.:* J. Chem. Soc. *A 1970*, 1919.
61. *Currie, M., Speakman, J. C.:* J. Chem. Soc. *A 1970*, 1923.
62. *Marshall, H., Cameron, A. T.:* Trans. Chem. Soc. *91*, 1519 (1907).
63. *McAdam, A.:* Ph. D. Thesis. Glasgow 1969. — *McAdam, A., Currie, M., Speakman, J. C.:* J. Chem. Soc. *A 1971*, 1994.
63a. — *Speakman, J. C.:* J. Chem. Soc. *A 1971*, 1997.
64. *Broadley, J. S., Cruickshank, D. W. J., Morrison, J. D., Robertson, J. M., Shearer, H. M. M.:* Proc. Roy. Soc. (London) *A 251*, 444 (1959).
65. *Macdonald, A. L.:* B. Sc. Thesis. Glasgow 1968. (Work to be published in J. Chem. Soc.).
66. — *Speakman, J. C.:* J. Cryst. Mol. Struct. *1*, 189 (1971).
66a. *Zachariasen, W. H.:* J. Chem. Phys. *1*, 634 (1933).
67. *Sharma, B. D.:* Acta Cryst. *18*, 818, (1965).
67a. *Gupta, M. P., Sahu, R. G.:* Acta Cryst. *B 26*, 1964 (1970).
68. *Zachariasen, W. H.:* Z. Krist. *89*, 442 (1934).
69. *Robertson, J. M., Woodward, I.:* J. Chem. Soc. *1936*, 1817.
70. *Delaplane, R. G., Ibers, J. A., Sabine, T. M., Cox, G. W., Craven, B. M., Coppens, P.:* Acta Cryst. *B 25*, 2423, 2437, 2442, 2451 (1969).
71. *Cox, E. G., Dougill, M. W., Jeffrey, G. A.:* J. Chem. Soc. *1952*, 4854.
71a. *e.g. Beagley, B., Small, R. W. H.:* Acta Cryst. *17*, 783 (1964).
72. *Ahmed, N. A. K., Liminga, R., Olovsson, I.:* Acta Chem. Scand. *22*, 88 (1968).
73. *Nilsson, A., Liminga, R., Olovsson, I.:* Acta Chem. Scand. *22*, 719 (1968).
74. *Pedersen, B. F.:* Acta Chem. Scand. *22*, 2953 (1968).
74a. *Einspahr, H., Marsh, R. E., Donohue, J.:* Results to be published, personally communicated by Prof. Donohue.
74b. *Tellgren, R., Olovsson, I.:* J. Chem. Phys., *54*, 127 (1971).
74c. *Follner, H., Kanters, J. A., Kroon, J.:* Z. Anorg. Allgem. Chem. *379*, 225 (1970).
74d. — Z. Anorg. Allgem. Chem. *373*, 198 (1970).
75. *Brown, C. J., Peiser, H. S., Turner-Jones, A.:* Acta Cryst. *2*, 167 (1949).
76. *Price, T. S., Brazier, S. A.:* Trans. Chem. Soc. *107*, 1367, 1713 (1915), and earlier papers.
77. *Adman, E., Margulis, T. N.:* J. Am. Chem. Soc. *90*, 4517 (1968).
78. *Haas, D. J.:* Acta Cryst. *17*, 1511 (1964).
79. *Currie, M., Speakman, J. C., Curry, N. A.:* J. Chem. Soc. *A 1967*, 1862.
79a. *Kroon, J., Kanters, J. A.:* personal communications. (See also *Kroon, J., Kanters, J. A., Peerdeman, A. F.:* Nature. Phys. Sciences. *229*, 121 (1971).
80. *Goedkoop, J. A., MacGillavry, C. H.:* Acta Cryst. *10*, 125 (1957).
81. *Nakahara, A., Saito, Y., Kuroya, H.:* Bull. Chem. Soc. Japan *25*, 331 (1952).
82. *Williams, J. M., Peterson, S. W.:* Inorg. Nucl. Chem. Letters *3*, 297 (1967).
83. — — J. Am. Chem. Soc. *91*, 776 (1969).

84. — — Acta Cryst. *A 25*, S. 113 (1969).
85. *Lundgren, J.-O., Olovsson, I.*: Acta Cryst. *23*, 966, 971 (1967).
86. *Beecham, A. F., Hurley, A. C., Mackay, M. F., Maslen, V. W., Mathieson, A. McL.*: J. Chem. Phys. *49*, 3312 (1968).
86a. *Derry, J. E., Hamor, T. A.*: Chem. Commun. *1970*, 1284.
87. *Hadži, D.* (editor): Hydrogen Bonding. Pergamon Press 1959.
88. *Pimentel, G. C., McClellan, A. L.*: The Hydrogen Bond. Freeman 1960.
89. *Hamilton, W. C., Ibers, J. A.*: Hydrogen Bonding in Solids. Benjamin 1968.
90. *Robertson, J. M.*: Trans. Faraday Soc. *36*, 913 (1940).
91. *Bellamy, L. J., Pace, R. J.*: Spectrochim. Acta *25 A*, 319 (1969).
91a. *Ganis, P., Dunitz, J. D.*: Helv. Chim. Acta *50*, 2370 (1967).
92. *Nakamoto, K., Margoshes, M., Rundle, R. E.*: J. Am. Chem. Soc. *77*, 6480 (1955).
93. *McGaw, B. L., Ibers, J. A.*: J. Chem. Phys. *39*, 2677 (1963).
94. *Claydon, M. F., Sheppard, N.*: Chem. Commun. *1969*, 1431.
95. *Hadži, D.*: Pure Appl. Chem. *11*, 435 (1965).
95a. *Golič, L., Hadži, D., Lazarini, F.*: Chem. Commun. *1971*, 860.
96. *Stirling, G. C., Ludman, C. J., Waddington, T. C.*: J. Chem. Phys. *52*, 2730 (1970). (See also *Collins, M. F., Haywood, B. C., Stirling, G. C.*: Neutron Beam and Magnetic Studies, p. 95. A.E.R.E., Harwell 1970.)
96a. *Ludman, C. J., Waddington, T. C., Salthouse, J. A., Lynch, R. J., Smith, J. A. S.*: Chem. Commun. *1970*, 405.
97. *Schroeder, L. W., Ibers, J. A.*: Inorg. Chem. *7*, 594 (1968).
97a. *Hadži, D.*: personal communication.
98. *Bratož, S.*: Advan. Quant. Chem. *3*, 208 (1967). (See also *Sheng Hsien Lin*: Physical Chemistry, Vol. V, Ch. 8, by *Eyring, H., Henderson, D., Jost, W.* New York: Academic Press 1970.
99. *Schlemper, E. O.*: Inorg. Chem. *7*, 1130 (1968).
100. *Hamilton, W. C., La Placa, S. J.*: Acta Cryst. *A 25*, 252 (1969). — *Schlemper, E. O., Hamilton, W. C., La Placa, S. J.*: J. Chem. Phys. *54*, 3990 (1971).
100a. *Dunitz, J. D.*: Pure Appl. Chem. *25*, 495 (1971).
101. *Cruickshank, D. W. J.*: Acta Cryst. *9*, 757 (1956).
102. *Snyder, R. G., Ibers, J. A.*: J. Chem. Phys. *36*, 1356 (1962).
103. *Hamilton, W. C., Ibers, J. A.*: Acta Cryst. *16*, 1209 (1963).
104. *Currie, M.*: Ph. D. Thesis. Glasgow 1969.
105. *Macdonald, A. L., Speakman, J, C.*: J. Chem. Soc., Perkin II, *1972*, 942.
106. *Rundle, R. E.*: J. Phys. *25*, 187 (1964).
107. *Skinner, J. M., Stewart, G. M. D., Speakman, J. C.*: J. Chem. Soc. *1954*, 180.
108. *Bryan, R. F., Mills, H. H., Speakman, J. C.*: J. Chem. Soc. *1963*, 4350.
109. *Mills, H. H., Speakman, J. C.*: J. Chem. Soc. *1963*, 4355.
110. *McGregor, D. R., Speakman, J. C.*: J. Chem. Soc. *A 1968*, 2106.
111. *Macdonald, A. L., Speakman, J. C., Hadži, D.*: J. Chem. Soc., Perkin II, *1972*, 825.
112. *Sinha, T. R., Wood, J. L.*: J. Chem. Phys. *48*, 4567 (1968).
113. *Currie, M.*: J. Chem. Soc., Perkin II, *1972*, 832.
114. *Kroon, J., Kanters, J. A.*: Acta Cryst., *B 28*, 714 (1972).

Received September 23, 1971.

Phase-fixed 3-Γ Symbols and Coupling Coefficients for the Point Groups

S. E. Harnung and C. E. Schäffer

Department I, (Inorganic Chemistry), The H. C. Ørsted Institute, University of Copenhagen, Universitetsparken 5, DK-2100 Copenhagen Ø, Denmark

Table of Contents

1. Introduction 202
a) The 3-Γ Symbol 202
b) The Importance of Standardizing the Basis Functions. Consequences of Schur's Lemma 204
c) The Ligand Field as an Impetus to Symmetry Standardization . 205

I. The 3-l Symbols of R_3, Generated by Use of Spherical Harmonics, and Those of R_{3i} 206

2. Solid Harmonic Bases and Their Symmetry Properties 206

3. Basis Function Operator Equivalents 211

4. The Three-Dimensional Rotation Group R_3 213
a) Even 3-l Symbols 213
b) Odd 3-l Symbols 216

5. The Three-Dimensional Rotation-Inversion Group R_{3i} 218

II. The Sub-Groups of R_3 and R_{3i} 219

6. Survey of the Crystallographic Point Groups 219
a) Introduction. Commutative Groups 219
b) Non-Commutative Groups 223

7. Non-Cubic Finite Point Groups as Sub-Groups of the Holohedric Infinite Group $D_{\infty h}$ 227

III. The 3-Γ Symbols for the Sub-Groups of R_3 and R_{3i} 232

8. The Non-Cubic Point Groups. Racah's Lemma 232
a) The Group D_∞ 232
b) The Sub-Groups of D_∞ 236
c) $D_{\infty h}$ and Its Sub-Groups 238

9. The 3-Γ Symbols for the Cubic Point Groups 239
a) The Octahedral Rotation Group, O 239
b) The Octahedral Group O_h and Its Sub-Groups 240

10. Conclusions 241

Appendix 1. The Evaluation of 3-l Symbols 241
Appendix 2. Racah's Lemma Applied to R_{3i} - $C_{\infty v}$ 243
Appendix 3. Tables 15—22 244

References 255

1. Introduction

a) The 3-Γ Symbol

The most general problem in quantum mechanics is the evaluation of integrals of the form

$$\int \Psi_1^* \mathfrak{O} \Psi_3 \, d\tau \tag{1}$$

where $\mathfrak{O}$ is some linear operator, Ψ_1 and Ψ_3 are functions characterizing the states with labels 1 and 3, and where τ represents all the variables involved in the problem.

The simplest situation, as far as symmetry is concerned, arises when $\mathfrak{O}$ represents a constant of motion, in which case the operator is invariant under the symmetry group of the system. However, $\mathfrak{O}$ may be any tensorial property and have any symmetry, so that selection rules and polarization properties are among the problems of the form of Eq. (1).

It is always possible to choose state functions as well as tensor operators to transform irreducibly under the symmetry group of the system. So Eq. (1) may be written in the form

$$<\alpha_1 \Gamma_1 \gamma_1 |\mathfrak{O}^{\Gamma_2}_{\gamma_2}| \alpha_3 \Gamma_3 \gamma_3> \tag{2}$$

where the Γ's refer to the irreducible representations and the γ's to their components. The α's specify additional quantum numbers and other properties of the functions which are necessary for the full characterization of the states.

Often basis functions are chosen which are bases for irreducible representations of the three-dimensional rotation-inversion group R_{3i}, even though the physical system has a sub-group symmetry. In this case the tensorial methods exibit their particular potency because the tensor operators — also those representing constants of motion — can be expanded into components of irreducible representations of R_{3i}.

In the present paper we shall be concerned with simply reducible groups (*1*). For these, the direct product of two irreducible representations contains no irreducible representation more than once. This implies that the direct product of three irreducible representations never contains the totally symmetric representation more than once, and has the consequence that the $[\Gamma_1, \Gamma_2, \Gamma_3]$ matrix elements[1]) embodied in

[1]) We use the symbol $[\Gamma_1, \Gamma_2, \ldots]$ for the product of the dimensions of the irreducible representations Γ_1, Γ_2, ...

Eq. (2), when the components are varied in all possible ways, can be factorized according to the Wigner-Eckart theorem,

$$<\alpha_1\,\Gamma_1\,\gamma_1|\mathfrak{O}^{\Gamma_2}_{\gamma_2}|\alpha_3\,\Gamma_3\,\gamma_3> = <\alpha_1\,\Gamma_1\|\mathfrak{O}^{\Gamma_2}\|\alpha_3\,\Gamma_3> \begin{pmatrix}\Gamma_1\,\Gamma_2\,\Gamma_3\\ \gamma_1\,\gamma_2\,\gamma_3\end{pmatrix} \quad (3)$$

which involves only one non-symmetry-determined parameter.

This parameter, $<\alpha_1\,\Gamma_1\|\mathfrak{O}^{\Gamma_2}\|\alpha_3\,\Gamma_3>$, which has the dimension of the operator, is generally called a "reduced matrix element". The term "reduced" means "having got rid of the three variables referring to the components of the irreducible representations" and has no numerical sense. In fact, the reduced matrix element has an absolute value which is always bigger than or equal to that of the matrix element itself. According to Eq. (4) the square of the reduced matrix element is equal to the sum of the squares of all the $[\Gamma_1, \Gamma_2, \Gamma_3]$ matrix elements comprised in Eq. (3).

The symbol $\begin{pmatrix}\Gamma_1 & \Gamma_2 & \Gamma_3\\ \gamma_1 & \gamma_2 & \gamma_3\end{pmatrix}$ that expresses the matrix element of Eq. (3) as a function of the components γ_1, γ_2, and γ_3 is a 3-Γ symbol which is analogous to Wigner's 3-j symbol. The 3-Γ symbols are dimensionless numbers and with the components in standard form their absolute values, but not necessarily their phases, are determined solely by the properties of the symmetry group of the system. For the 3-Γ symbols the normalization condition

$$\sum_{\gamma_1\,\gamma_2\,\gamma_3} \begin{pmatrix}\Gamma_1 & \Gamma_2 & \Gamma_3\\ \gamma_1 & \gamma_2 & \gamma_3\end{pmatrix}^2 = \delta(\Gamma_1\,\Gamma_2\,\Gamma_3) \quad (4)$$

is valid. Here the sum runs over all the $[\Gamma_1, \Gamma_2, \Gamma_3]$ components and the symbol $\delta(\Gamma_1\,\Gamma_2\,\Gamma_3)$ is one or zero, depending on whether or not the direct product of Γ_1, Γ_2, and Γ_3 contains the unit representation.

The 3-Γ symbols are invariant under an even permutation of the columns in the symbol and are multiplied by $(-1)^{\Gamma_1+\Gamma_2+\Gamma_3}$ under an odd permutation. Thus their behaviour under odd permutations classifies them as symmetrical (even 3-Γ symbols, Sect. 4a) or antisymmetrical (odd 3-Γ symbols, Sect. 4b). If a 3-Γ symbol belongs to the even or odd class, the exponent $\Gamma_1+\Gamma_2+\Gamma_3$ is an even or odd number, respectively. This classification, which will be discussed on p. 210 and treated in detail in Chapt. III, is associated with the relationship

$$\begin{pmatrix}\Gamma_1 & \Gamma_2 & \Gamma_3\\ \gamma_1 & \gamma_2 & \gamma_3\end{pmatrix} = (-1)^{\Gamma_1+\Gamma_2+\Gamma_3}\,[\Gamma_3]^{-1/2} <\Gamma_1\,\gamma_1\,\Gamma_2\,\gamma_2|\Gamma_3\,\gamma_3> \quad (5)$$

from which six coupling coefficients can be produced out of one 3-Γ symbol.

b) The Importance of Standardizing the Basis Functions.

Consequences of Schur's Lemma

When the direct product of two irreducible matrix representations of a group is reducible, it can be reduced to a direct sum of irreducible representations by an equivalent transformation with a constant matrix, i.e. the same matrix for all the matrix representatives of the symmetry operators of the group (*2*). We shall assume the irreducible representations in unitary form; then the constant matrix can be chosen as the real orthogonal matrix whose elements are the coupling coefficients occuring in Eq. (5). The orthogonality properties can be expressed as

$$\sum_{\gamma_1 \gamma_2} <\Gamma_3 \gamma_3 | \Gamma_1 \gamma_1 \Gamma_2 \gamma_2> <\Gamma_1 \gamma_1 \Gamma_2 \gamma_2 | \Gamma_3' \gamma_3'> = \delta(\Gamma_3 \Gamma_3')\, \delta(\gamma_3 \gamma_3')\, \delta(\Gamma_1 \Gamma_2 \Gamma_3) \qquad (6a)$$

$$\sum_{\Gamma_3 \gamma_3} <\Gamma_1 \gamma_1 \Gamma_2 \gamma_2 | \Gamma_3 \gamma_3> <\Gamma_3 \gamma_3 | \Gamma_1 \gamma_1' \Gamma_2 \gamma_2'> = \delta(\gamma_1 \gamma_1')\, \delta(\gamma_2 \gamma_2') \qquad (6b)$$

which, when combined with Eq. (5), leads to Eq. (4).

An important theorem which derives from Schur's lemma is the fundamental theorem for irreducible representations (*3*). This can be formulated as the Wigner-Eckart theorem, Eq. (3). In order to obtain this theorem in the present formulation, each irreducible representation must be chosen in identically the same form each time it occurs, rather than in an equivalent form. Therefore the irreducible representations are conveniently generated in standard form by applying the operators of the symmetry group to a properly chosen set of standard bases for the irreducible representations.

It can be shown also from Schur's lemma that a 3-Γ symbol can be defined by means of the coupling coefficients as in Eq. (5). Therefore the reduction of the irreducible product of the two sets $|\alpha_1 \Gamma_1\}$ and $|\alpha_2 \Gamma_2\}$, each forming a basis for an irreducible representation, can be written in terms of the 3-Γ symbol,

$$[|\alpha_1 \Gamma_1\} \times |\alpha_2 \Gamma_2\}]^{\Gamma_3}_{\gamma_3} = \sum_{\gamma_1 \gamma_2} (-1)^{\Gamma_1+\Gamma_2+\Gamma_3} [\Gamma_3]^{1/2} \begin{pmatrix} \Gamma_1 & \Gamma_2 & \Gamma_3 \\ \gamma_1 & \gamma_2 & \gamma_3 \end{pmatrix} |\alpha_1 \Gamma_1 \gamma_1> | \alpha_2 \Gamma_2 \gamma_2> \qquad (7)$$

Although the absolute values of the 3-Γ symbols are fully determined by the standard bases and the properties of the group, their phases are not. As has been pointed out by *J. S. Griffith*, who has previously introduced 3-Γ symbols for the point groups (*4*), there is an independent choice of phase for each ordered trio of Γ's in the symbol. In the present paper the spherical harmonics will be used as standard bases for the real representations of the point groups. The proposed method, by focusing on the analytical properties of these functions and keeping the group theoretical apparatus in the background, defines such symmetry quantities as the 3-Γ symbols, not only with respect to their absolute values, but also with respect to their signs.

A lemma by *Racah* (*5*), a consequence of Schur's lemma, provides a relationship between the coupling coefficients of a group and those of a sub-group. This relationship is here illustrated and standardized by introducing differential operator equivalents for the real standard bases of the three-dimensional rotation group R_3.

c) The Ligand Field as an Impetus to Symmetry Standardization

The authors have long been concerned with the application of the kind of ligand field models in which the concept of semi-empirical parameters is important. Whether the electrostatic model (EM) or the Angular Overlap Model (AOM) of the ligand field is used, the energies of the electrons of the partly filled shell are expressed as linear combinations of semi-empirical parameters whose coefficients are proportional to certain completely specified integrals over the angular part of hydrogen atom orbitals. The ligand field models generally assume the validity of the Laplace equation. This has the consequence that the potential can be expanded into spherical harmonics, each term being expressed as a product of three factors, a potential parameter $\mathfrak{A}_t^l$, a radial factor r^l, and an angular factor $\mathfrak{C}_t^l$ representing a surface harmonic, normalized to $[4\pi/(2l+1)]$ [see Eqs. (15)]. A general term in the matrix elements of the ligand field may then be written

$$\int \Psi_{t_1}^{l_1} (\mathfrak{A}_{t_2}^{l_2} r^{l_2} \mathfrak{C}_{t_2}^{l_2}) \Psi_{t_3}^{l_3} d\tau = <\alpha_1 l_1 t_1 | \mathfrak{O}_{t_2}^{l_2} | \alpha_3 l_3 t_3> \quad (8)$$

where the Ψ's represent central ion functions, which from now on will be assumed to be real. The validity of the Laplace equation also allows the functions to be expressed as one factor depending on the coordinate r only, and another factor depending only on the angular coordinates θ and φ. This is the reason why a function in Eq. (8) may be characterized by the quantum number l associating it with a particular irreducible representation of R_3, and by the t specifying the particular component of this representation.

By use of the factorization, Eq. (8) may be written

$$<\alpha_1 l_1 t_1 | \mathfrak{O}_{t_2}^{l_2} | \alpha_3 l_3 t_3> = (\mathfrak{A}_{t_2}^{l_2} <R_{\alpha_1} | r^{l_2} | R_{\alpha_3}>) <l_1 t_1 | \mathfrak{C}_{t_2}^{l_2} | l_3 t_3> \quad (9)$$

in striking analogy with Eq. (3). The factors in parenthesis on the right-hand side of Eq. (9) are the formal expressions of the semi-empirical ligand field parameters. The last factor, in which two hydrogen atom angular functions and a functionally analogous expression from the ligand field potential are integrated over the angular coordinates, may be further factorized according to Eq. (3). It is customary to include in the radial parameters certain constant factors (i.e. factors, independent of t_1, t_2, and t_3) from the angular integral, so as to obtain simple numbers for the coefficients to the radial parameters. Consequently the chemical literature offers many different choices of semi-empirical parameters, varying with respect both to sign and size, but representing the same observed quantity. This irrational state of affairs has prompted us previously (*6*) and now to go into the problem of standardization.

A more important aspect in this connection arises from the resemblance between Eqs. (3) and (9). Even 3-Γ symbols, where the Γ's refer to the irreducible representations of the group R_3, can be derived on the basis of this resemblance, as discussed in section 4a.

I. The 3-l Symbols of R_3, Generated by Use of Spherical Harmonics, and Those of R_{3i}

2. Solid Harmonic Bases and Their Symmetry Properties

Here we use as basis functions the spherical harmonics on a real solid harmonic form. They are eigenfunctions of the orbital angular momentum operator,

$$\mathfrak{l}^2 = \mathfrak{l}_Z^2 + \mathfrak{l}_X^2 + \mathfrak{l}_Y^2 \tag{10a}$$

$$= \left(x\frac{\partial}{\partial x} + y\frac{\partial}{\partial y} + z\frac{\partial}{\partial z}\right) + \left(x\frac{\partial}{\partial x} + y\frac{\partial}{\partial y} + z\frac{\partial}{\partial z}\right)^2 - r^2\nabla^2 \tag{10b}$$

$$= \mathfrak{s} + \mathfrak{s}^2 - r^2\nabla^2 \tag{10c}$$

where $r^2 = x^2 + y^2 + z^2$,and ∇^2 is the Laplacian in Cartesian coordinates. The operator $\mathfrak{s} = x\frac{\partial}{\partial x} + y\frac{\partial}{\partial y} + z\frac{\partial}{\partial z}$ is totally symmetric in the three-dimensional rotation group R_3 as well as in the three-dimensional rotation-inversion group R_{3i}. Homogeneous polynomials f_l of the degree l are eigenfunctions of the operator $\mathfrak{s}$ with the eigenvalue l so that the two first terms of $\mathfrak{l}^2$ acting on f_l give

$$(\mathfrak{s} + \mathfrak{s}^2)\,f_l = l(l+1)\,f_l \tag{11}$$

When f_l is a solid harmonic h_l, it is subject to the condition

$$\nabla^2 h_l = 0 \tag{12}$$

so one has

$$\mathfrak{l}^2 h_l = l(l+1) h_l \tag{13}$$

and h_l belongs to an irreducible tensorial set $|\alpha\, l\}$ of R_3. Since the parity of h_l is well defined, h_l also belongs to an irreducible tensorial set of R_{3i}.

We shall define sets of real solid harmonics which generally are not eigenfunctions of the operator

$$\mathfrak{l}_Z = -i\left(x\frac{\partial}{\partial y} - y\frac{\partial}{\partial x}\right) \tag{14}$$

These real solid spherical harmonics may be given explicitly by one of the alternative expressions, normalized to $4\pi/(2l+1)$ over the surface of the unit sphere (*11*, sec. 3d and 3e),

$$\begin{aligned}\mathfrak{H}^l_{\lambda\varsigma} &= r^l\,\mathfrak{C}^l_{\lambda\varsigma}\\ &= [\sqrt{2}\,(1-\delta(\lambda,0)) + \delta(\lambda,0)]\,[(l+\lambda)!]^{1/2}\,[(l-\lambda)!]^{-1/2}\,2^{-\lambda}(\lambda!)^{-1}\\ &\quad\times\left[z^{l-\lambda} - \frac{(l-\lambda)\,(l-\lambda-1)}{2\times(2\lambda+2)}\,z^{l-\lambda-2}\,(x^2+y^2)\right.\\ &\quad\left.+\frac{(l-\lambda)(l-\lambda-1)(l-\lambda-2)\,(l-\lambda-3)}{2\times4\times(2\lambda+2)\,(2\lambda+4)}\,z^{l-\lambda-4}\,(x^2+y^2)^2 - \cdots\right]\\ &\quad\times r^\lambda \sin^\lambda\theta\,\varsigma(\lambda\varphi)\end{aligned} \tag{15a}$$

or

$$\begin{aligned}\mathfrak{H}^l_{\lambda\varsigma} &= [\sqrt{2}(1-\delta(\lambda,0)) + \delta(\lambda,0)]\,[(l+\lambda)!\,(l-\lambda)!]^{-1/2}\,[(2l)!]\,(l!)^{-1}\,2^{-l}\\ &\quad\times\left[z^{l-\lambda} - \frac{(l-\lambda)\,(l-\lambda-1)}{2\times(2l-1)}\,z^{l-\lambda-2}\,r^2\right.\\ &\quad\left.+\frac{(l-\lambda)\,(l-\lambda-1)\,(l-\lambda-2)\,(l-\lambda-3)}{2\times4\times(2l-1)\,(2l-3)}\,z^{l-\lambda-4}\,r^4 - \cdots\right]\\ &\quad\times r^\lambda \sin^\lambda\theta\,\varsigma(\lambda\varphi)\end{aligned} \tag{15b}$$

Here ς stands for either s (for sine) or c (for cosine), but for a particular application of a given formula ς refers to a definite choice. In the two cases the factors $r^\lambda \sin^\lambda\theta\,\varsigma(\lambda\varphi)$ of Eqs. (15) are

$$r^\lambda \sin^\lambda\theta \sin(\lambda\varphi) = [\tbinom{\lambda}{1}x^{\lambda-1}y - \tbinom{\lambda}{3}x^{\lambda-3}y^3 + \cdots] \tag{16a}$$

and

$$r^\lambda \sin^\lambda\theta \cos(\lambda\varphi) = [\tbinom{\lambda}{0}x^\lambda - \tbinom{\lambda}{2}x^{\lambda-2}y^2 + \cdots] \tag{16b}$$

where the condition $0 \leqq \lambda \leqq l$ is valid for the integer λ. The Z-axis is the unique axis and the spherical polar coordinates are chosen in the usual way,

$$\begin{aligned} z &= r \cos\theta \\ x &= r \sin\theta \cos\varphi \\ y &= r \sin\theta \sin\varphi \end{aligned} \tag{17}$$

The functions $\mathfrak{H}^l_{\lambda\varsigma}$ are given for $0 \leqq l \leqq 9$ in appendix 3, Table 15.

Each function $\mathfrak{H}^l_{\lambda\varsigma}$ is fully characterized in the group R_3 by the three quantum numbers l, λ, and ς, where l characterizes the irreducible representation and $\lambda\varsigma$ its component.

The quantum number l which is common to the $2\,l + 1$ functions of the set $\mathfrak{H}^l$ is accordingly determining for the trace $\chi[\mathsf{R}(\alpha)]$ of the $(2\,l + 1)$-dimensional matrix representative of R_3 that corresponds to a rotation by the angle α around an arbitrary axis.

$$\chi[\mathsf{R}(\alpha)] = 1 + \sum_{\lambda=1}^{l} 2\cos(\lambda\alpha) = \left(\sin\frac{(2l+1)\,\alpha}{2}\right) / \sin\frac{\alpha}{2} \tag{18}$$

At the same time, l by its even- or oddness defines the parity g(erade) or u(ngerade) of the set $\mathfrak{H}^l$ under the inversion operator I of R_{3i}.

The quantum number λ is for $\lambda > 0$ common to the two functions of the set $\mathfrak{H}^l_\lambda$ which forms a basis for the two-dimensional irreducible representations Λ and E_λ of the groups $C_{\infty v}$ and $D_{\infty h}$, respectively. The trace of the 2×2 matrix representative corresponding to the rotation by α around the infinity axis, called the Z-axis, contains λ as a parameter,

$$\chi[\mathsf{R}^{\mathrm{Z}}(\alpha)] = 2\cos(\lambda\alpha) \tag{19}$$

The two functions[2]) $\mathfrak{H}^l_{\lambda c}$ and $\mathfrak{H}^l_{\lambda s}$ are not eigenfunctions of the operator l_{Z} although they are eigenfunctions of

$$\mathsf{l}_{\mathrm{Z}}^2 = x\frac{\partial}{\partial x} + y\frac{\partial}{\partial y} + 2\,xy\frac{\partial^2}{\partial x\,\partial y} - x^2\frac{\partial^2}{\partial y^2} - y^2\frac{\partial^2}{\partial x^2} \tag{20}$$

---

[2]) For the three-dimensional rotation group R_3 one obtains the highest symmetry of the irreducible matrix representations (7) by taking the λ-functions in the order sine before cosine. For the two-dimensional dihedral group D_∞ it is more natural (vide p. 230) to chose the opposite order of the λ-functions.

with the eigenvalues λ^2. They may be characterized by either of the relations

$$\mathsf{l_Z}\left[\mathfrak{H}_{\lambda c}^{l}\ \mathfrak{H}_{\lambda s}^{l}\right] = \left[\mathfrak{H}_{\lambda c}^{l}\ \mathfrak{H}_{\lambda s}^{l}\right]\begin{bmatrix} 0 & -i\lambda \\ +i\lambda & 0 \end{bmatrix} \tag{21}$$

or

$$\mathsf{R^Z}(\alpha)\left[\mathfrak{H}_{\lambda c}^{l}\ \mathfrak{H}_{\lambda s}^{l}\right] = \left[\mathfrak{H}_{\lambda c}^{l}\ \mathfrak{H}_{\lambda s}^{l}\right]\begin{bmatrix} \cos\lambda\alpha & -\sin\lambda\alpha \\ \sin\lambda\alpha & \cos\lambda\alpha \end{bmatrix} \tag{22}$$

where $\mathsf{R^Z}(\alpha)$ is the operator which rotates the contour of the functions by the angle α around the Z-axis, so that (7)

$$\mathsf{R^Z}(\alpha) = \exp\left(-\,i\,\alpha\,\mathsf{l_Z}\right) \tag{23}$$

The solid harmonics $\mathfrak{H}_{\lambda\varsigma}^{l}$ or the corresponding surface harmonics

$$\mathfrak{C}_{\lambda\varsigma}^{l} = r^{-l}\,\mathfrak{H}_{\lambda\varsigma}^{l} \tag{24}$$

are taken as the real standard basis functions for the irreducible representations of the group R_3. As mentioned above, the functions are also basis functions for the group R_{3i}, but in this case they only span half the irreducible representations

$$s_g,\ p_u,\ d_g,\ f_u,\ \ldots \tag{25}$$

This does not complicate our issue since R_{3i} is the direct product group of R_3 with the inversion group S_2, which is commutative and of the order two, and therefore trivial to account for (see Sect. 5).

The functions $\mathfrak{H}_{\lambda\varsigma}^{l}$, summarized in Eq. (25), generate the real standard irreducible representations of R_3 uniquely, and in order to obtain the standard matrices of R_{3i} we make the following definitions: A matrix representative associated with a proper rotation R, is identically the same in the two groups for corresponding gerade and ungerade representations. A representative associated with an improper rotation S of R_{3i}, corresponding to a proper one, R, through

$$\mathsf{S} = \mathsf{RI} = \mathsf{IR} \tag{26}$$

is identically the same as that of R for gerade irreducible representations and identically the same as that of R, except for a sign change of all matrix elements, for ungerade representations. By these definitions it is immediately obvious that the real irreducible tensorial sets[3]) which generate the irreducible representations

$$s_u,\ p_g,\ d_u,\ f_g,\ \ldots \tag{27}$$

not appearing in Eq. (25), are completely defined, except for a common sign within each l set. It will be shown below (p. 218), however, that in fact only one choice of sign, common to all the l sets, is arbitrary.

It is valuable to consider the direct product of the two sets $|\alpha_1 l\}$ and $|\alpha_2 l\}$ which form the bases for the same irreducible representation of R_3. One obtains in the direct sum symmetrical and skew-symmetrical irreducible tensorial sets

$$|\alpha_1 l\} \times |\alpha_2 l\} = |\alpha\mathrm{S}\} + \underline{|\alpha\mathrm{P}\}} + |\alpha\mathrm{D}\} + \underline{|\alpha\mathrm{F}\}} + |\alpha\mathrm{G}\} + \cdots + |\alpha\mathrm{L}(=2\,l)\} \tag{28}$$

where the skew terms are underlined. It appears that the skew terms are all associated with odd l-values and on this basis the basic rule for the choice of standard basis functions for any of the sub-groups of R_3 is suggested.

Basic rule: Standard basis functions for irreducible representations that occur in antisymmetrized (skew) direct products are chosen out of the spherical harmonics with l odd and the remaining ones out of those with l even. This rule is immediately applicable to R_3 and without difficulty to all non-abelian point groups except the icosahedral group which, however, is not a simply reducible group[4]).

We shall anticipate some consequences of this choice of basis functions for the irreducible representations of the point groups.

When a 3-Γ symbol contains an odd Γ (i.e. a Γ chosen from an odd l-value) an even number of times, this 3-Γ symbol will, apart from a (by definition positive) constant, be represented by an integral over a product

---

[3]) The basis sets of Eq. (27) are the axial tensorial sets. It should be noted that the concept ungerade has a significance in an absolute sense as discussed in connection with Eq. (26), but this is not true of the concept axial. Axial means having the same transformation properties as a given (polar) standard set of functions under proper rotations and the same, except for a sign change, under improper rotations.

[4]) Here a given trio of irreducible representations will sometimes give rise to two independent sets of coupling coefficients. The analysis along the lines of the present paper requires for the four and five-dimensional irreducible representations of the icosahedron in each case two independent sets of basis functions. For the four-dimensional representation, one basis function out of the f set and one out of the g set, and for the five-dimensional representation one out of the d set and one out of the g set form the simplest choice.

of three spherical harmonics. So the sign of the symbol is independent of the order of the Γ's and we have an even 3-Γ symbol.

When a 3-Γ symbol contains an odd Γ an odd number of times the corresponding integral vanishes because of its ungerade character. So in this case one of the Γ's must be represented by a corresponding axial operator (and the others still by their spherical harmonic standard basis functions) in order that the 3-Γ symbol can be represented by an integral. The 3-Γ symbol defined in this way will then automatically change sign under odd permutations of its columns, so we have an odd 3-Γ symbol.

3. Basis Function Operator Equivalents

The three differential operators $\frac{\partial}{\partial x}$, $\frac{\partial}{\partial y}$, and $\frac{\partial}{\partial z}$ and the three functions x, y, and z transform under proper and improper rotations by identically the same matrices as do the Cartesian unit base vectors $\mathbf{i}$, $\mathbf{j}$, and $\mathbf{k}$. The products of the differential operators and linear combinations of them also have the same transformation properties as the corresponding Cartesian unit base tensors or their linear combinations. The same holds true for the functions x, y, and z. For example,

$$\frac{\partial^2}{\partial z^2} - \frac{1}{2}\frac{\partial^2}{\partial x^2} - \frac{1}{2}\frac{\partial^2}{\partial y^2} = \frac{3}{2}\frac{\partial^2}{\partial z^2} - \frac{1}{2}\nabla^2 \tag{29}$$

and

$$\mathfrak{H}_\sigma^2 = z^2 - \frac{1}{2}x^2 - \frac{1}{2}y^2 \tag{30}$$

transform as the unit base tensor

$$(2\,\mathbf{k}\,\mathbf{k} - \mathbf{i}\,\mathbf{i} - \mathbf{j}\,\mathbf{j})/\sqrt{6} \tag{31}$$

It is useful to define in general differential operators from the solid harmonics $\mathfrak{H}_{\lambda\varsigma}^l$ by performing in Eqs. (15a) or (15b) the substitutions

$$\begin{aligned} &x \to \frac{\partial}{\partial x};\ y \to \frac{\partial}{\partial y};\ z \to \frac{\partial}{\partial z} \\ &r^2 \to \nabla^2 = \frac{\partial^2}{\partial x^2} + \frac{\partial^2}{\partial y^2} + \frac{\partial^2}{\partial z^2} \end{aligned} \tag{32}$$

where the last expression is a consequence of the first three.

These differential operators will be denoted $\mathfrak{D}_{\lambda\varsigma}^l$; they have the following property (*8*)

$$\mathfrak{D}^l_{\lambda\varsigma}\,\mathfrak{H}^l_{\lambda'\varsigma'} = \frac{(2l)!}{2^l l!}\,\delta(\lambda\lambda')\,\delta(\varsigma\varsigma')$$

$$= 1 \times 3 \times 5 \times \cdots \times (2l-1)\,\delta(\lambda\lambda')\,\delta(\varsigma\varsigma') \qquad (33)$$

where $\lambda\varsigma$ and $\lambda'\varsigma'$ both refer to the standards of Eq. (15), using the same coordinate axes of quantization. This simple result obtains because the operator $\mathfrak{D}^l_{\lambda\varsigma}$ operates on a solid harmonic of the same degree, l. It is advantageous to define an operator, $\mathfrak{N}^l_{\lambda\varsigma}$,

$$\mathfrak{N}^l_{\lambda\varsigma} \equiv \frac{2^l l!}{(2l)!}\,\mathfrak{D}^l_{\lambda\varsigma}, \qquad (34\,\mathrm{a})$$

so that

$$\mathfrak{N}^l_{\lambda\varsigma}\,\mathfrak{H}^l_{\lambda\varsigma} = 1 \qquad (34\,\mathrm{b})$$

The operator $\mathfrak{N}^l_{\lambda\varsigma}$ will be called a unification operator. On the basis of Eq. (34) we further define projection operators $\mathsf{O}^{(l)}_{\lambda\varsigma}$ by

$$\mathsf{O}^{(l)}_{\lambda\varsigma} = \mathfrak{H}^l_{\lambda\varsigma}\,\mathfrak{N}^l_{\lambda\varsigma}, \qquad (35\,\mathrm{a})$$

$$\mathsf{O}^{(l)}_{\lambda\varsigma}\,\mathfrak{H}^l_{\lambda\varsigma} = \mathfrak{H}^l_{\lambda\varsigma} \qquad (35\,\mathrm{b})$$

It is noted that the projection operator is an element of a reducible tensorial set and that its action as a projection operator is limited to homogeneous polynomials and spherical harmonics of the degree l.

To illustrate the properties of these operators it is observed that any homogeneous polynomial f_l of the degree l can be written

$$f_l = h_l + r^2 f_{l-2} \qquad (36)$$

where the spherical harmonic h_l is a linear combination of the elements of the set $\mathfrak{H}^l$,

$$h_l = \sum_{\lambda\varsigma} k(\lambda\varsigma)\,\mathfrak{H}^l_{\lambda\varsigma} \qquad (37)$$

Now, the projection operator $\mathsf{O}^{(l)}_{\lambda\varsigma}$ extracts (projects) the term $k(\lambda\varsigma)\mathfrak{H}^l_{\lambda\varsigma}$ out of f_l, and analogously the unification operator $\mathfrak{N}^l_{\lambda\varsigma}$ yields the coefficient $k(\lambda\varsigma)$ to $\mathfrak{H}^l_{\lambda\varsigma}$ [normalized to $4\pi/(2l+1)$] when operating on f_l. The analytical form of the operator $\mathfrak{D}^l_{\lambda\varsigma}$ may be obtained by using either Eq. (15a) or Eq. (15b), giving the two forms which appear in Eq. (29). The latter form is particularly useful when the operators $\mathfrak{N}^l_{\lambda\varsigma}$ and $\mathsf{O}^{(l)}_{\lambda\varsigma}$ are to be used on polynomials which are spherical harmonics

[$f_{l-2} = 0$ in Eq. (36)]. In this case all the terms containing ∇^2 or powers of ∇^2 will give vanishing contributions.

It is sometimes necessary to obtain the content of a harmonic $\mathfrak{H}^l_{\lambda\varsigma}$ of the degree l in a homogeneous polynomial f_{l+2n} of the degree $l+2n$. In this case one has to annihilate the terms of higher degree than l by operating on the polynomial f_{l+2n} with the totally symmetric operator ∇^{2n}. This can be done because

$$\nabla^2(r^q h_l) = q[q + (2l+1)]r^{q-2}h_l \tag{38}$$

Further applications of Eq. (38) will be shown elsewhere (*9*). Foı our purpose the important fact is that the coefficients to h_l on the right hand side of Eq. (38) are positive.

4. The Three-Dimensional Rotation Group R_3

a) Even 3-l Symbols

We now have the necessary preɪequisites to obtain the 3-Γ symbols for the group R_3. In this group they will, however, be termed 3-l symbols. We consider an integral over three basis functions of Eq. (15) and use the Wigner-Eckart theorem:

$$\frac{1}{4\pi}\int \mathfrak{H}^{l_1}_{t_1}\,\mathfrak{H}^{l_2}_{t_2}\,\mathfrak{H}^{l_3}_{t_3}\,dS = k(l_1\,l_2\,l_3)\begin{pmatrix} l_1 & l_2 & l_3 \\ t_1 & t_2 & t_3 \end{pmatrix} \tag{39}$$

Here t is an abbreviation for $\lambda\varsigma$ and we shall make the convention that such an integration is performed over the surface of the unit sphere, i.e. with the condition $r^2 = x^2 + y^2 + z^2 = 1$. Thus we have

$$\frac{1}{4\pi}\int dS = 1 \tag{40}$$

Firstly, it is noted that the integral of Eq. (39) is independent of the order of the functions. Secondly, the integral can only be non-vanishing if the degree, $l_1 + l_2 + l_3$, of the integrand is even. We define the constant $k(l_1\,l_2\,l_3)$ to be positive and independent of the order of the functions and thereby ensure that the 3-l symbols are unchanged under even as well as odd permutations of their columns. 3-l symbols with these properties will be called even 3-l symbols.

In Eq. (39) one can replace the functions $\mathfrak{H}^{l_1}_{t_1}$ or $\mathfrak{H}^{l_2}_{t_2}$ or both by their differential operator equivalents of $\mathfrak{D}$ type or $\mathfrak{N}$ type. This will alter the magnitude of the constant $k(l_1\,l_2\,l_3)$, which depends on the choice of operators, but not its sign (*10*). Therefore, using Eq. (3) and its subsequent discussion, the constant can in each case be obtained as the *positive* square root of the sum of the squares of the matrix elements, the sum ranging over t_1, t_2, and t_3.

An important example of this, is the case of Eq. (39), where the first function is the *s*-function, $\mathfrak{H}^0_\sigma = 1$, and where the second function is replaced by $\mathfrak{N}^{l_2}_{t_2}$

$$\frac{1}{4\pi}\int \mathfrak{N}^{l_2}_{t_2}\,\mathfrak{H}^{l_3}_{t_3}\,dS = \delta(l_2\,l_3)\,\delta(t_2\,t_3) \qquad (41)$$
$$= k'(0\,l_2\,l_3)\begin{pmatrix}0 & l_2 & l_3\\ \sigma & t_2 & t_3\end{pmatrix}$$

The Kronecker deltas arise for the following reasons. For $l_2 < l_3$ the integrand is necessarily a solid harmonic and its integral is accordingly vanishing. For $l_2 = l_3$ and $t_2 \neq t_3$, $\mathfrak{N}^{l_2}_{t_2}$ annihilates [Eq. (33)] $\mathfrak{H}^{l_3}_{t_3}$, and for $l_2 = l_3$ and $t_2 = t_3$, $\mathfrak{H}^{l_3}_{t_3}$ is unified. For $l_2 > l_3$ the integrand vanishes. By squaring Eq. (41) and taking the sum of the $(2\,l_2 + 1)$ non-vanishing terms, application of Eq. (4) yields $k'^2 = 2\,l_2 + 1$, so that $k' = +\,(2\,l_2 + 1)^{\frac{1}{2}}$ and as a consequence

$$\begin{pmatrix}s & l_2 & l_3\\ \sigma & t_2 & t_3\end{pmatrix} = \begin{pmatrix}0 & l_2 & l_3\\ \sigma & t_2 & t_3\end{pmatrix} = (2\,l_2 + 1)^{-\frac{1}{2}}\,\delta(l_2\,l_3)\,\delta(t_2\,t_3) \qquad (42)$$

The next example shows the procedure carried out in different ways with the values $l_1 = 1$, $l_2 = 1$, and $l_3 = 2$ fixed (Appendix 3, Table 15). It is convenient, however, to use the orthogonality conditions, Eq. (6), for the coupling coefficients together with the relation, Eq. (5), which connects them with the 3-*l* symbols.

Firstly, the two harmonics $\mathfrak{H}^{l_1}_{t_1}$ and $\mathfrak{H}^{l_2}_{t_2}$ of Eq. (39) are replaced by their differential operator equivalents. Then with $t_3 = \sigma$ the only non-vanishing integrals are

$$\frac{1}{4\pi}\int \frac{\partial}{\partial z}\frac{\partial}{\partial z}\left(z^2 - \tfrac{1}{2}x^2 - \tfrac{1}{2}y^2\right)dS = 2$$
$$\frac{1}{4\pi}\int \frac{\partial}{\partial x}\frac{\partial}{\partial x}\left(z^2 - \tfrac{1}{2}x^2 - \tfrac{1}{2}y^2\right)dS = -1 \qquad (43)$$
$$\frac{1}{4\pi}\int \frac{\partial}{\partial y}\frac{\partial}{\partial y}\left(z^2 - \tfrac{1}{2}x^2 - \tfrac{1}{2}y^2\right)dS = -1$$

The corresponding coupling coefficients $\langle pt\,pt|d\sigma\rangle$ are obtained simply by multiplication of the numbers on the right-hand side by the normalization factor $[2^2 + (-1)^2 + (-1)^2]^{-\frac{1}{2}} = 1/\sqrt{6}$ and the 3-*l* symbols by the normalization factor $1/\sqrt{5\cdot 6} = 1/\sqrt{30}$.

Secondly, only the first function in Eq. (39) is replaced by its differential operator equivalent. The product $\mathfrak{H}^1_{t_2}\,\mathfrak{H}^2_{t_3}$ is a homogeneous polynomial of the third degree which could be decomposed into a solid harmonic of f-type, and one of p-type multiplied by r^2. Since only the p-type part will give a non-vanishing contribution to the integral of Eq. (39), we replace the first function by $\mathfrak{D}^1_{t_1}\,\nabla^2$ so that ∇^2 annihilates the f-term and, according to Eq. (38), leaves the p-type term multiplied by some positive constant. Analogously with Eq. (43) one obtains

$$\frac{1}{4\pi}\int\frac{\partial}{\partial z}\,\nabla^2[z^3-\tfrac{1}{2}\,zx^2-\tfrac{1}{2}\,zy^2]\,dS=\frac{1}{4\pi}\int\frac{\partial}{\partial z}\,[6z-z-z]\,dS=4 \tag{44}$$

and corresponding expressions for the remaining integrals. This leads to the same result as before, except that this time the normalization constants are $1/(2\sqrt{6})$ and $1/(2\sqrt{30})$, respectively.

Thirdly, only the second harmonic of Eq. (39) is replaced by its differential operator equivalent giving

$$\frac{1}{4\pi}\int z\,\frac{\partial}{\partial z}\,[z^2-\tfrac{1}{2}\,x^2-\tfrac{1}{2}\,y^2]\,dS=2/3 \tag{45}$$

and analogous expressions. All 3-l symbols of ppd-type are given explicitly in Appendix 3, Table 16.

Finally, it is noted that the results obtained, by means of Eq. (7) can be used to derive that irreducible product of $\mathfrak{H}^1$ and $\mathfrak{D}^1$, which is of d-type, i.e. of the degree 2,

$$\begin{aligned}
[\mathfrak{H}^1\times\mathfrak{D}^1]^2_{\sigma}&=\frac{2}{\sqrt{6}}\,z\frac{\partial}{\partial z}-\frac{1}{\sqrt{6}}\,x\frac{\partial}{\partial x}-\frac{1}{\sqrt{6}}\,y\frac{\partial}{\partial y}\\
[\mathfrak{H}^1\times\mathfrak{D}^1]^2_{\pi s}&=\frac{1}{\sqrt{2}}\,y\frac{\partial}{\partial z}+\frac{1}{\sqrt{2}}\,z\frac{\partial}{\partial y}\\
[\mathfrak{H}^1\times\mathfrak{D}^1]^2_{\pi c}&=\frac{1}{\sqrt{2}}\,z\frac{\partial}{\partial x}+\frac{1}{\sqrt{2}}\,x\frac{\partial}{\partial z}\\
[\mathfrak{H}^1\times\mathfrak{D}^1]^2_{\delta s}&=\frac{1}{\sqrt{2}}\,x\frac{\partial}{\partial y}+\frac{1}{\sqrt{2}}\,y\frac{\partial}{\partial x}\\
[\mathfrak{H}^1\times\mathfrak{D}^1]^2_{\delta c}&=\frac{1}{\sqrt{2}}\,x\frac{\partial}{\partial x}-\frac{1}{\sqrt{2}}\,y\frac{\partial}{\partial y}
\end{aligned} \tag{46}$$

These operators are determined with sign and can be used to find the 3-l symbols of the type $\begin{pmatrix} l & 2 & l\\ t_1 & t_2 & t_3\end{pmatrix}$. In the case where $l=1$, results equivalent to those of Eq. (43) will be generated. It should be noted that when the operators of Eq. (46) are introduced instead of the function $\mathfrak{H}^l_{t_2}$ of

Eq. (39), they have an effective degree of zero; i.e. the degree of the integrand becomes $l_1 + l_3$. However, it is not required that $l_1 = l_3$ in order to have non-vanishing matrix elements of the operatoıs, Eq. (46). This is in contradistinction to the operators of Eq. (47), which, since they are proportional to the angular momentum operator, Eq. (14), are diagonal in l.

b) Odd 3-l Symbols

In the previous sub-section it was seen that the principle of associating with a basis function its differential operator equivalent, defined also with respect to sign, has many facets. However, even though the standard transformation properties of the basis functions of R_3 and R_{3i} are fully defined, 3-l symbols with $l_1 + l_2 + l_3$ equal to an odd number cannot be handled in the same direct way. As mentioned at the end of Sect. 2, one must in this case replace one of the spherical harmonic basis functions, which themselves form polar tensorial sets, by a corresponding axial tensoı. This is done here by considering first 3-l symbols of the type $\begin{pmatrix} l & 1 & l \\ t_1 & t_2 & t_3 \end{pmatrix}$ in which case the axial tensor $\mathfrak{K}^1_{t_2}$, which corresponds to the polar $\mathfrak{H}^1_{t_2}$, is skew-symmetric.

A convenient choice of a skew tensorial set of operators consists of the components of the angular momentum operator $\mathfrak{l}$ [Eq. (14)], which transform according to the irreducible representation p_g of R_{3i}. When these are divided by $\sqrt{-2}$, one obtains a set of real operators whose elements are the irreducible products of the degree one (effective degree zero) of the sets $\mathfrak{H}^1$ and $\mathfrak{D}^1$,

$$\begin{aligned}
[\mathfrak{H}^1 \times \mathfrak{D}^1]^1_\sigma &= -\frac{1}{\sqrt{2}}\, x \frac{\partial}{\partial y} + \frac{1}{\sqrt{2}}\, y \frac{\partial}{\partial x} \\
[\mathfrak{H}^1 \times \mathfrak{D}^1]^1_{\pi c} &= -\frac{1}{\sqrt{2}}\, y \frac{\partial}{\partial z} + \frac{1}{\sqrt{2}}\, z \frac{\partial}{\partial y} \qquad (47) \\
[\mathfrak{H}^1 \times \mathfrak{D}^1]^1_{\pi s} &= -\frac{1}{\sqrt{2}}\, z \frac{\partial}{\partial x} + \frac{1}{\sqrt{2}}\, x \frac{\partial}{\partial z}
\end{aligned}$$

In Eq. (47) we have made a definite phase choice which is actually the opposite of that occurring in the vector product of $x\mathbf{i} + y\mathbf{j} + z\mathbf{k}$ and $\frac{\partial}{\partial x}\,\mathbf{i} + \frac{\partial}{\partial y}\,\mathbf{j} + \frac{\partial}{\partial z}\,\mathbf{k}$. This particular choice has the consequence that reduced matrices derived by means of *Wigner's* 3-j symbols are identical to those derived in the present scheme (*10*). As was mentioned in connection with Eq. (27) and as will appear in this sub-section, this choice of phase determines the phases of all odd 3-l symbols.

By means of the operators of Eq. (47) we shall now derive odd 3-l symbols by a procedure similar to that of Sect. 4a.

Consider first the integral

$$\frac{1}{4\pi}\int \mathfrak{H}^1_{\pi s}[\mathfrak{H}^1 \times \mathfrak{D}^1]^1_\sigma \mathfrak{H}^1_{\pi c}\, dS = \frac{1}{4\pi}\int y\left(-\frac{1}{\sqrt{2}}x\frac{\partial}{\partial y}+\frac{1}{\sqrt{2}}y\frac{\partial}{\partial x}\right)x\, dS$$

$$= \frac{1}{3\sqrt{2}} = k\begin{pmatrix} p & p & p \\ \pi s & \sigma & \pi c\end{pmatrix} \tag{48}$$

Again the constant k is taken as the *positive* square root of the sum of the squares of all integrals of this kind [Eq. (4)]. There are six non-vanishing integrals of the same absolute value, and one obtains $k^2 = \frac{6}{18}$, $k = +1/\sqrt{3}$ and

$$\begin{pmatrix} p & p & p \\ \pi s & \sigma & \pi c\end{pmatrix} = \frac{+1}{\sqrt{6}} \tag{49}$$

This procedure can be carried out quite generally for 3-l symbols of the type $\begin{pmatrix} l & p & l \\ t_1 & t_2 & t_3\end{pmatrix}$. From Eq. (21) it follows, that

$$\frac{1}{4\pi}\int \mathfrak{H}^l_{\lambda c}\, \mathrm{l_Z}\, \mathfrak{H}^l_{\lambda s}\, dS = \frac{-i\lambda}{2l+1} \text{ and } \frac{1}{4\pi}\int \mathfrak{H}^l_{\lambda s}\, \mathrm{l_Z}\, \mathfrak{H}^l_{\lambda c}\, dS = \frac{i\lambda}{2l+1}$$

$$\lambda = 1, 2, \ldots, l \tag{50}$$

are the only non-vanishing integrals that can occur with the operator $\mathrm{l_Z}$. Therefore the sum of the squares of the matrix elements of the operator $[\mathfrak{H}^1 \times \mathfrak{D}^1]^1_\sigma$ is

$$2\sum_{\lambda=1}^{l}\frac{\lambda^2}{2(2l+1)^2} = \frac{1}{(2l+1)^2}\sum_{\lambda=1}^{l}\lambda^2$$

$$= \frac{1}{(2l+1)^2}\times\frac{l(l+1)\,(2l+1)}{6} = \frac{l(l+1)}{6(2l+1)} \tag{51}$$

or, in other words

$$\sum_{t_1 t_3}\left[\frac{1}{4\pi}\int \mathfrak{H}^l_{t_1}[\mathfrak{H}^1 \times \mathfrak{D}^1]^1_\sigma\, \mathfrak{H}^l_{t_3} dS\right]^2 = \frac{l(l+1)}{6(2l+1)}$$

$$= k^2 \sum_{t_1 t_3}\begin{pmatrix} l & p & l \\ t_1 & \sigma & t_3\end{pmatrix}^2 = k^2 \times \frac{1}{3} \tag{52}$$

where in the last step the Eqs. (5) and (6a) have been used. Thus for the constant k one has

$$k = + \sqrt{\frac{l(l+1)}{2(2l+1)}}$$

generally applicable for these 3-l symbols, which can be found from the relation

$$\frac{1}{4\pi} \int \mathfrak{H}^l_{t_1} [\mathfrak{H}^1 \times \mathfrak{D}^1]^1_{t_2} \mathfrak{H}^l_{t_3} dS = \sqrt{\frac{l(l+1)}{2(2l+1)}} \begin{pmatrix} l & p & l \\ t_1 & t_2 & t_3 \end{pmatrix} \tag{53}$$

As an example of this, we calculate $\begin{pmatrix} d & p & d \\ \delta s & \pi c & \pi c \end{pmatrix}$,

$$\frac{1}{4\pi} \int \sqrt{3} xy \left(-\frac{1}{\sqrt{2}} y \frac{\partial}{\partial z} + \frac{1}{\sqrt{2}} z \frac{\partial}{\partial y} \right) \sqrt{3}\, zx dS = \frac{-1}{5\sqrt{2}} = \sqrt{\frac{2 \cdot 3}{2 \cdot 5}} \begin{pmatrix} 2 & 1 & 2 \\ 2s & 1c & 1c \end{pmatrix}$$

$$\Rightarrow \begin{pmatrix} d & p & d \\ \delta s & \pi c & \pi c \end{pmatrix} = \frac{-1}{\sqrt{30}} \tag{54}$$

All the 3-l symbols of *dpd* type are given in Appendix 3, Table 17.

By means of the 3-l symbols evaluated through Eq. (53) one can now calculate any odd 3-l symbol. To illustrate this, it is noted that if $l_1 + l_2 + l_3$ is odd, then the integral whose integrand is of the degree $l_1 + l_2 + l_3 - 1$ can be calculated by the methods of Sect. 4a. Therefore, in the integrand of Eq. (39), we replace the middle function $\mathfrak{H}^{l_2}_{t_2}$, which is a component of a polar tensorial set, by that of its operator equivalents in the rotation group R_3, which is a component of an axial tensorial set and at the same time has an effective degree of $l_2 - 1$. This is

$$[\mathfrak{H}^{l_2} \times \mathfrak{D}^1]^{l_2}_{t_2} \tag{55}$$

where the irreducible product is expanded by means of the 3-l symbols of Eq. (53). Then Eq. (38) may be used to reduce the degree of $[\mathfrak{H}^{l_2} \times \mathfrak{D}^1]^{l_2}_{t_2} \mathfrak{H}^{l_3}_{t_3}$ down to the degree l_1 of the first factor in the integrand, and finally the projection operator $\mathsf{O}^{(l_1)}_{t_1}$ can be used, if desired.

The important thing to note is that, since the constant of Eq. (38) is positive, the phase convention made in Eq. (47) together with the decision to choose the constant k in Eq. (39) positive, determine the phases of all 3-l symbols.

5. The Three-Dimensional Rotation-Inversion Group R_{3i}

It was mentioned in connection with Eq. (25) that the spherical harmonics which span the irreducible representations of the group R_3 span only half the irreducible representations of the group R_{3i}. Consequently it is not possible to define the 3-l symbols of R_{3i} on the same basis as was used in Sect. 4.

However, we shall make use of the fact that the group R_{3i} is the direct product group of the group R_3 with the inversion group S_2. For a 3-Γ symbol of a direct product group the relation (4)

$$\begin{pmatrix} \Gamma_1 \Gamma_1' \ \Gamma_2 \Gamma_2' \ \Gamma_3 \Gamma_3' \\ \gamma_1 \gamma_1' \ \gamma_2 \gamma_2' \ \gamma_3 \gamma_3' \end{pmatrix} = \begin{pmatrix} \Gamma_1 \Gamma_2 \Gamma_3 \\ \gamma_1 \gamma_2 \gamma_3 \end{pmatrix} \begin{pmatrix} \Gamma_1' \Gamma_2' \Gamma_3' \\ \gamma_1' \gamma_2' \gamma_3' \end{pmatrix} \tag{56}$$

is valid. Here the unprimed and primed symbols refer to the irreducible representations of the constituent groups.

The 3-Γ symbols of the group S_2 are most simply defined by

$$\begin{pmatrix} \Gamma_1 \Gamma_2 \Gamma_3 \\ \gamma_1 \gamma_2 \gamma_3 \end{pmatrix} = \delta(\Gamma_1 \Gamma_2 \Gamma_3) \tag{57}$$

where the Γ's take on the values A_g or A_u, so that the 3-Γ symbol vanishes whenever A_u occurs an odd number of times. We conclude that the 3-l symbols of the group R_{3i} are equal to the corresponding ones of the group R_3 whenever they do not vanish for parity reasons.

The relations Eqs. (56) and (57) will be used similarly in Sect. 8c to generate the 3-Γ symbols of those point groups which can be described as direct product groups of a rotation group and either the inversion group S_2 or the reflexion group C_{1h}.

II. The Sub-Groups of R_3 and R_{3i}

6. Survey of the Crystallographic Point Groups

a) Introduction. Commutative Groups

The symmetry displayed by chemical entities such as molecules or ions is known as point symmetry* because the symmetry elements intersect in a specific point. To put this in another way: all symmetry operations which bring such an entity into a configuration equivalent to the original one leave one point unchanged. The assemblage of symmetry operators characterizing the geometry of an entity constitutes its point group.

---

*) For the groups C_n and C_{nv} the intersection takes place in a line, also for $n = \infty$. For the group $C_{1h} = C_s$ it takes place in a plane.

The point groups were first studied by crystallographers, who used them to characterize the macroscopic properties of crystals. An ideal crystal must belong to one of the thirty-two crystallographic point groups. (This limited number arises from the fact that only one-, two-, three-, four-, and six-fold axes can occur in crystals because of space-filling requirements.) These point groups are also the point groups of most importance for chemistry; however, a few others may have to be added.

Before giving a general survey of point groups, we study three simple ones, each of the order two: the inversion group S_2 (or C_i), the reflexion group C_{1h} (or C_s), whose only symmetry element is a mirror plane, and the group C_2, which has one two-fold axis. Although they are isomorphic, they represent quite different physical systems. Their character table is displayed in Table 1.

Table 1. Characters for the rotation group C_2 and the isomorphic groups S_2 and C_{1h}, which are hemihedrics of the holohedric C_{2h}. S_2 is the holohedric triclinic group

$S_2 = C_i$			E	$\mathsf{I} = \mathsf{C}_2^{\mathsf{Z}}\,\sigma_h = \mathsf{S}_2$
	$C_{1h} = C_s$		E	$\sigma_h = \mathsf{I}\,\mathsf{C}_2^{\mathsf{Z}} = \mathsf{S}_2\,\mathsf{C}_2^{\mathsf{Z}}$
		C_2	E	$\mathsf{C}_2^{\mathsf{Z}} = \mathsf{I}\,\sigma_h = \mathsf{S}_2\,\sigma_h$
A_g	A′	A	1	1
A_u	A″	B	1	−1

The operators $\mathsf{I} = \mathsf{S}_2$ and $\sigma_h = \mathsf{I}\mathsf{C}_2^{\mathsf{Z}}$ commute with all other symmetry operators. Therefore the direct product group of a group G of the order g with the inversion group gives a new group of the order $2g$ such that each of the irreducible representations of G occurs twice with the further labels g and u. Similarly, if a group does not contain the operator $\mathsf{I}\mathsf{C}_2^{\mathsf{Z}}$, then the direct product of this group with C_{1h} gives rise to a new group with twice the number of elements and of irreducible representations. In this case the irreducible representations are labeled with a prime when symmetrical with respect to $\mathsf{I}\mathsf{C}_2^{\mathsf{Z}}$ and with a double prime when anti-symmetrical.

Irreducible representations of dimension one are generally termed A if they are symmetrical with respect to the main axis operation and are termed B if antisymmetrical.

Using Table 1, one can form the direct product group of any pair of the three groups mentioned. As an example we take the direct product

Table 2. *Characters for the holohedric monoclinic group C_{2h}*

$C_{2h}=C_2\times S_2$	E	C_2^Z	S_2	$\sigma_h=S_2C_2^Z$
A_g	1	1	1	1
B_g	1	−1	1	−1
A_u	1	1	−1	−1
B_u	1	−1	−1	1

group C_{2h} of the groups C_2 and S_2, Table 2. It follows from the remarks in the previous paragraph that the four irreducible representations are labelled with A and B, and with g and u. It is seen that the character table for the group C_2 occurs four times, three times multiplied by $+1$ and the fourth time multiplied by -1. The same group C_{2h} could equally well have been obtained as the direct product group of the groups C_2 and C_{1h}. However, when the same group can be generated as the direct product group from either the group S_2 or the group C_{1h}, the inversion group is the perferred one by convention, as far as labeling of the resulting irreducible representations are concerned. Further, it is a useful convention to state the operations of the group with the rotations first and then the corresponding improper rotations in the same order.

One can obtain a group D_2 which is isomorphic with the group C_{2h} by using the generators C_2^Z and C_2^Y. Obviously these generators imply that the non-trivial operation C_2^X is in the group. The character table for the group D_2 is given in Table 3 and it is seen that with the irreducible representations of C_{2h} taken in the order A_g, A_u, B_g, B_u the two character tables are identical. In the group D_2 none of the three axes is a priori the principal one and the symbols B_1, B_2, and B_3 are used for anti-symmetrical about two axes. This implies symmetrical about one axis, the Z, Y, and X axes, respectively.

Table 3. *Characters for the dihedral orthorhombic rotation group D_2 and the hemihedral orthorhombic group C_{2v}*

C_{2v}		E	C_2^Z	$\sigma_v^{ZX}=S_2C_2^Y$	$\sigma_v^{YZ}=S_2C_2^X$
	D_2	E	C_2^Z	C_2^Y	C_2^X
A_1	A	1	1	1	1
A_2	B_1	1	1	−1	−1
B_1	B_2	1	−1	1	−1
B_2	B_3	1	−1	−1	1

It is a general feature that a *c*yclic group (C_n) of the order n with the Z-axis as symmetry axis becomes *d*ihedral (D_n) by addition of the generator C_2^Y, and thereby embraces twice as many elements.

The direct product group of the groups D_2 and S_2 gives the group D_{2h} with eight elements and with the irreducible representations of D_2 further characterized by g and u (Table 4). Besides the groups D_2 and S_2 also the previously mentioned groups C_2, C_{1h}, and C_{2h} are sub-groups of D_{2h}. Further the group C_{2v} (Table 3), consisting of the first proper rotation and the last two improper ones, is a sub-group of D_{2h} which is isomorphic with D_2.

Table 4. *Characters for the holohedric orthorhombic group* D_{2h}

$D_{2h}=D_2\times S_2$	E	C_2^Z	C_2^Y	C_2^X	S_2	$\sigma_h^{XY}=S_2C_2^Z$	$\sigma_v^{ZX}=S_2C_2^Y$	$\sigma_v^{YZ}=S_2C_2^X$
A_g	1	1	1	1	1	1	1	1
B_{1g}	1	1	−1	−1	1	1	−1	−1
B_{2g}	1	−1	1	−1	1	−1	1	−1
B_{3g}	1	−1	−1	1	1	−1	−1	1
A_u	1	1	1	1	−1	−1	−1	−1
B_{1u}	1	1	−1	−1	−1	−1	1	1
B_{2u}	1	−1	1	−1	−1	1	−1	1
B_{3u}	1	−1	−1	1	−1	1	1	−1

The point groups considered till now correspond to the three crystal systems which are optically biaxial. These are the triclinic system, the monoclinic system, and the orthorhombic system. As illustrated in Table 9, the groups S_2, C_{2h}, and D_{2h} may be called the holohedric triclinic, monoclinic, and orthorhombic groups. The other groups within each crystal system (Table 9) are sub-groups of the holohedric ones, and when they are isomorphic with the pure rotation group which makes up the first half of the elements of the holohedric group, they are called hemihedric groups.

The point groups discussed in this section are commutative (Abelian) groups. For these groups the number of irreducible representations equals the number of elements and all the irreducible representations have the dimension one. It is generally not the case, however, that the irreducible representations are real. Thus for n even and greater than two, all but two of the irreducible representations of the groups C_n and S_n fall in pairs whose characters generally are complex, and conjugate to one another. A similar statement holds true of all but one of the irreducible representations in the case of C_n groups with n odd and greater than one. For these groups we shall employ a real basis which connects such pairs into pseudo two-dimensional representations, which we shall call the *most reduced representations with a real basis*. This is somewhat artificial from a group theoretical point of view but has compensating advantages (Sect. 8). We shall mention that with a real basis the cyclic groups C_n ($n > 2$) and the alternating groups S_n (n even and > 2) are not simply reducible groups.

b) Non-Commutative Groups

The holohedric point groups that correspond to the optically uniaxial crystal systems are non-commutative groups. In such groups some of the symmetry operators fall in the same class, and at least one of the irreducible representations will have its dimension greater than one. We shall discuss these groups in the following general manner. From the pure rotation group with the highest number of elements the direct product group with the inversion group is formed, giving the corresponding holohedric group. Next the invariant sub-groups of the holohedric groups will be considered.

The lowest order possible for a non-commutative group is six. This follows from the basic group theoretic proposition, that the number of elements equals the sum of the squares of the dimensions of the irreducible representations. The pure rotation group D_3 which is of the order six and has three irreducible representations, is obtained by combining the generators $\mathsf{C}_3^{\mathrm{Z}}$ and $\mathsf{C}_2^{\mathrm{Y}}$. This group is characterized in Table 5 by its standard irreducible representations together with the basis functions which have been used to generate these representations through relations analogous to Eq. (22). In this way it is always possible to generate the matrix representatives whenever standard basis functions have been chosen.

In order to discuss the rest of the crystallographic point groups, one further has to consider the dihedral rotation groups D_4 and D_6, and the cubic rotation group O. Their character tables, standard basis functions, and a useful choice of group generators are displayed in Tables 6, 7, and 8. In this way the material required for symmetry considerations is directly available.

In the same way as the holohedric orthorhombic group D_{2h} was obtained as the direct product group of the rotation group D_2 with the inversion group S_2, the direct product groups of D_3, D_4, D_6, and O with S_2 are formed. As illustrated in Table 9, the groups D_{3d}, D_{4h}, D_{6h}, and O_h may be called the holohedric trigonal (or rhombohedral), tetragonal, hexagonal, and cubic groups, respectively. The corresponding crystal

Table 5. *Irreducible standard matrix representations of the point group D_3*

	D_3	E	$2\,\mathsf{C}_3$		$3\,\mathsf{C}_2$		
			$\mathsf{C}_3^{\mathrm{Z}}$	$(\mathsf{C}_3^{\mathrm{Z}})^2$	$\mathsf{C}_2^{\mathrm{Y}}$	$\mathsf{C}_2^{y=x/\sqrt{3}\,;\,z=0}$	$\mathsf{C}_2^{y=-x/\sqrt{3}\,;\,z=0}$
$\mathfrak{H}_\sigma^s$	A_1	1	1	1	1	1	1
$\mathfrak{H}_\sigma^p$	A_2	1	1	1	−1	−1	−1
$[\mathfrak{H}_{\pi c}^d, \mathfrak{H}_{\pi s}^d]$	E	$\begin{bmatrix} 1 & 0 \\ 0 & 1 \end{bmatrix}$	$\begin{bmatrix} -\frac{1}{2} & -\frac{\sqrt{3}}{2} \\ \frac{\sqrt{3}}{2} & -\frac{1}{2} \end{bmatrix}$	$\begin{bmatrix} -\frac{1}{2} & \frac{\sqrt{3}}{2} \\ -\frac{\sqrt{3}}{2} & -\frac{1}{2} \end{bmatrix}$	$\begin{bmatrix} 1 & 0 \\ 0 & -1 \end{bmatrix}$	$\begin{bmatrix} -\frac{1}{2} & -\frac{\sqrt{3}}{2} \\ -\frac{\sqrt{3}}{2} & +\frac{1}{2} \end{bmatrix}$	$\begin{bmatrix} -\frac{1}{2} & +\frac{\sqrt{3}}{2} \\ +\frac{\sqrt{3}}{2} & +\frac{1}{2} \end{bmatrix}$

Table 6. *Characters of the dihedral tetragonal rotation group D_4. The basis functions for standard irreducible representations are also given together with their matrix representatives of the standard generators of the group*

	D_4	E	$2\,C_4^Z$	C_2^Z	$2\,C_2'$	$2C_2''$
$\mathfrak{H}_\sigma^s$	A_1	1	1	1	1	1
$\mathfrak{H}_\sigma^p$	A_2	1	1	1	−1	−1
$\mathfrak{H}_{\delta c}^d$	B_1	1	−1	1	1	−1
$\mathfrak{H}_{\delta s}^d$	B_2	1	−1	1	−1	1
$\mathfrak{H}_{\pi c}^d$, $\mathfrak{H}_{\pi s}^d$	E	2	0	−2	0	0

Generators		C_4^Z	C_2^Y
Standard basis functions			
E	$[\mathfrak{H}_{\pi c}^d, \mathfrak{H}_{\pi s}^d]$	$\begin{bmatrix} 0 & -1 \\ 1 & 0 \end{bmatrix}$	$\begin{bmatrix} 1 & 0 \\ 0 & -1 \end{bmatrix}$

Table 7. *Characters of the dihedral hexagonal rotation group D_6. The basis functions for standard irreducible representations are also given, together with their matrix representatives of the standard generators of the group*

	D_6	E	$2\,C_6^Z$	$2\,C_3^Z$	C_2^Z	$3C_2'$	$3\,C_2''$
$\mathfrak{H}_\sigma^s$	A_1	1	1	1	1	1	1
$\mathfrak{H}_\sigma^p$	A_2	1	1	1	1	− 1	− 1
$\mathfrak{H}_{\varphi c}^g$	B_1	1	− 1	1	− 1	1	− 1
$\mathfrak{H}_{\varphi s}^g$	B_2	1	− 1	1	− 1	− 1	1
$\mathfrak{H}_{\pi c}^d$, $\mathfrak{H}_{\pi s}^d$	E_1	2	1	− 1	− 2	0	0
$\mathfrak{H}_{\delta c}^d$, $\mathfrak{H}_{\delta s}^d$	E_2	2	− 1	− 1	2	0	0

Generators		C_6^Z	C_2^Y
Standard basis functions			
E_1	$[\mathfrak{H}_{\pi c}^d, \mathfrak{H}_{\pi s}^d]$	$\begin{bmatrix} \frac{1}{2} & -\frac{\sqrt{3}}{2} \\ \frac{\sqrt{3}}{2} & \frac{1}{2} \end{bmatrix}$	$\begin{bmatrix} 1 & 0 \\ 0 & -1 \end{bmatrix}$
E_2	$[\mathfrak{H}_{\delta c}^d, \mathfrak{H}_{\delta s}^d]$	$\begin{bmatrix} -\frac{1}{2} & -\frac{\sqrt{3}}{2} \\ \frac{\sqrt{3}}{2} & -\frac{1}{2} \end{bmatrix}$	$\begin{bmatrix} 1 & 0 \\ 0 & -1 \end{bmatrix}$

Table 8. *Characters of the octahedral rotation group O. The basis functions for standard irreducible representations are also given, together with their matrix representatives of two generators of the group*

	O	E	8 C_3	3 C_2	6 C_4	6 C_2'
$\mathfrak{H}^s_\sigma$	A_1	1	1	1	1	1
$\mathfrak{H}^f_{\delta s}$	A_2	1	1	1	−1	−1
$\mathfrak{H}^d_\sigma, \mathfrak{H}^d_{\delta c}$	E	2	−1	2	0	0
$\mathfrak{H}^p_\sigma, \mathfrak{H}^p_{\pi s}, \mathfrak{H}^p_{\pi c}$	T_1	3	0	−1	1	−1
$\mathfrak{H}^d_{\pi s}, \mathfrak{H}^d_{\pi c}, \mathfrak{H}^d_{\delta s}$	T_2	3	0	−1	−1	1

Generators		C_3^{XYZ}	C_4^Z
Standard basis functions			
E	$[\mathfrak{H}^d_\sigma, \mathfrak{H}^d_{\delta c}]$	$\begin{bmatrix} -\frac{1}{2} & -\frac{\sqrt{3}}{2} \\ \frac{\sqrt{3}}{2} & -\frac{1}{2} \end{bmatrix}$	$\begin{bmatrix} 1 & 0 \\ 0 & -1 \end{bmatrix}$
T_1	$[\mathfrak{H}^p_\sigma, \mathfrak{H}^p_{\pi s}, \mathfrak{H}^p_{\pi c}]$	$\begin{bmatrix} 0 & 1 & 0 \\ 0 & 0 & 1 \\ 1 & 0 & 0 \end{bmatrix}$	$\begin{bmatrix} 1 & 0 & 0 \\ 0 & 0 & 1 \\ 0 & -1 & 0 \end{bmatrix}$
T_2	$[\mathfrak{H}^d_{\pi s}, \mathfrak{H}^d_{\pi c}, \mathfrak{H}^d_{\delta s}]$	$\begin{bmatrix} 0 & 0 & 1 \\ 1 & 0 & 0 \\ 0 & 1 & 0 \end{bmatrix}$	$\begin{bmatrix} 0 & 1 & 0 \\ -1 & 0 & 0 \\ 0 & 0 & -1 \end{bmatrix}$

systems are given in Table 9, and in Table 10 they are characterized by their symmetry operators collected into classes.

The invariant sub-groups of the non-cubic holohedric groups fall into two equally large collections: the first one contains the non-commutative groups and the second one the Abelian or commutative ones. For example, the holohedric group D_{4h} contains the non-commutative, isomorphic, and invariant sub-groups D_4, C_{4v}, and D_{2d}, and the Abelian invariant sub-groups C_{4h}, C_4, and S_4. For these latter groups, the operations which belong to the same class of the holohedric group will now fall into classes of their own.

It may be noted that the holohedric group D_{3d} is an invariant sub-group of the holohedric group D_{6h}. This fact has led some workers to eliminate the trigonal system and classify all the crystallographic groups

Table 9.

1. Triclinic crystal system: the holohedric group S_2, and its sub-group (no specific direction is defined)	S_2 C_1
2. Monoclinic crystal system: the holohedric group C_{2h}, and those sub-groups which do not belong to class 1 (one specific direction is defined)	C_{2h} C_2, C_s
3. Orthorhombic crystal system: the holohedric group D_{2h}, and those sub-groups which do not belong to classes 1 and 2 (three specific directions are defined)	D_{2h} D_2, C_{2v}
4. Trigonal crystal system: the holohedric group D_{3d}, and those sub-groups which do not belong to classes 1 and 2	D_{3d} D_3, C_{3v} S_6, C_3
5. Tetragonal crystal system: the holohedric group D_{4h}, and those sub-groups which do not belong to classes 1, 2, and 3	D_{4h} D_4, C_{4v}, D_{2d} C_{4h}, C_4, S_4
6. Hexagonal crystal system: the holohedric group D_{6h}, and those sub-groups which do not belong to classes 1, 2, 3, and 4	D_{6h} D_6, C_{6v}, D_{3h} C_{6h}, C_6, C_{3h}
7. Cubic crystal system: the holohedric group O_h, and those sub-groups which do not belong to classes 1, 2, 3, 4, and 5	O_h O, T_d T_h, T

In this table all the crystallographic point groups have been collected together. They can be briefly characterized as *all* the sub-groups of O_h plus the sub-groups of D_{6h} of class 6 of this table.

which are sub-groups of the group D_{6h} as belonging to the hexagonal crystal system except those belonging to the crystal systems of classes 1, 2, and 3 of Table 9.

For the cubic groups T_h and T the operators of the class 8 $\mathbf{C}_3$ of O_h fall into two classes, one corresponding to a rotation of $+\frac{2\pi}{3}$ and another one corresponding to a rotation of $-\frac{2\pi}{3}$.

7. Non-Cubic Finite Point Groups as Sub-Groups of the Holohedric Infinite Group $D_{\infty h}$

Before the holohedric group $D_{\infty h}$ is generated as the direct product group of the infinite dihedral group D_∞ with the inversion group S_2, the former group and its sub-groups will be studied in some detail.

The dihedral group D_∞ which may be generated by adding the operator $\mathbf{C}_2^Y$ to the two-dimensional rotation group C_∞ is characterized in

Table 10. *Review of symmetry classes of the holohedric crystallographic point groups and their sub-groups*

$D_3 \times S_2 =$	D_{3d}	**E**	**2C₃**	**3C₂**	**S₂**	**2S₆**	$3\sigma_d$						
	D_3	○	○	○									
	C_{3v}	○	○				○						
$C_3 \times S_2 =$	S_6	○	○		○	○							
	C_3	○	○										
$D_4 \times S_2 = D_4 \times C_{1h} =$	D_{4h}	**E**	$2C_4$	C_2	$2C_2'$	$2C_2''$	S_2	$2S_4$	σ_h	$2\sigma_v$	$2\sigma_d$		
	D_4	○	○	○	○	○							
	C_{4v}	○	○	○						○	○		
	D_{2d}	○		○	○			○			○		
$C_4 \times S_2 = C_4 \times C_{1h} =$	C_{4h}	○	○	○			○	○	○				
	C_4	○	○	○									
	S_4	○		○				○					
$D_6 \times S_2 = D_6 \times C_{1h} =$	D_{6h}	**E**	$2C_6$	$2C_3$	C_2	$3C_2'$	$3C_2''$	S_2	$2S_3$	$2S_6$	σ_h	$3\sigma_v$	$3\sigma_d$
	D_6	○	○	○	○	○	○						
	C_{6v}	○	○	○	○							○	○
$D_3 \times C_{1h} =$	D_{3h}	○		○			○		○		○	○	
$C_6 \times S_2 = C_6 \times C_{1h} =$	C_{6h}	○	○	○	○			○	○	○	○		
$C_3 \times C_{1h} =$	C_{3h}	○		○					○		○		
	C_6	○	○	○	○								
$O \times S_2 =$	O_h	**E**	$8C_3$	$3C_2$	$6C_4$	$6C_2$	S_2	$8S_6$	$3\sigma_h$	$6S_4$	$6\sigma_d$		
	O	○	○	○	○	○							
	T_d	○	○	○						○	○		
$T \times S_2 =$	T_h	○	○	○			○	○	○				
	T	○	○	○									

Table 11. *Characters of the infinite dihedral rotation group D_∞. The basis functions for standard irreducible representations are also given, together with their matrix representatives of the standard generators of the group*

	D_∞	E	2 R(φ)	C_2
$\mathfrak{H}_\sigma^s$	A_1	1	1	1
$\mathfrak{H}_\sigma^p$	A_2	1	1	−1
$[\mathfrak{H}_{\pi c}^d, \mathfrak{H}_{\pi s}^d]$	$E_1[\Pi]$	2	$2\cos\varphi$	0
$[\mathfrak{H}_{\delta c}^d, \mathfrak{H}_{\delta s}^d]$	$E_2[\Delta]$	2	$2\cos 2\varphi$	0
$[\mathfrak{H}_{\varphi c}^g, \mathfrak{H}_{\varphi s}^g]$	$E_3[\Phi]$	2	$2\cos 3\varphi$	0
.	.	.	.	.
.	.	.	.	.
.	.	.	.	.
$[\mathfrak{H}_{(2\lambda-1)c}^l, \mathfrak{H}_{(2\lambda-1)s}^l]$ $l=2\lambda$	$E_{(2\lambda-1)}$	2	$2\cos[(2\lambda-1)\varphi]$	0
$[\mathfrak{H}_{(2\lambda)c}^l, \mathfrak{H}_{(2\lambda)s}^l]$ $l=2\lambda$	$E_{(2\lambda)}$	2	$2\cos[(2\lambda)\varphi]$	0
.	.	.	.	.
.	.	.	.	.
.	.	.	.	.

Generators		$R^Z(\varphi)$	C_2^Y
Standard basis functions			
E_λ	$[\mathfrak{H}_{\lambda c}^l, \mathfrak{H}_{\lambda s}^l]$ (l even)	$\begin{bmatrix} \cos\lambda\varphi & -\sin\lambda\varphi \\ \sin\lambda\varphi & \cos\lambda\varphi \end{bmatrix}$	$\begin{bmatrix} 1 & 0 \\ 0 & -1 \end{bmatrix}$

Table **11**. The group D_∞ has all the finite axial pure rotation groups, i.e. the cyclic groups C_n as well as the dihedral groups D_n as its sub-groups. The structures of these groups depend on whether n is an even or an odd integer.

When n is odd, the C_n groups have one one-dimensional irreducible representation A, and the D_n groups have two such representations, A_1, the totally symmetric representation, and A_2 which is antisymmetrical under C_2^Y. The number of two-dimensional irreducible representations (or, for C_n, the number of most reduced representations with a real basis) is $\left(\frac{n}{2}-\frac{1}{2}\right)$.

When n is even, the C_n groups have two one-dimensional irreducible representations A and B. The representation B shows the transformation properties of a spherical harmonic with $\lambda=\frac{n}{2}$, and it is noted that both the cosine and the sine function belonging to this λ-value transform in this way. For even n, the D_n groups exhibit four one-dimensional irreducible representations A_1, A_2, B_1, and B_2, arising from A and B of C_n, each of which splits into two representations, one symmetrical and the other antisymmetrical with respect to C_2^Y. The number of two-dimensional irreducible representations (the most reduced ones for C_n) is in this case $\left(\frac{n}{2} - 1\right)$.

The antisymmetrized direct product of the two-dimensional representations with themselves is A for C_n and A_2 for D_n, independent of whether n is even or odd. The basic rule (p. 210) for the choice of standard basis functions for the irreducible representations naturally proposes the function $\mathfrak{H}_\sigma^p$ as a standard function for this representation. It is noted that C_n in this way requires two standard basis functions for A, namely further the function $\mathfrak{H}_\sigma^s$ representing the symmetrized direct product of the most reduced representations with themselves. This is due to the fact that the group C_n with a real basis is not a simply reducible group since in

$$E_\lambda \times E_\lambda = A\,[\text{symm}] + A\,(\text{skew}) + E_{2\lambda} \tag{58}$$

A occurs twice. One A is symmetric and the other A skew. These are different in the group D_n (which is simply reducible), namely A_1 and A_2, respectively.

The basic rule for the choice of standard basis functions requires that the basis functions for all the two-dimensional irreducible representations have l even. The simplest choice for D_∞ therefore is that given in table 11. This choice is also valid for C_∞ with real bases, and generally for the groups D_n and C_n, provided that the sequence E_1, E_2, ... in the table is truncated from below according to the number of irreducible representations (or most reduced representations for C_n) of the group in question. This choice of l even makes it natural to choose the operator C_2^Y as one of the standard generators for all the dihedral groups, since by this choice $\mathfrak{H}_{\lambda c}^l$ is always symmetrical and $\mathfrak{H}_{\lambda s}^l$ always antisymmetrical under the C_2^Y operation. If C_2^X had been chosen, the behaviour of $\mathfrak{H}_{\lambda c}^l$ functions (l even) under the standard twofold operator would depend on their λ values. Thus, for $(l - \lambda)$ even, the set $[\mathfrak{H}_{\lambda c}^l, \mathfrak{H}_{\lambda s}^l]$ transforms under C_2^X with the same matrix as under C_2^Y, whereas for $(l - \lambda)$ odd the matrix representatives differ by a factor of -1.

With the standardization mentioned in the previous paragraph, the standard basis functions for the B_1 and B_2 representations of the D_n groups with n even are $\mathfrak{H}_{\frac{n}{2}c}^l$ and $\mathfrak{H}_{\frac{n}{2}s}^l$, respectively, where $l = \frac{n}{2}$ when $\frac{n}{2}$ is even and $l = \frac{n}{2} + 1$ when $\frac{n}{2}$ is odd and greater than one.

The set of standard basis functions for the two-dimensional irreducible representations of D_∞ and D_n commonly used in the literature, is the set [$\cos(\lambda\varphi)$, $\sin(\lambda\varphi)$] whose elements are not simultaneously basis functions for R_3. It may be noted that if the standard generator C_2^X is chosen for D_∞, this set of functions spans the E_λ representations with identically the same matrices as do the present choice of standard bases with C_2^Y as the standard generator. This has the important consequence that the conventional symbols for the irreducible representations of the groups D_∞ and D_n agree with those of the present work. For the group D_6 (Table 7), for example, E_1 has the set $\mathfrak{H}^d_\pi$, E_2 the set $\mathfrak{H}^d_\delta$, B_1 the function $\mathfrak{H}^g_{\varphi c}$, and B_2 the function $\mathfrak{H}^g_{\varphi s}$ as generating functions. The last two functions mentioned have $\lambda = \frac{n}{2} = 3$ and $l = \frac{n}{2} + 1 = 4$.

The holohedric group $D_{\infty h}$ is obtained as the direct product group of the groups D_∞ and S_2. Its sub-groups are all the non-cubic point groups, including also the groups which do not belong to the crystal systems.

In the classification of the point groups we shall for our purpose be particularly interested in their relationship with the dihedral groups D_n for which the 3-Γ symbols can be uniquely defined. The groups C_n will always be described in terms of D_n symbols, whereby A [symm] and A (skew) of Eq. (58) become distinguishable. For isomorphic groups the same 3-Γ symbols can be chosen, and by establishing a one-to-one correspondence between the components of the irreducible representations of the group D_n and those of the isomorphic ones, the relation between these groups and R_{3i} becomes unique. Further 3-Γ symbols of direct product groups can be evaluated using Eq. (56). For these reasons, we note the following relations and isomorphisms (indicated by the symbol ∞)

$$\begin{array}{lll}
D_{nh} = D_n \times C_{1h} & (n \text{ odd}) & \\
D_{nh} = D_n \times S_2 & (n \text{ even}) & \\
D_{nd} = D_n \times S_2 & (n \text{ odd}) & \qquad (59)\\
D_{nd} \infty D_{2n} & (n \text{ even}) & \\
C_{nv} \infty D_n & (n \text{ odd and even}) &
\end{array}$$

We add the following[5])

$$\begin{array}{lll}
C_{nh} = C_n \times C_{1h} & (n \text{ odd}) & \\
C_{nh} = C_n \times S_2 & (n \text{ even}) & \qquad (60)\\
C_n \infty S_n & (n \text{ even}) & \\
C_{nh} \infty S_{2n} & (n \text{ odd}) &
\end{array}$$

which, however, are of less importance.

---

[5]) It may be noted that the group S_n (n odd) $= C_{nh}$ is of the order $2n$. Therefore the group symbol S_n is only used for n even, in which case the order of the group is n. Similarly the relation $C_\infty \times S_2 = C_\infty \times C_{1h} = S_\infty$ is valid.

III. 3-Γ Symbols for the Sub-Groups of R_3 and R_{3i}

8. The Non-Cubic Point Groups. Racah's Lemma

a. The Group D_∞

Once the choice of basis functions for the irreducible representations of D_∞ has been made, the 3-Γ symbols can be calculated as integrals analogously to the 3-l symbols (Sect. 4). Therefore, they may also be calculated from the corresponding 3-l symbols of R_3 using a positive constant of renormalization. The scheme for direct multiplication of the irreducible representations of D_∞ is given in Table 12.

Table 12.

D_∞	A_1	A_2	E_λ	$E_{\lambda'}$
A_1	A_1	A_2	E_λ	$E_{\lambda'}$
A_2	A_2	A_1	E_λ	$E_{\lambda'}$
E_λ	E_λ	E_λ	$A_1+A_2+E_{(2\lambda)}$	$E_{(\lambda'-\lambda)}+E_{(\lambda'+\lambda)}$
$E_{\lambda'}$	$E_{\lambda'}$	$E_{\lambda'}$	$E_{(\lambda'-\lambda)}+E_{(\lambda'+\lambda)}$	$A_1+A_2+E_{(2\lambda')}$

Scheme of direct multiplication of the irreducible representations of D_∞, where $\lambda' > \lambda$. From this table the multiplication schemes for the other point groups can be derived using the relationships discussed in Sect. 7. The antisymmetrized direct product $(E_\lambda^{\{2\}})$ is always A_2 and this is true also of all the dihedral sub-groups of D_∞

An alternative way of generating the 3-Γ symbols of D_∞ is to consider the transformation properties of the standard functions under the symmetry operator C_∞, and in this way Table 13 has been obtained.

The method of constructing the 3-Γ symbols by means of the standard basis functions implies that another choice of basis functions generally yields a set of 3-Γ symbols where some phases are changed. To discuss this in greater detail, we shall consider a lemma by *Racah* (*5*). This lemma describes a relationship between the coupling coefficients of a group and those of a sub-group and shows the value of the choice of basis functions for irreducible representations as specified by the basic rule, p. 210.

The irreducible representations of a group are generally reducible as representations of a sub-group. In Racah's lemma this statement is considered by introducing the concept of a group which is reduced with respect to a sub-group. By this is meant that the irreducible representations of the group are in a form where the representatives of the elements of the sub-group appear in block-diagonal form, each block being chosen so as to be standard for the sub-group. The lemma states that the coupling coefficients for the irreducible representations of the group in this form are proportional to the coupling coefficients for the corresponding irreducible representations of the sub-group. One has, for example, taking R_3 as the group,

$$\begin{aligned} &< l_1\, c_1\, \Gamma_1\, \gamma_1 \quad l_2\, c_2\, \Gamma_2\, \gamma_2 \,|\, l_3\, c_3\, \Gamma_3\, \gamma_3 > \\ &= (\underline{l_1}\, c_1\, \Gamma_1 \quad \underline{l_2}\, c_2\, \Gamma_2 \,|\, l_3\, c_3\, \underline{\Gamma_3}) \times < \Gamma_1\, \gamma_1 \quad \Gamma_2\, \gamma_2 \,|\, \Gamma_2\, \gamma_3 > \end{aligned} \tag{61}$$

where the labels c_i may be introduced if more than one irreducible sub-group representation of the type Γ_i arises from the irreducible representation l_i of the group. The constants of proportionality, which are independent of the components γ_i, form orthogonal matrices,

$$\begin{aligned} &\sum_{l_3\, c_3} (l_1\, c_1\, \Gamma_1 \; l_2\, c_2\, \Gamma_2 \,|\, l_3\, c_3 \Gamma_3)\, (l_3\, c_3\, \Gamma_3 \,|\, l_1\, c_1'\, \Gamma_1' \; l_2\, c_2'\, \Gamma_2') \\ &\qquad = \delta(c_1\, \Gamma_1\,,\, c_1'\, \Gamma_1')\; \delta(c_2\, \Gamma_2\,,\, c_2'\, \Gamma_2') \end{aligned} \tag{62a}$$

$$\begin{aligned} &\sum_{c_1\, \Gamma_1\, c_2\, \Gamma_2} (l_3\, c_3\, \Gamma_3 \,|\, l_1\, c_1\, \Gamma_1 \; l_2\, c_2\, \Gamma_2)\, (l_1\, c_1\, \Gamma_1 \; l_2\, c_2\, \Gamma_2 \,|\, l_3'\, c_3'\, \Gamma_3) \\ &\qquad = \delta(l_3\,,\, l_3')\; \delta(c_3\,,\, c_3') \end{aligned} \tag{62b}$$

These orthogonal matrices, which we shall call Racah lemma matrices, are characterized by two irreducible representations of the group and one irreducible representation of the sub-group [underlined in Eqs. (61), (63), and (64)]. The number of times this sub-group representation occurs in the direct product of the two irreducible representations of the group is equal to the order of the Racah lemma matrix.

As an example, we consider the direct product of the p and d representations of R_3 and the resultant π representation of D_∞. Each of the terms in the direct sum, $p \times d = p + d + f$ contributes π once, so that the Racah lemma matrix is three by three. By use of the Tables 11, 16, 17, and 18 and Eq. (5) one obtains

$\underline{p}$	$\underline{d}$	$f\, \underline{E_\pi}$	$d\, \underline{E_\pi}$	$p\, \underline{E_\pi}$	
A_2	E_π	$-\frac{4}{\sqrt{30}}$	$+\frac{1}{\sqrt{6}}$	$-\sqrt{\frac{3}{10}}$	
E_π	A_1	$+\frac{2}{\sqrt{10}}$	$+\frac{1}{\sqrt{2}}$	$-\frac{1}{\sqrt{10}}$	(63)
E_π	E_δ	$+\frac{1}{\sqrt{15}}$	$-\frac{1}{\sqrt{3}}$	$-\sqrt{\frac{3}{5}}$	

and with the R_3 basis functions taken in the opposite order the matrix changes to

$\underline{d}$	$\underline{p}$	$f\,\underline{E_\pi}$	$d\,\underline{E_\pi}$	$p\,\underline{E_\pi}$
E_π	A_2	$+\frac{4}{\sqrt{30}}$	$+\frac{1}{\sqrt{6}}$	$+\sqrt{\frac{3}{10}}$
A_1	E_π	$+\frac{2}{\sqrt{10}}$	$-\frac{1}{\sqrt{2}}$	$-\frac{1}{\sqrt{10}}$
E_δ	E_π	$+\frac{1}{\sqrt{15}}$	$+\frac{1}{\sqrt{3}}$	$-\sqrt{\frac{3}{5}}$

(64)

where the first row and the second column have had sign changes because of the odd character of the 3-Γ symbols involving the trios ($E_\pi\ A_2\ E_\pi$) and ($d\ p\ d$), respectively.

To illustrate the Racah lemma matrices in more detail, the middle element of Eq. (64) is calculated. According to Eq. (61) it is expressed by

$$\frac{\langle d\,A_1\,a_1\ p\,E_\pi\,\pi_1|d\,E_\pi\,\pi_1\rangle}{\langle A_1\,a_1\ E_\pi\,\pi_1|E_\pi\,\pi_1\rangle} = \frac{\langle d\,A_1\,a_1\ p\,E_\pi\,\pi_2|d\,E_\pi\,\pi_2\rangle}{\langle A_1\,a_1\ E_\pi\,\pi_2|E_\pi\,\pi_2\rangle} \tag{65}$$

It is noted that among the R_3 bases used in Eq. (65) the function $\mathfrak{H}^d_\sigma$ and the set ($\mathfrak{H}^d_{\pi c}$, $\mathfrak{H}^d_{\pi s}$) transform in the standard way according to the A_1 and $E_1 = E_\pi$ representations of D_∞ (Table 11), whereas the set ($\mathfrak{H}^p_{\pi c}$, $\mathfrak{H}^p_{\pi s}$) is not standard under C_2^Y. The function $\mathfrak{H}^p_{\pi s}$ is symmetrical under C_2^Y and the function $\mathfrak{H}^p_{\pi c}$ antisymmetrical. However, in order to transform by the standard matrices, i.e. by those of [$\mathfrak{H}^d_{\pi c}$, $\mathfrak{H}^d_{\pi s}$], one of the p-functions will have to change its sign. We take the set [$-\mathfrak{H}^p_{\pi s}$, $\mathfrak{H}^p_{\pi c}$] as the standard choice so that the set ($\mathfrak{H}^p_\sigma$, $-\mathfrak{H}^p_{\pi s}$, $\mathfrak{H}^p_{\pi c}$) = (z, $-y$, x) forms a right-handed coordinate system. By this choice, the two last columns of the Racah lemma matrix are completely specified. One has, returning to the middle element,

$$\langle d\,A_1\,a_1\ p\,E_\pi\pi_1|d\,E_\pi\pi_1\rangle = -\langle d\,\sigma\ \ p\pi\,s|d\pi\,c\rangle \tag{66}$$

and

$$\langle d\,A_1\,a_1\ p\,E_\pi\pi_2|d\,E_\pi\pi_2\rangle = \langle d\,\sigma\ \ p\pi\,c|d\pi\,s\rangle \tag{67}$$

which may be evaluated from Table 17 and Eq. (5),

$$\langle d\,A_1\,a_1\ p\,E_\pi\pi_1|d\,E_\pi\pi_1\rangle = \langle d\,A_1\,a_1\ p\,E_\pi\pi_2|d\,E_\pi\pi_2\rangle$$
$$= -\frac{1}{\sqrt{10}} \times \sqrt{5} = -\frac{1}{\sqrt{2}} \tag{68}$$

The coupling coefficients for D_∞ are found from Table 13 and Eq. (5),

$$<\mathrm{A}_1\, a_1\, \mathrm{E}_\pi \pi_1 | \mathrm{E}_\pi \pi_1> = <\mathrm{A}_1\, a_1\, \mathrm{E}_\pi \pi_2 | \mathrm{E}_\pi \pi_2> = \frac{1}{\sqrt{2}} \times \times \sqrt{2} = 1 \tag{69}$$

Combination of Eqs. (68) and (69) according to Eq. (65) then gives the matrix element desired.

It is noted that the adaptation of $\mathfrak{H}^l$ functions to D_∞ can always be done analogously to the case of d and p functions. For l even the set $[\mathfrak{H}^l_{\lambda c}, \mathfrak{H}^l_{\lambda s}]$ and for l odd the set $[-\mathfrak{H}^l_{\lambda s}, \mathfrak{H}^l_{\lambda c}]$ transform in the standard way under D_∞. By means of this convention, the first column of the Racah lemma matrices of Eqs. (63) and (64) is completely specified.

In order to apply Racah's lemma to the group R_3 reduced with respect to C_∞ (most reduced representations with a real basis), one simply uses the classification of irreducible representations according to D_∞ as explained in Table 13. As the two sigma terms in $\Lambda \times \Lambda$ cannot be distinguished in the group C_∞, the Racah lemma matrices whose elements are characterized by $(\underline{l\Lambda}\, \underline{l\Lambda} \mid l' \underline{\Sigma})$, will appear in block-diagonal form, since the Σ can be either Σ_1 or Σ_2 (Table 13).

Table 13. *Non-vanishing 3-Γ symbols for C_∞, based upon the most reduced representations with a real basis*

$$\begin{pmatrix} \Sigma_1 & \Sigma_1 & \Sigma_1 \\ \sigma_1 & \sigma_1 & \sigma_1 \end{pmatrix} = \begin{pmatrix} \Sigma_2 & \Sigma_1 & \Sigma_2 \\ \sigma_2 & \sigma_1 & \sigma_2 \end{pmatrix} = 1$$

$$\begin{pmatrix} \Lambda & \Sigma_1 & \Lambda' \\ c & \sigma_1 & c \end{pmatrix} = \begin{pmatrix} \Lambda & \Sigma_1 & \Lambda' \\ s & \sigma_1 & s \end{pmatrix} = (1/\sqrt{2})\, \delta(\Lambda, \Lambda')$$

$$\begin{pmatrix} \Lambda & \Sigma_2 & \Lambda' \\ s & \sigma_2 & c \end{pmatrix} = - \begin{pmatrix} \Lambda & \Sigma_2 & \Lambda' \\ c & \sigma_2 & s \end{pmatrix} = (1/\sqrt{2})\, \delta(\Lambda, \Lambda') \quad \text{(odd)}$$

$$\begin{pmatrix} \Lambda & \Lambda' & \Lambda+\Lambda' \\ s & c & s \end{pmatrix} = \begin{pmatrix} \Lambda & \Lambda' & \Lambda+\Lambda' \\ c & s & s \end{pmatrix}$$

$$= \begin{pmatrix} \Lambda & \Lambda' & \Lambda+\Lambda' \\ c & c & c \end{pmatrix} = - \begin{pmatrix} \Lambda & \Lambda' & \Lambda+\Lambda' \\ s & s & c \end{pmatrix} = (1/2)$$

The group C_∞ with a real basis is not a simply reducible group, because $\Lambda \times \Lambda = \Sigma + \Sigma + (2\Lambda)$ where (2Λ) stands for the representation with $\lambda = 2\Lambda$. The 3-Γ symbols containing Σ fall into an even class called $(\Gamma\, \Sigma_1\, \Gamma)$ and an odd class called $(\Gamma\, \Sigma_2\, \Gamma)$ with reference to the group D_∞. All other 3-Γ symbols are even. When in D_∞ the Y-axis is used to standardize the components (sub-indices 1 and 2 for symmetrical and antisymmetrical under C_2^Y) and when all standard basis functions are chosen with even l-values, all cosine functions are symmetrical and all sine functions antisymmetrical under C_2^Y and the table is valid for D_∞ by the substitution, $\Sigma_1\, \sigma_1 \to \mathrm{A}_1\, a_1$, $\Sigma_2\, \sigma_2 \to \mathrm{A}_2\, a_2$, $\Lambda c \to \mathrm{E}_\lambda\, \lambda_1$, and $\Lambda s \to \mathrm{E}_\lambda\, \lambda_2$.

b) The Sub-Groups of D_∞

As mentioned in Sect. 7, the structure of the groups D_n and C_n is different for n odd and n even. The groups D_3 and D_4 are used to exemplify this.

The group D_3 has the irreducible representations A_1, A_2, and E (Table 5), where E is the conventional abbreviation for E_1. Racah's lemma is fulfilled for these representations when they are associated with the first three irreducible representations of D_∞ (Table 11, compare Table 5) and the 3-Γ symbols can be directly read out of Table 13, except for the EEE symbols, which require additional considerations. The basis functions for E_2 of D_∞ span the E representation of D_3 in the standard manner when taken as $[\mathfrak{H}^d_{\delta c}, -\mathfrak{H}^d_{\delta s}]$. Hence the 3-$\Gamma$ symbols follow from Table 17 (using a positive renormalization constant) or from Table 13. From Table 13 one obtains

$$-\begin{pmatrix} \mathrm{E} & \mathrm{E} & \mathrm{E} \\ e_2 & e_1 & e_2 \end{pmatrix} = -\begin{pmatrix} \mathrm{E} & \mathrm{E} & \mathrm{E} \\ e_1 & e_2 & e_2 \end{pmatrix} = +\begin{pmatrix} \mathrm{E} & \mathrm{E} & \mathrm{E} \\ e_1 & e_1 & e_1 \end{pmatrix} = -\begin{pmatrix} \mathrm{E} & \mathrm{E} & \mathrm{E} \\ e_2 & e_2 & e_1 \end{pmatrix} = \frac{1}{2} \tag{70}$$

where the first two symbols and the last one are the same, as they should be, since this 3-Γ symbol is even.

The irreducible representations of D_4 are A_1, A_2, B_1, B_2, and E (Table 6) where again E means E_1. As discussed in Sect. 7, the standard basis functions are $\mathfrak{H}^s_\sigma$, $\mathfrak{H}^p_\sigma$, $\mathfrak{H}^d_{\delta c}$, $\mathfrak{H}^d_{\delta s}$, and $[\mathfrak{H}^d_{\pi c}, \mathfrak{H}^d_{\pi s}]$, respectively. The 3-$\Gamma$ symbols follow from Table 13 using a positive constant of renormalization for those involving the δ-functions. One has

$$\begin{aligned}
&\begin{pmatrix} \mathrm{B}_1 & \mathrm{A}_1 & \mathrm{B}_1 \\ b_1 & a_1 & b_1 \end{pmatrix} = \begin{pmatrix} \mathrm{B}_2 & \mathrm{A}_1 & \mathrm{B}_2 \\ b_2 & a_1 & b_2 \end{pmatrix} = 1 \\
&\begin{pmatrix} \mathrm{B}_2 & \mathrm{A}_2 & \mathrm{B}_1 \\ b_2 & a_2 & b_1 \end{pmatrix} = -\begin{pmatrix} \mathrm{B}_1 & \mathrm{A}_2 & \mathrm{B}_2 \\ b_1 & a_2 & b_2 \end{pmatrix} = 1 \qquad \text{(odd)} \\
&\begin{pmatrix} \mathrm{E} & \mathrm{E} & \mathrm{B}_2 \\ e_2 & e_1 & b_2 \end{pmatrix} = \begin{pmatrix} \mathrm{E} & \mathrm{E} & \mathrm{B}_2 \\ e_1 & e_2 & b_2 \end{pmatrix} = \begin{pmatrix} \mathrm{E} & \mathrm{E} & \mathrm{B}_1 \\ e_1 & e_1 & b_1 \end{pmatrix} = -\begin{pmatrix} \mathrm{E} & \mathrm{E} & \mathrm{B}_1 \\ e_2 & e_2 & b_1 \end{pmatrix} = \frac{1}{\sqrt{2}}
\end{aligned} \tag{71}$$

In order to obtain general expressions for the 3-Γ symbols of the D_n groups it is noted (Table 11) that for E_λ the standard representative of the $\mathbf{C_n}$ operator is

$$\begin{bmatrix} \cos\dfrac{2\pi\lambda}{n} & -\sin\dfrac{2\pi\lambda}{n} \\[2ex] \sin\dfrac{2\pi\lambda}{n} & \cos\dfrac{2\pi\lambda}{n} \end{bmatrix} \tag{72}$$

where the standard order of the basis functions is $[\mathfrak{H}^l_{\lambda c}, \mathfrak{H}^l_{\lambda s}]$ and l even. For $\lambda = n$, this is a unit matrix and the functions span the A_1 and A_2 representations. For $\lambda = \frac{n}{2}$ (n even), the matrix is minus one times the unit matrix and the functions span B_1 and B_2. The $E_{n-\lambda}$ representative of the C_n operation is the same as that of E_λ except for a sign change on the non-diagonal elements in Eq. (72). Therefore, in order to fulfil the conditions for Racah's lemma, one of the basis functions of $E_{n-\lambda}$ has to change sign so that the set transforms with the standard E_λ matrices. We choose, as above with D_3, the standard $[\mathfrak{H}^l_{(n-\lambda)c}, -\mathfrak{H}^l_{(n-\lambda)s}]$ for the representations $E_{n-\lambda}$. For the D_n groups the E_λ representations all have $\lambda < \frac{n}{2}$ so that when the direct product $E_\lambda \times E_{\lambda'}$ contains $E_{\lambda+\lambda'}$ with $\lambda + \lambda' > \frac{n}{2}$ we have, for example, using Table 13,

$$\begin{pmatrix} E_\lambda & E_{\lambda'} & E_{\lambda+\lambda'} \\ e_2 & e_1 & e_2 \end{pmatrix} = -\begin{pmatrix} E_\lambda & E_{\lambda'} & E_{n-\lambda-\lambda'} \\ e_2 & e_1 & e_2 \end{pmatrix} = \frac{1}{2} \tag{73}$$

Table 14. *Non-vanishing 3-Γ symbols for the D_n group. For n odd, all those containing the B representations vanish*

$\begin{pmatrix} A_1 & A_1 & A_1 \\ a_1 & a_1 & a_1 \end{pmatrix}$	$=$	$\begin{pmatrix} A_2 & A_1 & A_2 \\ a_2 & a_1 & a_2 \end{pmatrix}$	$= 1$					
$\begin{pmatrix} E_\lambda & A_1 & E_{\lambda'} \\ e_1 & a_1 & e_1 \end{pmatrix}$	$=$	$\begin{pmatrix} E_\lambda & A_1 & E_{\lambda'} \\ e_2 & a_1 & e_2 \end{pmatrix}$	$= (1/\sqrt{2})\, \delta(\lambda, \lambda')$					
$\begin{pmatrix} E_\lambda & A_2 & E_{\lambda'} \\ e_2 & a_2 & e_1 \end{pmatrix}$	$=-$	$\begin{pmatrix} E_\lambda & A_2 & E_{\lambda'} \\ e_1 & a_2 & e_2 \end{pmatrix}$	$= (1/\sqrt{2})\, \delta(\lambda, \lambda')$	(odd)				
$\begin{pmatrix} B_1 & A_1 & B_1 \\ b_1 & a_1 & b_1 \end{pmatrix}$	$=$	$\begin{pmatrix} B_2 & A_1 & B_2 \\ b_2 & a_1 & b_2 \end{pmatrix}$	$= 1$					
$\begin{pmatrix} B_2 & A_2 & B_1 \\ b_2 & a_2 & b_1 \end{pmatrix}$	$=-$	$\begin{pmatrix} B_1 & A_2 & B_2 \\ b_1 & a_2 & b_2 \end{pmatrix}$	$= 1$	(odd)				
$\begin{pmatrix} E_\lambda & E_{\lambda'} & E_{\lambda+\lambda'} \\ e_2 & e_1 & e_2 \end{pmatrix}$	$=$	$\begin{pmatrix} E_\lambda & E_{\lambda'} & E_{\lambda+\lambda'} \\ e_1 & e_2 & e_2 \end{pmatrix}$	$=$	$\begin{pmatrix} E_\lambda & E_{\lambda'} & E_{\lambda+\lambda'} \\ e_1 & e_1 & e_1 \end{pmatrix}$	$=-$	$\begin{pmatrix} E_\lambda & E_{\lambda'} & E_{\lambda+\lambda'} \\ e_2 & e_2 & e_1 \end{pmatrix}$	$=(1/2)$	
$-\begin{pmatrix} E_\lambda & E_{\lambda'} & E_{n-\lambda-\lambda'} \\ e_2 & e_1 & e_2 \end{pmatrix}$	$=-$	$\begin{pmatrix} E_\lambda & E_{\lambda'} & E_{n-\lambda-\lambda'} \\ e_1 & e_2 & e_2 \end{pmatrix}$	$=$	$\begin{pmatrix} E_\lambda & E_{\lambda'} & E_{n-\lambda-\lambda'} \\ e_1 & e_1 & e_1 \end{pmatrix}$	$=-$	$\begin{pmatrix} E_\lambda & E_{\lambda'} & E_{n-\lambda-\lambda'} \\ e_2 & e_2 & e_1 \end{pmatrix}$	$=(1/2)$	
$\begin{pmatrix} E_\lambda & E_{\frac{1}{2}n-\lambda} & B_2 \\ e_2 & e_1 & b_2 \end{pmatrix}$	$=$	$\begin{pmatrix} E_\lambda & E_{\frac{1}{2}n-\lambda} & B_2 \\ e_1 & e_2 & b_2 \end{pmatrix}$	$=$	$\begin{pmatrix} E_\lambda & E_{\frac{1}{2}n-\lambda} & B_1 \\ e_1 & e_1 & b_1 \end{pmatrix}$	$=-$	$\begin{pmatrix} E_\lambda & E_{\frac{1}{2}n-\lambda} & B_1 \\ e_2 & e_2 & b_1 \end{pmatrix}$	$= 1/\sqrt{2}$	

This table can be used for all the other finite non-cubic point groups as well, as discussed in Sect. 8. Similar tables have been given by *Griffith* (*4*), but his signs are in some cases different from ours.

where the minus sign arises because it is minus one times the sine component of $E_{\lambda+\lambda'}$, which gives the standard form of $E_{n-\lambda-\lambda'}$. Generally, whenever one has the sine component of $E_{\lambda+\lambda'}$ and $\lambda + \lambda' > \frac{n}{2}$, the 3-$\Gamma$ symbol changes sign relative to Table 13.

When $\lambda + \lambda' = n/2$ (n even), the cosine component transforms as B_1 and the sine component as B_2, and the four last 3-Γ symbols of Table 13 fall into two groups; both are even and keep the sign of Table 13. The generalized results for D_n groups are collected in Table 14.

As has been mentioned (Sect. 7 and Table 13), the real basis choice makes it possible to classify a cyclic sub-group C_n by its corresponding dihedral group D_n and to use its 3-Γ symbols. This can be done in all cases where the degeneracies within the C_n groups are not lifted, as is the case with electric fields, for example, in contrast to magnetic fields, which will lift the degeneracies and thereby most conveniently require a complex basis for the description.

c) $D_{\infty h}$ and Its Sub-Groups

The 3-Γ symbols of $D_{\infty h}$ may be obtained through the relation $D_{\infty h} = D_\infty \times S_2$ in an exactly analogous way to that used for $R_{3i} = R_3 \times S_2$ in Sect. 5. For the groups D_{nh} similar relations apply [Eq. (59)] and the same is true of the groups D_{nd}, so that the 3-Γ symbols for the holohedric groups can be obtained from those of the corresponding pure rotation groups.

For the group $C_{\infty v}$ the isomorphism $D_\infty \infty C_{\infty v}$ makes it possible to use the 3-Γ symbols of the group D_∞ for the group $C_{\infty v}$. However, in order to give such a choice a practical meaning, one has to associate the irreducible representations and components of the two groups in a unique way. This is done by the following relations

$$\begin{aligned} &A_1\, a_1 \leftrightarrow \Sigma^+\, \sigma^+; \quad A_2\, a_2 \leftrightarrow \Sigma^-\, \sigma^- \\ &E_\lambda\, \lambda_1 \leftrightarrow \Lambda\, \lambda^+; \quad E_\lambda\, \lambda_2 \leftrightarrow \Lambda\, \lambda^- \end{aligned} \tag{74}$$

Here, the components, which in D_∞ are labeled 1 and 2 for symmetrical and antisymmetrical under the $\mathsf{C}_2^{\mathrm{Y}}$ operation, are in $C_{\infty v}$ labeled $+$ and $-$ for symmetrical and antisymmetrical under the operation $\sigma^{\mathrm{ZX}} = \mathsf{IC}_2^{\mathrm{Y}}$.

In order to use Racah's lemma to the group R_{3i} reduced with respect to $C_{\infty v}$, it is noted that the standard basis functions of the irreducible representations in Eq. (25) [which are the functions of Eq. (15)] are symmetry-adapted to $C_{\infty v}$ in such a way that sigma and cosine functions are symmetrical under σ^{ZX}, and sine functions antisymmetrical. The basis functions of the irreducible representations in Eq. (27) have under the proper rotations $R^{Z}(\varphi)$ the same behaviour as the corresponding ones of Eq. (25) but transform with the opposite signs under the improper rotation, σ^{ZX}. In analogy with previous choices[6]) (p. 234 and p. 236), we make the choice of standard basis functions $[-\mathfrak{K}^{lp}_{\lambda s}, \mathfrak{K}^{lp}_{\lambda c}]$ for those irreducible representations of R_{3i} whose bases are not represented by spherical harmonics, i.e. the axial tensorial sets which have l even, parity p ungerade or l odd, parity p gerade. Thereby the Racah lemma matrices for R_{3i} reduced with respect to $C_{\infty v}$ are fully defined.

For the groups C_{nv}, the isomorphism $D_n \infty C_{nv}$ makes an association with the corresponding D_n groups possible in the same way as for the infinite groups. However, here the conventional labeling of the irreducible representations is a different one in that the subindices 1 and 2 in the irreducible representations of C_{nv} mean symmetrical and antisymmetrical with respect to σ^{ZX}. So, for the groups D_n and C_{nv}, one associates those irreducible representations and components which have the same set of labels, and one has only to remember the different meanings of the sub-indices in the two groups.

For the group D_{nd} (with n even) the isomorphism $D_{nd} \infty D_{2n}$ may be used to define the 3-Γ symbols. Again the irreducible representations and components that have the same set of labels will be associated. One has to remember that A and B in D_{2n} refer to the symmetry operator $C_{(2n)}$ and in D_{nd} to the operator $S_{(2n)}$. In this case, however, the sub-indices 1 and 2 refer to the C_2^Y operator in both groups.

Finally, in the commutative point groups C_{nh} and S_n one can apply the 3-Γ symbols of D_{nh} and D_n, using the relations of Eq. (60).

9. The 3-Γ Symbols for the Cubic Point Groups

a) The Octahedral Rotation Group, O

In the three-dimensional rotation group R_3 every other irreducible representation occurs in the antisymmetrized direct products [Eq. (28)], whereas in the dihedral group D_∞ only one particular irreducible representation occurs in these direct products [Table 12]. For the octahedral rotation group O, one has an intermediate situation.

---

[6]) We note that whenever in this paper the sets $[\mathfrak{H}^l_{\lambda c}, \mathfrak{H}^l_{\lambda s}]$ or $[\mathfrak{K}^l_{\lambda c}, \mathfrak{K}^l_{\lambda s}]$ cannot be taken to transform standard, we have always, independent of the standard order within a particular set, achieved the standard behaviour by allotting a minus sign to the sine component.

In the group O the irreducible representations A_2 and T_1 occur in the antisymmetrized direct products only, while A_1, E, and T_2 occur in the symmetrized direct products only. According to the basic rule (p. 210), the spherical harmonic basis functions are chosen so that odd l-values are associated with A_2 and T_1 and even l-values with A_1, E, and T_2. Using for the group O always the lowest possible l-value, the sets of standard basis functions[7]) are as stated in Table 8.

With this choice of standard basis functions the **3-Γ** symbols of the group O are completely defined from the **3-l** symbols, provided that a positive constant of renormalization is used for each trio of irreducible representations of O. The trivial 3-Γ symbols are the even symbols,

$$\begin{pmatrix} A_1 & A_1 & A_1 \\ a_1 & a_1 & a_1 \end{pmatrix} = \begin{pmatrix} A_2 & A_1 & A_2 \\ a_2 & a_1 & a_2 \end{pmatrix} = 1 \tag{75}$$

The non-trivial ones may be classified as follows. *1.* Those containing the E or the T_2 representation twice. They may be found by renormalization of the **3-l** symbols of Table **17** and are given in the Tables **19** and 21. *2.* The **3-Γ** symbols that contain T_1 twice. They may be found from Table 16 and are given in Table 20. *3.* The **3-Γ** symbols of the form $\begin{pmatrix} T_2 & \Gamma & T_1 \\ \gamma_1 & \gamma_2 & \gamma_3 \end{pmatrix}$. They can be found from Table **18** and are given in Table 22.

The **3-Γ** symbols defined in this way have the same character of even- and oddness under permutations of their columns as have their parent **3-l** symbols. Thus, when A_2 and T_1 occur an odd number of times, one has an odd **3-Γ** symbol. By use of the permutation properties it appears that the 3-Γ symbols that involve the trios ($T_2T_1T_1$) and ($T_2T_2T_1$) in Table 22 also occur in Tables 20 and 21, respectively.

b) The Octahedral Group O_h and Its Sub-Groups

The **3-Γ** symbols of the group O can be taken as **3-Γ** symbols also for the isomorphic group T_d, and these symbols may again be used for the tetrahedral rotation group T when a real basis is sufficient.

The **3-Γ** symbols of the group of the octahedron O_h can be found through the relation $O_h = O \times S_2$ using Eq. (56) and these **3-Γ** symbols may again be applied to T_h, provided that the most reduced representations with a real basis are sufficient for the description of this group.

---

[7]) This choice is the same as that of *Griffith* (*4*), to whom we are greatly indebted for the idea of the basic rule. Apparently, he himself did not realize its scope, since he did not apply it to associate the groups R_3 and O and did not use it for the D_n groups at all.

10. Conclusions

A standard choice of real basis functions for the three-dimensional rotation group R_3 and its sub-groups has been used to generate phase-fixed even 3-l symbols for R_3 and even 3-Γ symbols for the sub-groups, in each case together with the corresponding coupling coefficients.

The odd 3-l and 3-Γ symbols for the pure rotation groups have similarly been defined through the introduction of axial operators involving partial differentiation. When three given functions (or one operator and two functions) are used to define a 3-Γ symbol (or a 3-l symbol) for more than one group, the sign is fixed by the functions and independent of the group.

For the holohedric groups which arise as the direct product of the pure rotation groups and the inversion group S_2, the 3-Γ symbols (or for $R_{3i} = R_3 \times S_2$ the 3-l symbols) have been defined on the basis of those for the rotation groups themselves.

For the hemihedric groups the 3-Γ symbols have been chosen to be equal to those of their isomorphic pure rotation groups.

Through these simple conventions, plus a few other cases of choice of basis functions transforming standard, the relation between the coupling coefficients of the group R_{3i} and those of its sub-groups has been completely specified.

As a consequence of a lemma by *Racah*, it has been necessary to distinguish between standard basis sets which are used to generate with sign the 3-Γ symbols of the rotation groups, and basis sets transforming standard. The latter sets would in general have generated the 3-Γ symbols with different phases.

Appendix 1

The Evaluation of 3-l Symbols.

The even 3-l symbols can be found directly from Eq. (39) by integration, provided that the constant, $k(l_1 l_2 l_3)$ is known. As it will be shown in the following paper (*10*), Eq. (39) may be written

$$\frac{1}{4\pi}\int \mathfrak{H}_{t_1}^{l_1}\mathfrak{H}_{t_2}^{l_2}\mathfrak{H}_{t_3}^{l_3}\,dS = \begin{pmatrix} l_1 & l_2 & l_3 \\ \sigma & \sigma & \sigma \end{pmatrix}\begin{pmatrix} l_1 & l_2 & l_3 \\ t_1 & t_2 & t_3 \end{pmatrix} \tag{76}$$

where $l_1 + l_2 + l_3 = \mathrm{L}$ is even, and

$$\begin{pmatrix} l_1 & l_2 & l_3 \\ \sigma & \sigma & \sigma \end{pmatrix} = \frac{\left(\frac{\mathrm{L}}{2}\right)!}{\left(\frac{\mathrm{L}}{2}-l_1\right)!\left(\frac{\mathrm{L}}{2}-l_2\right)!\left(\frac{\mathrm{L}}{2}-l_3\right)!}\sqrt{\frac{(\mathrm{L}-2l_1)!(\mathrm{L}-2l_2)!(\mathrm{L}-2l_3)!}{(\mathrm{L}+1)!}} \tag{77}$$

The integrand may be found by multiplication of the spherical harmonics, Eq. (15a) or Table 15, in which they are listed up to $l = 9$. The integration is conveniently carried out by use of the expression

$$\frac{1}{4\pi}\int x^{2p} y^{2q} z^{2r}\, dS = \frac{(2p)!\,(2q)!\,(2r)!\,(p+q+r)!}{p!\,q!\,r!\,(2p+2q+2r+1)!} \tag{78}$$

which we have derived by the trigonometric substitution, Eq. (17), and the integrals

$$\begin{aligned}\int_0^{\pi} \cos^m\alpha \sin^n\alpha\, d\alpha &= \frac{m-1}{m+n}\int_0^{\pi} \cos^{m-2}\alpha \sin^n\alpha\, d\alpha \\ &= \frac{n-1}{m+n}\int_0^{\pi} \cos^m\alpha \sin^{n-2}\alpha\, d\alpha\end{aligned} \tag{79}$$

$$\begin{aligned}\int_0^{2\pi} \cos^m\alpha \sin^n\alpha\, d\alpha &= \frac{m-1}{m+n}\int_0^{2\pi} \cos^{m-2}\alpha \sin^n\alpha\, d\alpha \\ &= \frac{n-1}{m+1}\int_0^{2\pi} \cos^m\alpha \sin^{n-2}\alpha\, d\alpha\end{aligned} \tag{80}$$

In the special case of an even 3-l symbol which has $l_1 = l_3 - l_2$ one can use Eq. (39) in the form

$$\frac{1}{4\pi}\int \mathfrak{H}^{l_1}_{t_1} \mathfrak{N}^{l_2}_{t_2} \mathfrak{H}^{l_3}_{t_3}\, dS = k'(l_1\, l_2\, l_3) \begin{pmatrix} l_1 & l_2 & l_3 \\ t_1 & t_2 & t_3 \end{pmatrix},\; l_1 = l_3 - l_2 \tag{81}$$

where the constant k' is given [(*10*), Eq. (48)] by

$$k'(l_1\, l_2\, l_3) = (2\, l_1 + 1)^{-\frac{1}{2}} \binom{2 l_3 + 1}{2 l_2}^{\frac{1}{2}} \tag{82}$$

The odd 3-l symbols can be evaluated as outlined in Sect. 4b by an operator equivalent method which ultimately brings the calculation into the form of Eq. (76). However, although this method has considerable theoretical interest, it is laborious and not feasible for high l-values. We refer to the following paper (*10*) for a general and useful way to calculate any 3-l symbol in terms of a 3-j symbol.

Generally, any 3-l symbol can be expressed as the integral

$$\frac{1}{4\pi}\int \mathfrak{H}^{l_1}_{t_1} \mathfrak{V}^{l_2}_{t_2} \mathfrak{H}^{l_3}_{t_3}\, dS = \langle l_1 \,\|\, \mathfrak{V}^{l_2} \,\|\, l_3 \rangle \begin{pmatrix} l_1 & l_2 & l_3 \\ t_1 & t_2 & t_3 \end{pmatrix} \tag{83}$$

where the reduced matrix element [see Eq. (3) and comments on Eq. (4)] is given by

$$< l_1 \| \mathfrak{V}^{l_2} \| l_3 > = \delta(l_1\, l_2\, l_3) \tag{84}$$

Eq. (84) is a general form of the set $\mathfrak{U}^k$ of unit tensorial operators, whose reduced matrices are

$$< l_1 \| \mathfrak{U}^k \| l_3 > = \delta(l_1\, l_3)\; \delta(l_1\, k\, l_3) \tag{85}$$

and the operators $\mathfrak{V}^{l_2}_{t_2}$, of which there is an infinite number, may be found from Ref. (*10*), Eq. (33), and operator equivalents for $\mathfrak{V}^{l_2}_{t_2}$ may be found from Ref. (*10*), Eq. (57). In Tables 16–22 we have represented a few 3-l symbols of R_3 and 3-Γ symbols of O in terms of the general unit tensorial operators $\mathfrak{V}^l_t$ and $\mathfrak{V}^\Gamma_\gamma$, respectively.

Appendix 2

Racah's Lemma Applied to the Group R_{3i} Reduced with Respect to the Group $C_{\infty v}$.

In Sect. 8a Racah's lemma was illustrated for the group R_3 reduced with respect to D_∞. It is valuable also to consider the group R_{3i} reduced with respect to $C_{\infty v}$, since this will exemplify the standardizations made in section 8c.

We shall use as an example one of the coupling coefficients of R_3 which corresponds to the middle element of Eq. (64),

$$< d\sigma\, p\pi c \mid d\pi s > = -\frac{1}{\sqrt{2}} \tag{86}$$

By use of Eqs. (56) and (57) one can generate for R_{3i} the four coupling coefficients

$$\begin{aligned} &< d_g\sigma\, p_g\pi c \mid d_g\pi s > = d_g\sigma\, p_u\pi c \mid d_u\pi s > \\ &= < d_u\sigma\, p_g\pi c \mid d_u\pi s > = < d_u\sigma\, p_u\pi c \mid d_g\pi s > = -\frac{1}{\sqrt{2}} \end{aligned} \tag{87}$$

Each of these yields one Racah lemma constant after R_{3i} has been reduced with respect to $C_{\infty v}$. Since the basis set of the irreducible representation p_g is an axial tensorial set $\mathfrak{R}^{p_g}$ (p. 210), and the basis set of d_g is a polar tensorial set $\mathfrak{H}^{d_g}$, the first of the coefficients of Eq. (87) becomes

$$< d_g\, \Sigma^+\sigma^+\, p_g\, \Pi\pi^- \mid d_g\, \Pi\pi^- > \, = -\frac{1}{\sqrt{2}}$$

$$= (\, \underline{d_g}\, \Sigma^+\, \underline{p_g}\, \Pi \mid d_g\, \underline{\Pi}\,) < \Sigma^+\sigma^+\, \Pi\pi^- \mid \Pi\pi^- > \, = \left(-\frac{1}{\sqrt{2}}\right)\left(+\,1\right) \quad (88)$$

Here, the coupling coefficient of $C_{\infty v}$ follows from Eq. (74), Table 13, and Eq. (5), and the value $+1$ implies that the Racah lemma constant is $-\frac{1}{\sqrt{2}}$. Similarly the remaining three coupling coefficients of Eq. (87) become

$$< d_g\, \Sigma^+\sigma^+\, p_u\, \Pi\pi^+ \mid d_u\, \Pi\pi^+ > \, = +\frac{1}{\sqrt{2}}$$

$$= (\underline{d_g}\, \Sigma^+\, \underline{p_u}\, \Pi \mid d_u\, \underline{\Pi}) < \Sigma^+\sigma^+\, \Pi\pi^+ \mid \Pi\pi^+ > \, = \left(+\frac{1}{\sqrt{2}}\right)\left(+\,1\right)$$

$$< d_u\, \Sigma^-\sigma^-\, p_g\, \Pi\pi^- \mid d_u\, \Pi\pi^+ > \, = +\frac{1}{\sqrt{2}}$$

$$= (\underline{d_u}\, \Sigma^-\, \underline{p_g}\, \Pi \mid d_u\, \underline{\Pi}) < \Sigma^-\sigma^-\, \Pi\pi^- \mid \Pi\pi^+ > \, = \left(+\frac{1}{\sqrt{2}}\right)\left(+\,1\right),$$

$$< d_u\, \Sigma^-\sigma^-\, p_u\, \Pi\pi^+ \mid d_g\, \Pi\pi^- > \, = -\frac{1}{\sqrt{2}}$$

$$= (\underline{d_u}\, \Sigma^-\, \underline{p_u}\, \Pi \mid d_g\, \underline{\Pi}) < \Sigma^-\sigma^-\, \Pi\pi^+ \mid \Pi\pi^- > \, = \left(+\frac{1}{\sqrt{2}}\right)\left(-\,1\right) \quad (89)$$

when reduced to $C_{\infty v}$.

Appendix 3

(Tables 15—22)

Table 15. *Solid Harmonics* $\mathfrak{H}^l_{\lambda\varsigma}$, *normalized to* $4\pi/(2l+1)$ *over the Unit Sphere. When normalized to unity they are the usual real hydrogen angular functions*

l	$=$	0
$\mathfrak{H}^s_\sigma$	$=$	1
l	$=$	1
$\mathfrak{H}^p_\sigma$	$=$	z
$\mathfrak{H}^p_{\pi s}$	$=$	y
$\mathfrak{H}^d_{\pi c}$	$=$	x
l	$=$	2
$\mathfrak{H}^d_\sigma$	$=$	$(1/2)\,(2z^2-x^2-y^2)$
$\mathfrak{H}^d_{\pi s}$	$=$	$\sqrt{3}\,zy$
$\mathfrak{H}^d_{\pi c}$	$=$	$\sqrt{3}\,zx$
$\mathfrak{H}^d_{\delta s}$	$=$	$(\sqrt{3}/2)\,(2\,xy)$
$\mathfrak{H}^d_{\delta c}$	$=$	$(\sqrt{3}/2)\,(x^2-y^2)$

Table 15 (continued)

l	$= 3$
$\mathfrak{H}^f_\sigma$	$= (1/2)\,(2z^2 - 3x^2 - 3y^2)\,z$
$\mathfrak{H}^f_{\pi s}$	$= (\sqrt{6}/4)\,(4z^2 - x^2 - y^2)\,y$
$\mathfrak{H}^f_{\pi c}$	$= (\sqrt{6}/4)\,(4z^2 - x^2 - y^2)\,x$
$\mathfrak{H}^f_{\delta s}$	$= (\sqrt{15}/2)\,z\,(2\,xy)$
$\mathfrak{H}^f_{\delta c}$	$= (\sqrt{15}/2)\,z\,(x^2 - y^2)$
$\mathfrak{H}^f_{\varphi s}$	$= (\sqrt{10}/4)\,(3x^2y - y^3)$
$\mathfrak{H}^f_{\varphi c}$	$= (\sqrt{10}/4)\,(x^3 - 3xy^2)$

l	$= 4$
$\mathfrak{H}^g_\sigma$	$= (1/8)\,(8z^4 + 3x^4 + 3y^4 - 24z^2x^2 - 24y^2z^2 + 6x^2y^2)$
$\mathfrak{H}^g_{\pi s}$	$= (\sqrt{10}/4)\,(4z^2 - 3x^2 - 3y^2)\,zy$
$\mathfrak{H}^g_{\pi c}$	$= (\sqrt{10}/4)\,(4z^2 - 3x^2 - 3y^2)\,zx$
$\mathfrak{H}^g_{\delta s}$	$= (\sqrt{5}/4)\,(6z^2 - x^2 - y^2)\,(2\,xy)$
$\mathfrak{H}^g_{\delta c}$	$= (\sqrt{5}/4)\,(6z^2 - x^2 - y^2)\,(x^2 - y^2)$
$\mathfrak{H}^g_{\varphi s}$	$= (\sqrt{70}/4)\,z\,(3x^2y - y^3)$
$\mathfrak{H}^g_{\varphi c}$	$= (\sqrt{70}/4)\,z\,(x^3 - 3xy^2)$
$\mathfrak{H}^g_{\gamma s}$	$= (\sqrt{35}/8)\,(4x^3y - 4xy^3)$
$\mathfrak{H}^g_{\gamma c}$	$= (\sqrt{35}/8)\,(x^4 - 6x^2y^2 + y^4)$

$l = 5$

$$\mathfrak{H}^h_\sigma = (1/8)\,(8z^5 - 40z^3x^2 - 40z^3y^2 + 15zx^4 + 15zy^4 + 30zy^2x^2)$$

$$\mathfrak{H}^h_{\pi s} = (\sqrt{15}/8)\,(8z^4 + x^4 + y^4 + 2x^2y^2 - 12z^2x^2 - 12z^2y^2)\,y$$

$$\mathfrak{H}^h_{\pi c} = (\sqrt{15}/8)\,(8z^4 + x^4 + y^4 + 2x^2y^2 - 12z^2x^2 - 12z^2y^2)\,x$$

$$\mathfrak{H}^h_{\delta s} = (\sqrt{105}/4)\,(2z^2 - x^2 - y^2)\,z\,(2\,xy)$$

$$\mathfrak{H}^h_{\delta c} = (\sqrt{105}/4)\,(2z^2 - x^2 - y^2)\,z\,(x^2 - y^2)$$

$$\mathfrak{H}^h_{\varphi s} = (\sqrt{70}/16)\,(8z^2 - x^2 - y^2)\,(3x^2y - y^3)$$

$$\mathfrak{H}^h_{\varphi c} = (\sqrt{70}/16)\,(8z^2 - x^2 - y^2)\,(x^3 - 3xy^2)$$

$$\mathfrak{H}^h_{\gamma s} = (3\sqrt{35}/8)\,z\,(4x^3y - 4xy^3)$$

$$\mathfrak{H}^h_{\gamma c} = (3\sqrt{35}/8)\,z\,(x^4 - 6x^2y^2 + y^4)$$

$$\mathfrak{H}^h_{\eta s} = (3\sqrt{14}/16)\,(5x^4y - 10x^2y^3 + y^5)$$

$$\mathfrak{H}^h_{\eta c} = (3\sqrt{14}/16)\,(x^5 - 10x^3y^2 + 5xy^4)$$

$l = 6$

$$\mathfrak{H}^i_\sigma = (1/16)\,(16z^6 - 120z^4x^2 - 120z^4y^2 + 90z^2x^4 + 90z^2y^4 + 180z^2x^2y^2 - 5x^6 - 5y^6 - 15x^4y^2 - 15x^2y^4)$$

$$\mathfrak{H}^i_{\pi s} = (\sqrt{21}/8)\,(8z^5 - 20z^3x^2 - 20z^3y^2 + 5zx^4 + 5zy^4 + 10zx^2y^2)\,y$$

$$\mathfrak{H}^i_{\pi c} = (\sqrt{21}/8)\,(8z^5 - 20z^3x^2 - 20z^3y^2 + 5zx^4 + 5zy^4 + 10zx^2y^2)\,x$$

$$\mathfrak{H}^i_{\delta s} = (\sqrt{210}/32)\,(16z^4 - 16z^2x^2 - 16z^2y^2 + x^4 + y^4 + 2x^2y^2)\,(2xy)$$

$$\mathfrak{H}^i_{\delta c} = (\sqrt{210}/32)\,(16z^4 - 16z^2x^2 - 16z^2y^2 + x^4 + y^4 + 2x^2y^2)\,(x^2 - y^2)$$

$$\mathfrak{H}^i_{\varphi s} = (\sqrt{210}/16)\,(8z^2 - 3x^2 - 3y^2)\,z\,(3x^2y - y^3)$$

$$\mathfrak{H}^i_{\varphi c} = (\sqrt{210}/16)\,(8z^2 - 3x^2 - 3y^2)\,z\,(x^3 - 3xy^2)$$

Table 15 (continued)

$\mathfrak{H}^i_{\gamma s} = (3\sqrt{7}/16)\,(10z^2 - x^2 - y^2)\,(4x^3y - 4xy^3)$

$\mathfrak{H}^i_{\gamma c} = (3\sqrt{7}/16)\,(10z^2 - x^2 - y^2)\,(x^4 - 6x^2y^2 + y^4)$

$\mathfrak{H}^i_{\eta s} = (3\sqrt{154}/16)\,z\,(5x^4y - 10x^2y^3 + y^5)$

$\mathfrak{H}^i_{\eta c} = (3\sqrt{154}/16)\,z\,(x^5 - 10x^3y^2 + 5xy^4)$

$\mathfrak{H}^i_{\iota s} = (\sqrt{462}/32)\,(6x^5y - 20x^3y^3 + 6xy^5)$

$\mathfrak{H}^i_{\iota c} = (\sqrt{462}/32)\,(x^6 - 15x^4y^2 + 15x^2y^4 - y^6)$

$l = 7$

$\mathfrak{H}^k_{\sigma} = 1/16\,(16z^7 - 168z^5x^2 - 168z^5y^2 + 210z^3x^4 + 210z^3y^4 + 420z^3x^2y^2 - 35zx^6 - 35zy^6 - 105zx^4y^2 - 105zx^2y^4)$

$\mathfrak{H}^k_{\pi s} = (\sqrt{7}/32)\,(64z^6 - 240z^4x^2 - 240z^4y^2 + 120z^2x^4 + 120z^2y^4 + 240z^2x^2y^2 - 5x^6 - 5y^6 - 15x^4y^2 - 15x^2y^4)\,y$

$\mathfrak{H}^k_{\pi c} = (\sqrt{7}/32)\,(64z^6 - 240z^4x^2 - 240z^4y^2 + 120z^2x^4 + 120z^2y^4 + 240z^2x^2y^2 - 5x^6 - 5y^6 - 15x^4y^2 - 15x^2y^4)\,x$

$\mathfrak{H}^k_{\delta s} = (\sqrt{42}/32)\,(48z^4 - 80z^2x^2 - 80z^2y^2 + 15x^4 + 15y^4 + 30x^2y^2)\,z\,(2xy)$

$\mathfrak{H}^k_{\delta c} = (\sqrt{42}/32)\,(48z^4 - 80z^2x^2 - 80z^2y^2 + 15x^4 + 15y^4 + 30x^2y^2)\,z\,(x^2 - y^2)$

$\mathfrak{H}^k_{\varphi s} = (\sqrt{21}/32)\,(80z^4 - 60z^2x^2 - 60z^2y^2 + 3x^4 + 3y^4 + 6x^2y^2)\,(3x^2y - y^3)$

$\mathfrak{H}^k_{\varphi c} = (\sqrt{21}/32)\,(80z^4 - 60z^2x^2 - 60z^2y^2 + 3x^4 + 3y^4 + 6x^2y^2)\,(x^3 - 3xy^2)$

$\mathfrak{H}^k_{\gamma s} = (\sqrt{231}/16)\,(10z^2 - 3x^2 - 3y^2)\,z\,(4x^3x - 4xy^3)$

$\mathfrak{H}^k_{\gamma c} = (\sqrt{231}/16)\,(10z^2 - 3x^2 - 3y^2)\,z\,(x^4 - 6x^2y^2 - y^4)$

$\mathfrak{H}^k_{\eta s} = (\sqrt{231}/32)\,(12z^2 - x^2 - y^2)\,(5x^4y - 10x^2y^3 + y^5)$

$\mathfrak{H}^k_{\eta c} = (\sqrt{231}/32)\,(12z^2 - x^2 - y^2)\,(x^5 - 10x^3y^2 + 5xy^4)$

$\mathfrak{H}^k_{\iota s} = (\sqrt{6006}/32)\,z\,(6x^5y - 20x^3y^3 + 6xy^5)$

$$\mathfrak{H}^k_{\iota c} = (\sqrt{6006}/32)\, z\,(x^6 - 15x^4y^2 + 15x^2y^4 - y^6)$$
$$\mathfrak{H}^k_{\varkappa s} = (\sqrt{429}/32)\,(7x^6y - 35x^4y^3 + 21x^2y^5 - y^7)$$
$$\mathfrak{H}^k_{\varkappa c} = (\sqrt{429}/32)\,(x^7 - 21x^7y^2 + 35x^3y^4 - 7xy^6)$$

$l = 8$

$$\mathfrak{H}^l_{\sigma} = 1/128\,(128z^8 - 1792z^6x^2 - 1792z^6y^2 + 3360z^4x^4 + 3360z^4y^4 + 6720z^4x^2y^2 - 1120z^2x^6 - 1120z^2y^6 - 3360z^2x^4y^2 - 3360z^2x^2y^4 + 35x^8 + 35y^8 + 140x^6y^2 + 140x^2y^6 + 210x^4y^4)$$
$$\mathfrak{H}^l_{\pi s} = (3/32)\,(64z^6 - 336z^4x^2 - 336z^4y^2 + 280z^2x^4 + 280z^2y^4 + 560z^2x^2y^2 - 35x^6 - 35y^6 - 105x^4y^2 - 105x^2y^4)\,zy$$
$$\mathfrak{H}^l_{\pi c} = (3/32)\,(64z^6 - 336z^4x^2 - 336z^4y^2 + 280z^2x^4 + 280z^2y^4 + 560z^2x^2y^2 - 35x^6 - 35y^6 - 105x^4y^2 - 105x^2y^4)\,zx$$
$$\mathfrak{H}^l_{\delta s} = (3\sqrt{70}/64)\,(32z^6 - 80z^4x^2 - 80z^4y^2 + 30z^2x^4 + 30z^2y^4 + 60z^2x^2y^2 - x^6 - y^6 - 3x^4y^2 - 3x^2y^4)\,(2\,xy)$$
$$\mathfrak{H}^l_{\delta c} = (3\sqrt{70}/64)\,(32z^6 - 80z^4x^2 - 80z^4y^2 + 30z^2x^4 + 30z^2y^4 + 60z^2x^2y^2 - x^6 - y^6 - 3x^4y^2 - 3x^2y^4)\,(x^2 - y^2)$$
$$\mathfrak{H}^l_{\varphi s} = (\sqrt{1155}/32)\,(16z^4 - 20z^2x^2 - 20z^2y^2 + 3x^4 + 3y^4 + 6x^2y^2)\,z\,(3x^2y - y^3)$$
$$\mathfrak{H}^l_{\varphi c} = (\sqrt{1155}/32)\,(16z^4 - 20z^2x^2 - 20z^2y^2 + 3x^4 + 3y^4 + 6x^2y^2)\,z\,(x^3 - 3xy^2)$$
$$\mathfrak{H}^l_{\gamma s} = (3\sqrt{77}/64)\,(40z^4 - 24z^2x^2 - 24z^2y^2 + x^4 + y^4 + 2x^2y^2)\,(4x^3y - 4xy^3)$$
$$\mathfrak{H}^l_{\gamma c} = (3\sqrt{77}/64)\,(40z^4 - 24z^2x^2 - 24z^2y^2 + x^4 + y^4 + 2x^2y^2)\,(x^4 - 6x^2y^2 - y^4)$$
$$\mathfrak{H}^l_{\eta s} = (3\sqrt{1001}/32)\,(4z^2 - x^2 - y^2)\,z\,(5x^4y - 10x^2y^3 + y^5)$$
$$\mathfrak{H}^l_{\eta c} = (3\sqrt{1001}/32)\,(4z^2 - x^2 - y^2)\,z\,(x^5 - 10x^3y^2 + 5xy^4)$$
$$\mathfrak{H}^l_{\iota s} = (\sqrt{858}/64)\,(14z^2 - x^2 - y^2)\,(6x^5y - 20x^3y^3 + 6xy^5)$$
$$\mathfrak{H}^l_{\iota c} = (\sqrt{858}/64)\,(14z^2 - x^2 - y^2)\,(x^6 - 15x^4y^2 + 15x^2y^4 - y^6)$$
$$\mathfrak{H}^l_{\varkappa s} = (3\sqrt{715}/32)\,z\,(7x^6y - 35x^4y^3 + 21x^2y^5 - y^7)$$
$$\mathfrak{H}^l_{\varkappa c} = (3\sqrt{715}/32)\,z\,(x^7 - 21x^5y^2 + 35x^3y^4 - 7xy^6)$$
$$\mathfrak{H}^l_{\lambda s} = (3\sqrt{715}/128)\,(8x^7y - 56x^5y^3 + 56x^3y^5 - 8xy^7)$$
$$\mathfrak{H}^l_{\lambda c} = (3\sqrt{715}/128)\,(x^8 - 28x^6y^2 + 70x^4y^4 - 28x^2y^6 + y^8)$$

Table 15 (continued)

$l = 9$

$$\mathfrak{H}_{\sigma}^{m} = 1/128\,(128z^9 - 2304z^7x^2 - 2304z^7y^2 + 6048z^5x^4 + 6048z^5y^4 + 12096z^5x^2y^2 - 3360z^3x^6 - 3360z^3y^6 - 10080z^3x^4y^2 - 10080z^3x^2y^4 + 315zx^8 + 315zy^8 + 1260zx^6y^2 + 1260x^2y^6 + 1890zx^4y^4)$$

$$\mathfrak{H}_{\pi s}^{m} = (3\sqrt{5}/128)\,(128z^8 - 896z^6x^2 - 896z^6y^2 + 1120z^4x^4 + 1120z^4y^4 + 2240z^4x^2y^2 - 280z^2x^6 - 280z^2y^6 - 840z^2x^4y^2 - 840z^2x^2y^4 + 7x^8 + 7y^8 + 28x^6y^2 + 28x^2y^6 + 42x^4y^4)\,y$$

$$\mathfrak{H}_{\pi c}^{m} = (3\sqrt{5}/128)\,(128z^8 - 896z^6x^2 - 896z^6y^2 + 1120z^4x^4 + 1120z^4y^4 + 2240z^4x^2y^2 - 280z^2x^6 - 280z^2y^6 - 840z^2x^4y^2 - 840z^2x^2y^4 + 7x^8 + 7y^8 + 28x^6y^2 + 28x^2y^6 + 42x^4y^4)\,x$$

$$\mathfrak{H}_{\delta s}^{m} = (3\sqrt{110}/64)\,(32z^6 - 112z^4x^2 - 112z^4y^2 + 70z^2x^4 + 70z^2y^4 + 140z^2x^2y^2 - 7x^6 - 7y^6 - 21x^4y^2 - 21x^2y^4)\,z\,(2xy)$$

$$\mathfrak{H}_{\delta c}^{m} = (3\sqrt{110}/64)\,(32z^6 - 112z^4x^2 - 112z^4y^2 + 70z^2x^4 + 70z^2y^4 + 140z^2x^2y^2 - 7x^6 - 7y^6 - 21x^4y^2 - 21x^2y^4)\,z\,(x^2 - y^2)$$

$$\mathfrak{H}_{\varphi s}^{m} = (\sqrt{2310}/128)\,(64z^6 - 120z^4x^2 - 120z^4y^2 + 36z^2x^4 + 36z^2y^4 + 72z^2x^2y^2 - x^6 - y^6 - 3x^4y^2 - 3x^2y^4)\,(3x^2y - y^3)$$

$$\mathfrak{H}_{\varphi c}^{m} = (\sqrt{2310}/128)\,(64z^6 - 120z^4x^2 - 120z^4y^2 + 36z^2x^4 + 36z^2y^4 + 72z^2x^2y^2 - x^6 - y^6 - 3x^4y^2 - 3x^2y^4)\,(x^3 - 3xy^2)$$

$$\mathfrak{H}_{\gamma s}^{m} = (3\sqrt{5005}/64)\,(8z^4 - 8z^2x^2 - 8z^2y^2 + x^4 + y^4 + 2x^2y^2)\,z\,(4x^3y - 4xy^3)$$

$$\mathfrak{H}_{\gamma c}^{m} = (3\sqrt{5005}/64)\,(8z^4 - 8z^2x^2 - 8z^2y^2 + x^4 + y^4 + 2x^2y^2)\,z\,(x^4 - 6x^2y^2 - y^4)$$

$$\mathfrak{H}_{\eta s}^{m} = (3\sqrt{286}/128)\,(56z^4 - 28z^2x^2 - 28z^2y^2 + x^4 + y^4 + 2x^2y^2)\,(5x^4 - 10x^2y^3 + y^5)$$

$$\mathfrak{H}_{\eta c}^{m} = (3\sqrt{286}/128)\,(56z^4 - 28z^2x^2 - 28z^2y^2 + x^4 + y^4 + 2x^2y^2)\,(x^5 - 10x^3y^2 + 5xy^4)$$

$$\mathfrak{H}_{\iota s}^{m} = (\sqrt{4290}/64)\,(14z^2 - 3x^2 - 3y^2)\,z\,(6x^5y - 20x^3y^3 + 6xy^5)$$

$$\mathfrak{H}_{\iota c}^{m} = (\sqrt{4290}/64)\,(14z^2 - 3x^2 - 3y^2)\,z\,(x^6 - 15x^4y^2 + 15x^2y^4 - y^6)$$

$$\mathfrak{H}_{\kappa s}^{m} = (3\sqrt{1430}/256)\,(16z^2 - x^2 - y^2)\,(7x^6y - 35x^4y^3 + 21x^2y^5 - y^5)$$

$$\mathfrak{H}_{\kappa c}^{m} = (3\sqrt{1430}/256)\,(16z^2 - x^2 - y^2)\,(x^5 - 21x^5y^2 + 35x^3y^4 - 7xy^6)$$

$$\mathfrak{H}_{\lambda s}^{m} = (3\sqrt{12155}/128)\,z\,(8x^7y - 56x^5y^3 + 56x^3y^5 - 8xy^7)$$

$$\mathfrak{H}_{\lambda c}^{m} = (3\sqrt{12155}/128)\,z\,(x^8 - 28x^6y^2 + 70x^4y^4 - 28x^2y^6 + y^8)$$

$$\mathfrak{H}_{\mu s}^{m} = (\sqrt{24310}/256)\,(9x^8y - 84x^6y^3 + 126x^4y^5 - 36x^2y^7 + y^9)$$

$$\mathfrak{H}_{\mu c}^{m} = (\sqrt{24310}/256)\,(x^9 - 36x^7y^2 + 126x^5y^4 - 84x^3y^6 + 9xy^8)$$

Table 16. *The 3-l symbols* $\begin{pmatrix} p & l & p \\ t_1 & t_2 & t_3 \end{pmatrix}$ *represented as matrix elements of the general unit tensorial operator* $\mathfrak{B}^l$ [*Eqs.* (83) *and* (84)]. *In the table, the entries concerned with* $l = s$ *must be divided by* $\sqrt{3}$, *those with* $l = p$ *divided by* $\sqrt{6}$, *and those with* $l = d$ *by* $\sqrt{30}$. *It is noted that except for* $l = s$ *the trace of the matrix is zero. This is known as the baricenter rule or the center of gravity rule and is a consequence of Eqs.* (6a) *and* (5),

$$\sum_{\gamma_1\gamma_2} \begin{pmatrix} \Gamma_1 & \Gamma_2 & \Gamma_3 \\ \gamma_1 & \gamma_2 & \gamma_3 \end{pmatrix} \begin{pmatrix} \Gamma_1 & \Gamma_2 & \Gamma_3' \\ \gamma_1 & \gamma_2 & \gamma_3' \end{pmatrix} = [\Gamma_3]^{-1}\, \delta(\Gamma_3\, \Gamma_3')\, \delta(\gamma_3\, \gamma_3')\, \delta(\Gamma_1\, \Gamma_2\, \Gamma_3).$$

Here, if $\Gamma_3' = s$, *then* $\Gamma_1 = \Gamma_2$, $\gamma_1 = \gamma_2$, *and* $\begin{pmatrix} \Gamma_1 & \Gamma_2 & s \\ \gamma_1 & \gamma_2 & \sigma \end{pmatrix} = 1/\sqrt{[\Gamma_1]}$, *and*

$$\sum_{\gamma_1} \begin{pmatrix} \Gamma_1 & \Gamma_1 & \Gamma_3 \\ \gamma_1 & \gamma_1 & \gamma_3 \end{pmatrix} = \sqrt{[\Gamma_1]}\, \delta(\Gamma_3, s)\, \delta(\gamma_3, \sigma).$$

The matrix is symmetrical for even 3-l symbols. Further, the square sum rules implied from Eqs. (4), (5), *and* (6) *may be noted*

$\mathfrak{B}^s + \mathfrak{B}^p + \mathfrak{B}^d$	$p\sigma$	$p\pi s$	$p\pi c$
$p\sigma$	1 $\quad 2$ s $\quad d\sigma$	1 $\quad \sqrt{3}$ $p\pi c$ $\quad d\pi s$	-1 $\quad \sqrt{3}$ $p\pi s$ $\quad d\pi c$
$p\pi s$	-1 $\quad \sqrt{3}$ $p\pi c$ $\quad d\pi s$	1 $\quad -1$ $\quad -\sqrt{3}$ s $\quad d\sigma$ $\quad d\delta c$	1 $\quad \sqrt{3}$ $p\sigma$ $\quad d\delta s$
$p\pi c$	1 $\quad \sqrt{3}$ $p\pi s$ $\quad d\pi c$	-1 $\quad \sqrt{3}$ $p\sigma$ $\quad d\delta s$	1 $\quad -1$ $\quad \sqrt{3}$ s $\quad d\sigma$ $\quad d\delta c$

Table 17. *The 3-l symbols* $\begin{pmatrix} d & l & d \\ t_1 & t_2 & t_3 \end{pmatrix}$ *represented as matrix elements of the general unit tensorial operator* $\mathfrak{B}^l$. *For* $l = s$, p, *and* d *the entries of the table must be divided by* $\sqrt{5}$, $\sqrt{30}$, *and* $\sqrt{70}$, *respectively. For* $l = f$ *only the elements of* $\mathfrak{B}^f_{\delta s}$ *are given; they must be divided by* $\sqrt{14}$. *The elements of* $\mathfrak{B}^g$ *are omitted*

$\mathfrak{B}^s + \mathfrak{B}^p + \mathfrak{B}^d + \mathfrak{B}^f_{\delta s}$	$d\sigma$	$d\pi s$	$d\pi c$	$d\delta s$	$d\delta c$
$d\sigma$	1 2 s $d\sigma$	$\sqrt{3}$ 1 $p\pi c$ $d\pi s$	$-\sqrt{3}$ 1 $p\pi s$ $d\pi c$	-2 $d\delta s$	-2 -1 $d\delta c$ $f\delta s$
$d\pi s$	$-\sqrt{3}$ 1 $p\pi c$ $d\pi s$	1 1 $-\sqrt{3}$ s $d\sigma$ $d\delta c$	1 $\sqrt{3}s$ $p\sigma$ $d\delta$	-1 $\sqrt{3}$ $p\pi s$ $d\pi c$	-1 $-\sqrt{3}$ $p\pi c$ $d\pi s$
$d\pi c$	$\sqrt{3}$ 1 $p\pi s$ $d\pi c$	-1 $\sqrt{3}$ $p\sigma$ $d\delta s$	1 1 $\sqrt{3}$ s $d\sigma$ $d\delta c$	1 $\sqrt{3}$ $p\pi c$ $d\pi s$	-1 $\sqrt{3}$ $p\pi s$ $d\pi c$
$d\delta s$	-2 $d\delta s$	1 $\sqrt{3}$ $p\pi s$ $d\pi c$	-1 $\sqrt{3}$ $p\pi c$ $d\pi s$	1 -2 s $d\sigma$	2 $p\sigma$
$d\delta c$	-2 1 $d\delta c$ $f\delta s$	1 $-\sqrt{3}$ $p\pi c$ $d\pi s$	1 $\sqrt{3}$ $p\pi s$ $d\pi c$	-2 $p\sigma$	1 -2 s $d\sigma$

Table 18. *The 3-l symbols* $\begin{pmatrix} d & l & p \\ t_1 & t_2 & t_3 \end{pmatrix}$ *represented as matrix elements of the general unit tensorial operator* $\mathfrak{V}^l$. *For* $l = p$, d, *and* f *the entries of the table must be divided by* $\sqrt{30}$, $\sqrt{30}$, *and* $\sqrt{210}$, *respectively*

$\mathfrak{V}^p + \mathfrak{V}^d + \mathfrak{V}^f$	$p\sigma$			$p\pi s$				$p\pi c$			
$d\sigma$	2 $p\sigma$		$3\sqrt{2}$ $f\sigma$	-1 $p\pi s$	$\sqrt{3}$ $d\pi c$	$2\sqrt{3}$ $f\pi s$		-1 $p\pi c$	$-\sqrt{3}$ $d\pi s$		$2\sqrt{3}$ $f\pi c$
$d\pi s$	$\sqrt{3}$ $p\pi s$	-1 $d\pi c$	4 $f\pi s$	$\sqrt{3}$ $p\sigma$	1 $d\delta s$	$-\sqrt{6}$ $f\sigma$	$-\sqrt{10}$ $f\delta c$		$\sqrt{3}$ $d\sigma$	1 $d\delta c$	$\sqrt{10}$ $f\delta s$
$d\pi c$	$\sqrt{3}$ $p\pi c$	1 $d\pi s$	4 $f\pi c$		$-\sqrt{3}$ $d\sigma$	1 $d\delta c$	$\sqrt{10}$ $f\delta s$	$\sqrt{3}$ $p\sigma$	-1 $d\delta s$	$-\sqrt{6}$ $f\sigma$	$\sqrt{10}$ $f\delta c$
$d\delta s$		-2 $d\delta c$	$\sqrt{10}$ $f\delta s$	$\sqrt{3}$ $p\pi c$	-1 $d\pi s$	-1 $f\pi c$	$-\sqrt{15}$ $f\varphi c$	$\sqrt{3}$ $p\pi s$	1 $d\pi c$	-1 $f\pi s$	$\sqrt{15}$ $f\varphi s$
$d\delta c$		2 $d\delta s$	$\sqrt{10}$ $f\delta c$	$-\sqrt{3}$ $p\pi s$	-1 $d\pi c$	1 $f\pi s$	$\sqrt{15}$ $f\varphi s$	$\sqrt{3}$ $p\pi c$	-1 $d\pi s$	-1 $f\pi c$	$\sqrt{15}$ $f\varphi c$

Table 19. *The 3-Γ symbols* $\begin{pmatrix} E & \Gamma & E \\ \gamma_1 & \gamma_1 & \gamma_3 \end{pmatrix}$ *of the octahedral rotation group O represented as matrix elements of the general set of unit tensorial operators* $\mathfrak{V}^{\Gamma}$, $\Gamma = A_1$, A_2, *and E*

$\mathfrak{V}^{A_1}+\mathfrak{V}^{A_2}+\mathfrak{V}^{E}$	$E\theta$	$E\varepsilon$
$E\theta$	$1/\sqrt{2}$ $1/2$ A_1 $E\theta$	$-1/\sqrt{2}$ $-1/2$ A_2 $E\varepsilon$
$E\varepsilon$	$1/\sqrt{2}$ $-1/2$ A_2 $E\varepsilon$	$1/\sqrt{2}$ $-1/2$ A_1 $E\theta$

Table 20. *The 3-Γ symbols* $\begin{pmatrix} T_1 & \Gamma & T_1 \\ \gamma_1 & \gamma_2 & \gamma_3 \end{pmatrix}$ *of the octahedral rotation group O represented as matrix elements of the general set of unit tensorial operators* $\mathfrak{V}^{\Gamma}$, $\Gamma = A_1$, E, T_1, *and* T_2

$\mathfrak{V}^{A_1}+\mathfrak{V}^{E}+\mathfrak{V}^{T_1}+\mathfrak{V}^{T_2}$	$T_1 z$	$T_1 y$	$T_1 x$
$T_1 z$	$1/\sqrt{3}$ $+1/\sqrt{3}$ A_1 $E\theta$	$1/\sqrt{6}$ $1/\sqrt{6}$ $T_1 x$ $T_2 \xi$	$-1/\sqrt{6}$ $1/\sqrt{6}$ $T_1 y$ $T_2 \eta$
$T_1 y$	$-1/\sqrt{6}$ $1/\sqrt{6}$ $T_1 x$ $T_2 \xi$	$1/\sqrt{3}$ $-1/\sqrt{12}$ $-1/2$ A_1 $E\theta$ $E\varepsilon$	$1/\sqrt{6}$ $1/\sqrt{6}$ $T_1 z$ $T_2 \zeta$
$T_1 x$	$1/\sqrt{6}$ $1/\sqrt{6}$ $T_1 y$ $T_2 \eta$	$-1/\sqrt{6}$ $1/\sqrt{6}$ $T_1 z$ $T_2 \zeta$	$1/\sqrt{3}$ $-1/\sqrt{12}$ $+1/2$ A_1 $E\theta$ $E\varepsilon$

Table 21. *The 3-Γ symbols* $\begin{pmatrix} T_2 & \Gamma & T_2 \\ \gamma_1 & \gamma_2 & \gamma_3 \end{pmatrix}$ *of the octahedral rotation group O represented as matrix elements of the general set of unit tensorial operators* $\mathfrak{V}^{\Gamma}$, $\Gamma = A_1$, E, T_1, *and* T_2

$\mathfrak{V}^{A_1}+\mathfrak{V}^{E}+\mathfrak{V}^{T_1}+\mathfrak{V}^{T_2}$	$T_2 \xi$	$T_2 \eta$	$T_2 \zeta$
$T_2 \xi$	$1/\sqrt{3}$ $1/\sqrt{12}$ $-1/2$ A_1 $E\theta$ $E\varepsilon$	$1/\sqrt{6}$ $1/\sqrt{6}$ $T_1 z$ $T_2 \zeta$	$-1/\sqrt{6}$ $1/\sqrt{6}$ $T_1 y$ $T_2 \eta$
$T_2 \eta$	$-1/\sqrt{6}$ $1/\sqrt{6}$ $T_1 z$ $T_2 \zeta$	$1/\sqrt{3}$ $1/\sqrt{12}$ $1/2$ A_1 $E\theta$ $E\varepsilon$	$1/\sqrt{6}$ $1/\sqrt{6}$ $T_1 x$ $T_2 \xi$
$T_2 \zeta$	$1/\sqrt{6}$ $1/\sqrt{6}$ $T_1 y$ $T_2 \eta$	$-1/\sqrt{6}$ $1/\sqrt{6}$ $T_1 x$ $T_2 \xi$	$1/\sqrt{3}$ $-1/\sqrt{3}$ A_1 $E\theta$

Table 22. *The 3-Γ symbols* $\begin{pmatrix} T_2 \, \Gamma \, T_1 \\ \gamma_1 \, \gamma_2 \, \gamma_3 \end{pmatrix}$ *of the octahedral rotation group O represented as matrix elements of the general set of unit tensorial operators* $\mathfrak{B}^{\Gamma}$, $\Gamma = A_2$, E, T_1, and T_2

$\mathfrak{B}^{A_2} + \mathfrak{B}^{E} + \mathfrak{B}^{T_1} + \mathfrak{B}^{T_2}$	$T_1 z$	$T_1 y$	$T_1 x$
$T_2 \xi$	$1/\sqrt{6}$ $-1/\sqrt{6}$ $T_1 y$ $T_2 \eta$	$1/\sqrt{6}$ $1/\sqrt{6}$ $T_1 z$ $T_2 \zeta$	$1/\sqrt{3}$ $1/2$ $1/\sqrt{12}$ A_2 $E\theta$ $E\varepsilon$
$T_2 \eta$	$1/\sqrt{6}$ $1/\sqrt{6}$ $T_1 x$ $T_2 \xi$	$1/\sqrt{3}$ $-1/2$ $1/\sqrt{12}$ A_2 $E\theta$ $E\varepsilon$	$1/\sqrt{6}$ $-1/\sqrt{6}$ $T_1 z$ $T_2 \zeta$
$T_2 \zeta$	$1/\sqrt{3}$ $-1/\sqrt{3}$ A_2 $E\varepsilon$	$1/\sqrt{6}$ $-1/\sqrt{6}$ $T_1 x$ $T_2 \xi$	$1/\sqrt{6}$ $1/\sqrt{6}$ $T_1 y$ $T_2 \eta$

References

1. *Wigner, E. P.:* Am. J. Math. *63,* 57 (1941).
2. — Group Theory. New York–London: Academic Press 1959.
3. *Löwdin, P. O.:* Rev. Mod. Phys. *39,* 259 (1967).
4. *Griffith, J. S.:* The Irreducible Tensor Method for Molecular Symmetry Groups. New Jersey: Prentice-Hall, Inc. 1962.
5. *Racah, G.:* Phys. Rev. *76,* 1352 (1949).
6. *Schäffer, C. E.*: Proc. Roy. Soc. (London) *A 297,* 96 (1967).
7. — Struct. Bonding *5,* 68 (1968).
8. — Intern. J. Quantum Chem. *5,* 379 (1971).
9. — Struct. Bonding *14,* inprint.
10. *Harnung,* S. E., *Schäffer, C. E.*: Struct. Bonding *12,* 257 (1972).
11. *Schäffer, C. E.:* Pure Appl. Chem. *24,* 361 (1970).

Real Irreducible Tensorial Sets and their Application to Ligand-Field Theory

S. E. Harnung and C. E. Schäffer

Department I, (Inorganic Chemistry), The H. C. Ørsted Institute, University of Copenhagen, Universitetsparken 5, DK-2100 Copenhagen Ø, Denmark

Table of Contents

Introduction 257

I. Irreducible Tensorial Sets with Real Bases 258

1. Standard Conventions 258
2. The 3-*l* Symbol 263
3. The 6-*l* and 9-*l* Symbols 265
4. The Wigner-Eckart Theorem 267
5. Reduced Matrices of Irreducible Products of Operator Sets 268
6. Irreducible Sets of Spherical Harmonic Operators 270

II. Application to Ligand-Field Matrices 275

1. The Ligand-Field Model 275
2. The Ligand-Field Operator and Its One-electron Matrix-Elements ... 277
3. Calculation of Ligand-Field Matrices 280
4. Conclusions 283

III. Appendices. 285

1. Alternative Evaluation of 3-*l* Symbols 285
2. A Short Table of 3-*l* Symbols 285
3. Notes on the 6-*l* Symbol, Eq. (23) 286
4. Calculation of Ligand-Field Matrices. An Example 288

References 294

Introduction

The application of irreducible tensorial sets to quantum mechanical problems provides a powerful tool for the calculation of observables. This was recognized in the early forties by *G. Racah* (*12—14*) who developed the tensor method and used it for an extension of the theory of atomic spectra.

In a later work (*4*) *Fano* and *Racah* have given a lucid account of the complex of ideas leading to this application of irreducible tensorial sets. They consider the three-dimensional rotation group and follow common practice in requiring that the basis functions correspond to definite projection quantum numbers. This implies that the irreducible representations generally have complex elements.

However, for many purposes it is desirable to extend the theory of irreducible tensorial sets to the real irreducible representations that are carried by sets of real spherical harmonics. In the present paper this has been put into effect by using the properties of the invariant irreducible products considered by *Fano* and *Racah*. In particular, the phase conventions proposed in that work have been preserved so that *invariants* (reduced matrices, $\overline{W}$- and X-coefficients) *are identical in the two representations*.

The real version of the irreducible tensor method, related to the complex representations as mentioned, is highly useful in the ligand-field theory, as will be shown in the second part of this paper. An additional reason for this is the fact that expansion theorems concerning functions and operators achieve apt forms when the tensor method is applied to real spherical harmonics.

I. Irreducible Tensorial Sets with Real Bases

1. Standard Conventions

The basic operation in tensor theory is the unitary transformation that reduces the direct product of two irreducible unitary representations and transforms it into a direct sum of irreducible representations. In the three-dimensional rotation group a term of a certain symmetry type (irreducible representation) occurs at most once (*23*) in the direct sum. Thus, the effect of the transformation on the bases for the initial representations to yield an element of one of the direct product sets may be expressed according to Wigner's formula

$$[|j_1\} \times |j_2\}]^{[j]}_m = \sum_{m_1, m_2} \langle j_1 m_1 j_2 m_2 | jm \rangle |j_1 m_1\rangle |j_2 m_2\rangle \qquad (1)$$

In this context the angular momentum quantum numbers are considered as sufficient labels for the functions. Sets denoted by the j-number in a superscript square bracket or in the symbol $|\}$ are contrastandard (*4*), and in this paper the $(2j+1)$ elements constituting such sets are arranged in a row matrix according to decreasing values of m.

The phase of the irreducible product, Eq. (1), is fixed by the commonly accepted convention (*4*, Appendix G) that the element

$$\langle j_1 j_1 \; j_2 j - j_1 | jj \rangle \geq 0 \qquad (2)$$

is real and positive.

The vector coupling coefficients of Eq. (1) are not symmetric in the j-numbers. Therefore, in order to stress the inherent symmetry, the highly symmetric function, $\bar{V}$ [here denoted $\bar{V}(j)$], was defined by *Fano* and *Racah*. It is related to the vector coupling coefficient by the expression

$$\langle j_1 m_1 j_2 m_2 | jm \rangle = (-1)^{2j_2+j-m} (2j+1)^{1/2} \bar{V} \begin{pmatrix} j_1 & j_2 & j \\ m_1 & m_2 & -m \end{pmatrix} \qquad (3)$$

The function $\bar{V}(j)$ is invariant under even permutations of its columns and has to be multiplied by $(-1)^{j_1+j_2+j}$ under odd permutations. $\bar{V}(j)$ differs from the 3-j symbol of *Wigner* (*15*) by the factor $(-1)^{j_1+j_2+j}$.

The irreducible representations may be classified according to whether j is an integer or half of an odd integer. We shall consider here the former, which are the potentially real representations (*22b*, p. 287). These representations in contrastandard form (*4*) can be transformed into real form by a constant unitary matrix, i.e. the same matrix for every element in the group. The elements of the constant matrix will be chosen such that the contrastandard, self-conjugate sets which form the bases for the potentially real representations in complex standard form are connected to the sets of the usual real spherical harmonics which form the bases for the real standard representations. From the constant matrix, the vector coupling coefficients pertaining to the real functions will be deduced.

Elements of contrastandard, self-conjugate sets have been chosen in Ref. (*4*). The basis functions $|jm\rangle$ are for integer j represented by the complex standard spherical harmonics, $\mathfrak{Y}_m^{[j]}$ which differ from the spherical harmonics $Y_{jm} = \Theta(jm)\,\Phi(m)$ used by *Condon* and *Shortley* (*2*) by the common phase $i^j = (-1)^{j/2}$:

$$\mathfrak{Y}_m^{[j]} = i^j\, Y_{jm}, \quad \text{integer } j. \qquad (4)$$

The real standard basis functions are chosen as the usual orthonormal spherical harmonics (*21*). They will be denoted $\mathfrak{Z}_{\lambda\varsigma}^l$ and in solid harmonic form[1]) they may be given by the analytical expressions

---

[1]) The functions $r^l\,\mathfrak{Z}_{\lambda\varsigma}^l$ which belong to the solutions (*6*) of the Laplace equation $\left(\frac{\partial^2}{\partial x^2} + \frac{\partial^2}{\partial y^2} + \frac{\partial^2}{\partial z^2}\right) V = \nabla^2 V = 0$ are called solid spherical harmonics or (in mathematical physics) just spherical harmonics. If the functions are further subject to the condition $x^2+y^2+z^2=1$, they are called surface harmonics. In the present paper real functions of this type will be denoted $|\alpha l \lambda \varsigma\rangle$ or $r^l\,\mathfrak{Z}_{\lambda\varsigma}^l$ when normalized to unity on the surface of the unit sphere. The solid harmonics occur in the solutions of the Schrödinger equation for hydrogen like atoms (*2*)

$$\psi_{nl} = -\left(\frac{2\,Z}{na_0}\right)^{\frac{3}{2}+l} \sqrt{\frac{(n-l-1)!}{2\,n\,(n+l)!}}\, e^{-\frac{Zr}{na_0}}\, L_{n+l}^{2l+1}\left(\frac{2\,Z}{na_0}\, r\right) \times r^l\,\mathfrak{Z}_{\lambda\varsigma}^l, \quad a_0 = \frac{\hbar^2}{m\,e^2}$$

but since the factor r^l belongs to the radial part one often uses the term "spherical harmonic" for the surface harmonic angular part of the solutions.

$$r^l \mathfrak{Z}^l_{\lambda c} = \sqrt{\frac{2l+1}{4\pi}} \left[\{1 - \delta(\lambda, 0)\} \sqrt{2} + \delta(\lambda, 0)\right] \frac{(2l)!}{2^l l! \sqrt{(l+\lambda)!\,(l-\lambda)!}}$$

$$\times \left[z^{l-\lambda} - \frac{(l-\lambda)\,(l-\lambda-1)}{2 \times (2l-1)} z^{l-\lambda-2}\, r^2 \right. \tag{5a}$$

$$\left. + \frac{(l-\lambda)\,(l-\lambda-1)\,(l-\lambda-2)\,(l-\lambda-3)}{2 \times 4 \times (2l-1)\,(2l-3)} z^{l-\lambda-4}\, r^4 - \ldots \right]$$

$$\times\, r^\lambda \sin^\lambda\theta\ \varsigma(\lambda\varphi) \qquad (0 \leqq \lambda \leqq l)$$

$$= \sqrt{\frac{2l+1}{4\pi}} \left[\{1 - \delta(\lambda, 0)\} \sqrt{2} + \delta(\lambda, 0)\right] \frac{\sqrt{(l+\lambda)!}}{2^\lambda \lambda! \sqrt{(l-\lambda)!}}$$

$$\times \left[z^{l-\lambda} - \frac{(l-\lambda)\,(l-\lambda-1)}{2 \times (2\lambda+2)} z^{l-\lambda-2}\,(x^2+y^2) \right. \tag{5b}$$

$$\left. + \frac{(l-\lambda)\,(l-\lambda-1)\,(l-\lambda-2)\,(l-\lambda-3)}{2 \times 4 \times (2\lambda+2)\,(2\lambda+4)} z^{l-\lambda-4}(x^2+y^2)^2 - \ldots \right]$$

$$\times\, r^\lambda \sin^\lambda\theta\ \varsigma(\lambda\varphi) \qquad (0 \leqq \lambda \leqq l)$$

Here ς means either of the trigonometric functions sine or cosine. When ς represents sine,

$$r^\lambda \sin^\lambda\theta\ \sin(\lambda\varphi) = \left[\tbinom{\lambda}{1} x^{\lambda-1} y - \tbinom{\lambda}{3} x^{\lambda-3} y^3 + \ldots\right] \tag{6a}$$

and when ς represents cosine,

$$r^\lambda \sin^\lambda\theta\ \cos(\lambda\varphi) = \left[\tbinom{\lambda}{0} x^{\lambda} - \tbinom{\lambda}{2} x^{\lambda-2} y^2 + \ldots\right] \tag{6b}$$

The usual spherical polar coordinate system (r, θ, φ) is defined with respect to the right-handed Cartesian coordinate system whose coordinates occur in the equations. The Z-axis is the unique axis around which σ-functions ($\lambda = 0$) have rotational symmetry. Functions with $0 < \lambda \leqq l$ are characterized (*19*) through λ by their standard transformation properties when they undergo rotations around the Z-axis:

$$\mathsf{R}^{\mathsf{Z}}(\psi)\,[\mathfrak{Z}^l_{\lambda s}\,\mathfrak{Z}^l_{\lambda c}] = [\mathfrak{Z}^l_{\lambda s}\,\mathfrak{Z}^l_{\lambda c}] \begin{bmatrix} \cos\lambda\psi & \sin\lambda\psi \\ -\sin\lambda\psi & \cos\lambda\psi \end{bmatrix} \tag{7}$$

The first square bracket of Eq. (5) contains Kronecker deltas that are necessary in order to obtain a single expression and which assure that all functions are normalized to unity.

The l is placed on $\mathfrak{Z}$ without any kind of bracket to indicate that in the real basis the transition to the contragredient representation is effected by the unit matrix. The functions $\mathfrak{Z}^{l}_{\lambda\varsigma}$ carry all the irreducible representations of the three-dimensional rotation group, R_3, and since their parity under the inversion operation is well defined, they also carry every other of the irreducible representations of the corresponding inversion group, R_{3i}. The bases of the pure rotation group will be denoted $|l\lambda\varsigma\rangle$ or just $|lt\rangle$ and for these bases the parity is not defined.

The constant unitary matrix that transforms the set $|j\}$ (integer j) of the complex representation into the corresponding set $|l\}$ of the real representation is denoted by $\{j|l\}$ so that

$$|l\} = |j\}\{j|l\} \tag{8}$$

where the standard sequence in $|l\}$ is $(|l\sigma\rangle|l1\ s\rangle|l1\ c\rangle|l2\ s\rangle\ldots|\,ll c\rangle)$. A matrix of this type has been given previously (*20*, Eq. 10). When that matrix is adjusted to the present phase choice, eq. (4), the elements of $\{j|l\}$ are generated by

$$(|l\sigma\rangle|l\lambda s\rangle|l\lambda c\rangle) = (|j\lambda\rangle|j0\rangle|j-\lambda\rangle)$$
$$\times\begin{bmatrix} 0 & (-1)^{\lambda}\dfrac{(-i)^{j+1}}{\sqrt{2}} & (-1)^{\lambda}\dfrac{(-i)^{j}}{\sqrt{2}} \\ (-i)^{j} & 0 & 0 \\ 0 & -\dfrac{(-i)^{j+1}}{\sqrt{2}} & \dfrac{(-i)^{j}}{\sqrt{2}} \end{bmatrix} \tag{9}$$

where $\lambda = |m|$ and $l = j$.

Elements of an irreducible product of sets whose elements transform under the pure rotation group in identically the same way as do the functions, Eq. (5), are defined through

$$[|l_1\} \times |l_2\}]^{l}_{\lambda\varsigma} = \sum_{\lambda_1\varsigma_1\lambda_2\varsigma_2} \langle l_1\lambda_1\varsigma_1\ l_2\lambda_2\varsigma_2|l\lambda\varsigma\rangle|l_1\lambda_1\varsigma_1\rangle|l_2\lambda_2\varsigma_2\rangle \tag{10}$$

and the expansion coefficients follow from Eqs. (1) and (8).

With the conventions of Eqs. (2) and (9) they are all real,

$$\begin{aligned}&\langle l_1\lambda_1\varsigma_1\ l_2\lambda_2\varsigma_2|l\lambda\varsigma\rangle\\ &= \sum_{m_1m_2m} \langle l_1\lambda_1\varsigma_1|j_1m_1\rangle\langle l_2\lambda_2\varsigma_2|j_2m_2\rangle\langle j_1m_1\ j_2m_2|jm\rangle\\ &\qquad\times\langle jm|l\lambda\varsigma\rangle\end{aligned} \tag{11}$$

The evaluation of Eq. (11) yields the general expression

$$<l_1\,\lambda_1\,\varsigma_1\;l_2\,\lambda_2\,\varsigma_2|l\,\lambda\,\varsigma> \\ = i^{l_1+l_2-l}\,(\varsigma_1\times\varsigma_2\times\bar{\varsigma}_3)\,2^{-1/2}\,<l_1\lambda_1\;l_2\lambda_2|l\,\lambda> \quad (12)$$

which is valid except when σ-functions occur. This restriction, which is associated with the first square bracket of Eqs. (5a) and (5b), emphasises the difference in the constant of normalization between σ-functions and functions with $\lambda\neq 0$. The phase before the square root of Eq. (12) is real. It is found by the substitution $\varsigma=1$ for cosine functions and $\varsigma=i$ for sine functions; the bar denotes complex conjugation. The coupling coefficient on the right-hand side belongs to those complex standard representations, Eq. (1), which have integer j and positive m values. When $\lambda_1=0$, or $\lambda_2=0$, the right-hand side of Eq. (12) must be multiplied by $\sqrt{2}$ and $\varsigma=1$ used for σ-functions. When $\lambda=0$ one finds

$$\begin{aligned} &<l_1\,\lambda_1\,c_1\;l_2\,\lambda_2\,c_2|l\sigma> \\ &= (-1)^{\lambda_1}\,i^{l_1+l_2-l}(1\times 1\times 1)<l_1-\lambda_1\,l_2\,\lambda_2|l0> \\ &<l_1\,\lambda_1\,s_1\;l_2\,\lambda_2\,s_2|l\sigma> \\ &= -(-1)^{\lambda_1}\,i^{l_1+l_2-l}(i\times i\times 1)<l_1-\lambda_1\,l_2\,\lambda_2|l0> \\ &<l_1\,\lambda_1\,c_1\;l_2\,\lambda_2\,s_2|l\sigma> \qquad (13) \\ &= (-1)^{\lambda_1}\,i^{l_1+l_2-l}(1\times i\times 1)<l_1-\lambda_1\,l_2\,\lambda_2|l0> \\ &<l_1\,\lambda_1\,s_1\;l_2\,\lambda_2\,c_2|l\sigma> \\ &= -(-1)^{\lambda_1}\,i^{l_1+l_2-l}(\mathrm{i}\times 1\times 1)<l_1-\lambda_1\,l_2\,\lambda_2|l0> \end{aligned}$$

With these findings it may be deduced from Eq. (10) that the irreducible product of the degree one of two sets of the degree one differs from the ordinary vector product by the factor $-1/\sqrt{2}$.

The same difference occurs in the complex representation and goes back to *Wigner's* (*22a*, p. 204; *22b*, p. 190) original condition, $<j_1\,j_1\;j_2\,j_2|j\,j_1-j_2>\;\geqq 0$, which is equivalent to Eq. (2). The sign difference is independent of the handedness of the coordinate system and would have disappeared if the signs of the irreducible products had been defined by $(-1)^{j_1+j_2+j}<j_1\,j_1\quad j_2\,j-j_1|jj>\;\geqq 0$ to be compared with Eq. (2). This definition would not have altered the relation between the $\bar{V}(j)$ and the 3-j but would have changed the phases of all odd symbols. However, in agreement with common usage, the convention of Eq. (2) will be maintained here.

2. The 3-l Symbols

Irreducible multiple products of several standard sets are constructed by applying Eq. (10) to two factors at a time. Generally, the associative rule of combination does not apply to multiple products which accordingly must be classified by the appropriate coupling scheme. It is of particular importance to consider irreducible multiple products of the degree zero, since their coefficients in an expansion by means of Eq. (10) have useful symmetry properties. When irreducible triple products of the degree zero are formed and expanded in terms of the constituent functions, the expansion coefficients are found to be related to the coupling coefficients of Eq. (11) in a simple way. The expansion coefficients, $\bar{V}(l)$ are analogous to the $\bar{V}(j)$ functions, and the 3-l symbols, which are the basic quantities in the applications, are related to $\bar{V}(l)$ in the same manner as the 3-j symbols are related to the $\bar{V}(j)$ functions.

With the three initial irreducible sets kept in the same order, two totally symmetric functions may be derived,

$$[[|l_1\} \times |l_2\}]^{l_{12}} \times |l_3\}]^0_\sigma \text{ and } [|l_1\} \times [|l_2\} \times |l_3\}]^{l_{23}}]^0_\sigma .$$

Provided that the triple products are non-vanishing, they differ at most by a sign since l_3 is contained just once among the values of l_{12} and similarly l_1 occurs once among l_{23}. Expansion of the two triple products in terms of the direct product functions yields

$$\begin{aligned}
&[[|l_1\} \times |l_2\}]^{l_{12}} \times |l_3\}]^0_\sigma \\
&= \sum_{t_1 t_2 t_3} \Big(\sum_{t_{12}} <l_1 t_1\ l_2 t_2 | l_{12} t_{12}> <l_{12} t_{12}\ l_3 t_3 | 0\, \sigma> \Big) |l_1 t_1>|l_2 t_2>|l_3 t_3> \\
&= \sum_{t_1 t_2 t_3} [l_3]^{-\frac{1}{2}} <l_1 t_1\ l_2 t_2 | l_3 t_3>\ \ |l_1 t_1>|l_2 t_2>|l_3 t_3> \qquad (14\text{a})
\end{aligned}$$

and, similarly,

$$\begin{aligned}
&[|l_1\} \times [|l_2\} \times |l_3\}]^{l_{23}}]^0_\sigma \\
&= \sum_{t_1 t_2 t_3} [l_1]^{-\frac{1}{2}} <l_2 t_2\ l_3 t_3 | l_1 t_1>\ \ |l_1 t_1>|l_2 t_2>|l_3 t_3> \qquad (14\text{b})
\end{aligned}$$

The abbreviations[2]) $[l_i] = (2l_i + 1)$ and $t = \lambda\varsigma$ have been used.

---

[2]) In general we shall use $[l_1, l_2, \ldots] = (2\,l_1 + 1)\,(2\,l_2 + 1)\,\ldots$

In the reduction use has been made of the coefficient

$$<l_1\, t_1\; l_2\, t_2 | 0\, \sigma> \; = + \, [l_1]^{-\frac{1}{2}} \, \delta(l_1\, l_2) \, \delta(t_1\, t_2) \tag{15}$$

which, except for a sign, can be derived from the fundamental theorem in the theory of irreducible representations (*10*) and the fact that the group considered is simply reducible (*23*). The positive sign derives from the expression, Eq. (11), in the form of Eq. (13). By means of Eq. (12) it can be verified that for all possible trios of irreducible representations the expansion coefficients of Eq. (14a) are identical to those of Eq. (14b). Further, an odd permutation of the initial sets changes the expansion coefficients by the factor $(-1)^{l_1+l_2+l_3}$.

Thus the triple product of the degree zero is unique and may be written without indication of coupling scheme. We define the $\bar{V}(l)$ functions as the expansion coefficients of the invariant triple product,

$$[|l_1\} \times |l_2\} \times |l_3\}]^0_\sigma = \sum_{t_1, t_2, t_3} \bar{V}\begin{pmatrix} l_1\, l_2\, l_3 \\ t_1\, t_2\, t_3 \end{pmatrix} |l_1\, t_1> |l_2\, t_2> |l_3\, t_3> \tag{16}$$

and have from Eq. (14),

$$\bar{V}\begin{pmatrix} l_1\, l_2\, l_3 \\ t_1\, t_2\, t_3 \end{pmatrix} = [l_3]^{-\frac{1}{2}} <l_1\, t_1\; l_2\, t_2 | l_3\, t_3> \tag{17}$$

The $\bar{V}(l)$ functions have the same behaviour under permutations of the columns as have the $\bar{V}(j)$ functions. They are invariant under even permutations and are multiplied by the factor $(-1)^{l_1+l_2+l_3}$ under odd permutations. In order to bring the present description into closer agreement with usual practice (*7, 24*), we define the **3-*l*** symbol. This symbol is different from the $\bar{V}(l)$ function of Eq. (17) by the phase $(-1)^{l_1+l_2+l_3}$,

$$\begin{pmatrix} l_1\, l_2\, l_3 \\ t_1\, t_2\, t_3 \end{pmatrix} = (-1)^{l_1+l_2+l_3} \, \bar{V}\begin{pmatrix} l_1\, l_2\, l_3 \\ t_1\, t_2\, t_3 \end{pmatrix} \tag{18}$$

As the $\bar{V}(l)$ derives from the construction of invariants, it is the more natural symbol. On the other hand, the **3-*l*** symbol is preferred in the applications since this corresponds to the generally accepted use of the **3-*j*** symbol (*vide infra*, Eqs. (27) and (28)).

The orthogonality properties of the Wigner coefficients, Eq. (10), carry over via Eqs. (17) and (18) to the **3-*l*** symbols in the form,

$$\sum_{t_1, t_2} \begin{pmatrix} l_1\, l_2\, l_3 \\ t_1\, t_2\, t_3 \end{pmatrix} \begin{pmatrix} l_1\, l_2\, l_3' \\ t_1\, t_2\, t_3' \end{pmatrix} = [l_3]^{-1} \, \delta(l_3\, l_3') \, \delta(t_3\, t_3') \, \delta(l_1\, l_2\, l_3) \tag{19a}$$

$$\sum_{l_3, t_3} [l_3] \begin{pmatrix} l_1 \, l_2 \, l_3 \\ t_1 \, t_2 \, t_3 \end{pmatrix} \begin{pmatrix} l_1 \, l_2 \, l_3 \\ t_1' \, t_2' \, t_3 \end{pmatrix} = \delta(t_1 \, t_1') \, \delta(t_2 \, t_2') \tag{19b}$$

$$\sum_{t_1, t_2, t_3} \begin{pmatrix} l_1 \, l_2 \, l_3 \\ t_1 \, t_2 \, t_3 \end{pmatrix}^2 = \delta(l_1 \, l_2 \, l_3) \tag{19c}$$

Eq. (19c) is simply the normalization property of the coefficients in the expansion, Eq. (16). The symbol $\delta(l_1 \, l_2 \, l_3)$ has the usual meaning (*4*) of being one or zero, depending on whether or not the triangular conditions for the angular momentum quantum numbers are fulfilled.

The values of the 3-l symbols may be found from the definitions, Eqs. (17), (18), (3), and Eq. (12),

$$\begin{pmatrix} l_1 & l_2 & l_3 \\ \lambda_1 \varsigma_1 & \lambda_2 \varsigma_2 & \lambda_3 \varsigma_3 \end{pmatrix} = i^{l_1 + l_2 - l_3} \, (\varsigma_1 \times \varsigma_2 \times \bar{\varsigma}_3) \, \frac{(-1)^{l_3 - \lambda_3}}{\sqrt{2}} \begin{pmatrix} l_1 & l_2 & l_3 \\ \lambda_1 & \lambda_2 & -\lambda_3 \end{pmatrix} \tag{20}$$

in cases where no σ-functions occur. The last symbol on the right-hand side is actually a 3-j symbol and may be looked up in tables (*15*). When $\lambda_1 = 0$, or $\lambda_2 = 0$, the $\sqrt{2}$ has to be removed from Eq. (20). Since one may use the permutation properties of the 3-l symbol, it is not necessary to complicate the formula, Eq. (20), by allowing for $\lambda_3 = 0$. It is interesting to note that for $\lambda_1 = \lambda_2 = \lambda_3 = 0$, the 3-$l$ symbol is positive,

$$\begin{pmatrix} l_1 \, l_2 \, l_3 \\ \sigma \, \sigma \, \sigma \end{pmatrix} \tag{21}$$

$$= \frac{\left(\frac{L}{2}\right)!}{\left(\frac{L}{2} - l_1\right)! \left(\frac{L}{2} - l_2\right)! \left(\frac{L}{2} - l_3\right)!} \sqrt{\frac{(L - 2\, l_1)! \, (L - 2\, l_2)! \, (L - 2\, l_3)!}{(L + 1)!}}$$

when $L = l_1 + l_2 + l_3$ is even, and vanishes for L odd.

3. The 6-l and 9-l Symbols

The definition of the invariant quantities, the 6-l and the 9-l symbols may be inferred from a consideration of the orthogonal recoupling transformation between the different coupling schemes of irreducible products of the same degree:

In virtue of Eq. (10) the connecting matrix between the two coupling schemes of the irreducible triple products indicated, is orthogonal and has elements given by

$$\begin{aligned} &< ((l_1\, l_2) l_{12}\, l_3)\; l \,|\, (l_1 (l_2\, l_3)\; l_{23}) l > \\ &= (-1)^{l_1+l_2+l_3+l} [l_{12}, l_{23}]^{\frac{1}{2}} \begin{pmatrix} l_1 & l_2 & l_{12} \\ l_3 & l & l_{23} \end{pmatrix} \end{aligned} \tag{22}$$

in terms of the 6-l symbol (see Appendix 3) defined by

$$\begin{pmatrix} l_1 & l_2 & l_{12} \\ l_3 & l & l_{23} \end{pmatrix} = \sum_{\text{All } t} \begin{pmatrix} l_1 & l_2 & l_{12} \\ t_1 & t_2 & t_{12} \end{pmatrix} \begin{pmatrix} l_1 & l & l_{23} \\ t_1 & t & t_{23} \end{pmatrix} \begin{pmatrix} l_2 & l_{23} & l_3 \\ t_2 & t_{23} & t_3 \end{pmatrix} \begin{pmatrix} l_{12} & l_3 & l \\ t_{12} & t_3 & t \end{pmatrix} \tag{23}$$

A similar calculation of an element of the orthogonal transformation matrix between irreducible quadruple products yields

$$\begin{aligned} &< ((l_1\, l_2) l_{12} (l_3\, l_4) l_{34})\; l \,|\, ((l_1\, l_3) l_{13} (l_2\, l_4) l_{24}) l > \\ &= [l_{12}, l_{34}, l_{13}, l_{24}]^{\frac{1}{2}} \begin{bmatrix} l_1 & l_2 & l_{12} \\ l_3 & l_4 & l_{34} \\ l_{13} & l_{24} & l \end{bmatrix} \end{aligned} \tag{24}$$

in terms of the 9-l symbol defined by

$$\begin{aligned} \begin{bmatrix} l_1 & l_2 & l_{12} \\ l_3 & l_4 & l_{34} \\ l_{13} & l_{24} & l \end{bmatrix} = \sum_{\text{All } t} & \begin{pmatrix} l_1 & l_2 & l_{12} \\ t_1 & t_2 & t_{12} \end{pmatrix} \begin{pmatrix} l_3 & l_4 & l_{34} \\ t_3 & t_4 & t_{34} \end{pmatrix} \begin{pmatrix} l_{13} & l_{24} & l \\ t_{13} & t_{24} & t \end{pmatrix} \\ \times & \begin{pmatrix} l_1 & l_3 & l_{13} \\ t_1 & t_3 & t_{13} \end{pmatrix} \begin{pmatrix} l_2 & l_4 & l_{24} \\ t_2 & t_4 & t_{24} \end{pmatrix} \begin{pmatrix} l_{12} & l_{34} & l \\ t_{12} & t_{34} & t \end{pmatrix} \end{aligned} \tag{25}$$

Once the definite sequence of all the l's occurring in the definitions of Eqs. (23) and (25) is established, the 6-l and the 9-l symbols are true invariants in the group, i.e. they do not depend on the form in which the irreducible representations are chosen. Since the present definitions follow generally accepted practice (*4, 15*), the 6-l symbol is identical to the $\overline{W}$ of *Racah* (*4*) and to the 6-j symbol of *Wigner* (*15*), and similarly the 9-l symbol is identical to the X of (*4*) and to the 9-j symbol of (*15*).

Due to the identity of the 6-l and the $\overline{W}$ functions and to the identity of the 9-l and the X functions, the symmetry properties and all algebraic relations which involve the $\overline{W}$ and the X functions are preserved; such relations have been presented elsewhere (*3, 4, 7, 15, 22*). In contrast to this, equations that contain 3-l symbols together with 6-l symbols appear in a simpler form (without phase factors), which may be derived as indicated in (*4*). As an example, a transscription of [(*4*), Eq. (11.19)] may be mentioned

$$\begin{pmatrix} l_1 \, l_2 \, l_{12} \\ l_3 \, l \;\; l_{23} \end{pmatrix} \begin{pmatrix} l_1 \, l_2 \, l_{12} \\ t_1 \, t_2 \, t_{12} \end{pmatrix} = \sum_{t\, t_{23}\, t_3} \begin{pmatrix} l_1 \, l \, l_{23} \\ t_1 \, t \, t_{23} \end{pmatrix} \begin{pmatrix} l_2 \, l_{23} \, l_3 \\ t_2 \, t_{23} \, t_3 \end{pmatrix} \begin{pmatrix} l_{12} \, l_3 \, l \\ t_{12} \, t_3 \, t \end{pmatrix} \tag{26}$$

This relation may be used as a recursion formula for the 3-l symbols in a manner indicated in Ref. (*3*, p. 42).

4. The Wigner-Eckart Theorem

The Wigner-Eckart theorem (*3, 4, 7, 15, 22, 24*) may be derived by means of the properties of the invariant triple product, Eq. (16). We adapt the definition

$$<\alpha_1 \, l_1 \, t_1 | \mathfrak{O}^{l_2}_{t_2} | \alpha_3 \, l_3 \, t_3> \; = \; <\alpha_1 \, l_1 \| \mathfrak{O}^{l_2} \| \alpha_3 \, l_3> \begin{pmatrix} l_1 \, l_2 \, l_3 \\ t_1 \, t_2 \, t_3 \end{pmatrix} \tag{27}$$

which we compare with the definitions in Refs. (*4*, p. 79) and (*15*, p. 6), written in their respective notations:

$$\begin{aligned} &<\alpha_1 \, j_1 \, m_1 | \mathfrak{O}^{[j_2]}_{m_2} | \alpha_3 \, j_3 \, m_3> \\ &= \; <\alpha_1 \, j_1 \| \mathfrak{O}^{[j_2]} \| \alpha_3 \, j_3>(-1)^{j_1-m_1} \, \overline{V} \begin{pmatrix} j_1 & j_3 & j_2 \\ -m_1 & m_3 & m_2 \end{pmatrix} \end{aligned} \tag{28a}$$

$$\begin{aligned} &<\alpha_1 \, j_1 \, m_1 | \mathfrak{O}^{j_2}_{m_2} | \alpha_3 \, j_3 \, m_3> \\ &= \; <\alpha_1 \, j_1 \| \mathfrak{O}^{j_2} \| \alpha_3 \, j_3> (-1)^{j_1-m_1} \begin{pmatrix} j_1 & j_2 & j_3 \\ -m_1 & m_2 & m_3 \end{pmatrix} \end{aligned} \tag{28b}$$

Since the left-hand sides are identical, it is seen that, due to the phase difference in the definition of the $\overline{V}(j)$ function and the 3-j symbol [analogously to Eq. (18)], the two reduced matrix elements in Eqs. (28a) and (28b) are identical. The operator set as well as the sets of functions are supposed to transform according to irreducible representations, and the labels α_1 and α_3 introduce extra parameters when necessary.

The reduced matrix of Eq. (27) is identical to those of Eqs. (28) provided that corresponding sets of functions and operators span the same space:

$$<\alpha_1 \, l_1 || \mathfrak{O}^{l_2} || \alpha_3 \, l_3> \; = + \; <\alpha_1 \, j_1 || \mathfrak{O}^{[j_2]} || \alpha_3 \, j_3> \tag{29}$$

This follows from an expansion of the left-hand side of Eq. (27),

$$\begin{aligned}
&<\alpha_1 l_1 t_1 | \mathfrak{O}^{l_2}_{t_2} | \alpha_3 l_3 t_3> \\
&= \sum_{m_1 m_2 m_3} <l_1 t_1 | j_1 m_1> <\alpha_1 j_1 m_1 | \mathfrak{O}^{[j_2]}_{m_2} | \alpha_3 j_3 m_3> <j_2 m_2 | l_2 t_2> \\
&\qquad \times <j_3 m_3 | l_3 t_3> \\
&= <\alpha_1 j_1 || \mathfrak{O}^{[j_2]} || \alpha_3 j_3> \frac{(-1)^{j_1+j_2+j_3}}{\sqrt{2 j_1 + 1}} \sum_{m_1 m_2 m_3} <l_1 t_1 | j_1 m_1> \\
&\qquad \times <j_1 m_1 | j_2 m_2 \; j_3 m_3> <j_2 m_2 | l_2 t_2> <j_3 m_3 | l_3 t_3> \\
&= <\alpha_1 j_1 || \mathfrak{O}^{[j_2]} || \alpha_3 j_3> \begin{pmatrix} l_1 \, l_2 \, l_3 \\ t_1 \, t_2 \, t_3 \end{pmatrix}
\end{aligned} \tag{30}$$

Here we have used the complex conjugate of Eq. (11) and the fact that the Wigner coefficients are real quantities in both representations. Then, by comparison with the right-hand side of Eq. (27) the identity of Eq. (29) follows.

5. Reduced Matrices of Irreducible Products of Operator Sets

Irreducible sets of operators are fully characterized by their reduced matrices and the function sets contained herein. This is seen by application of the identity operator on both sides of the operator $\mathfrak{O}^k_q$,

$$\begin{aligned}
\mathfrak{O}^k_q &= \sum_{\substack{\alpha l t \\ \alpha' l' t'}} |\alpha l t> <\alpha l t | \mathfrak{O}^k_q | \alpha' l' t'> <\alpha' l' t'| \\
&= \sum_{\alpha l \alpha' l'} <\alpha \, l || \mathfrak{O}^k || \alpha' \, l'> \sum_{t t'} \begin{pmatrix} l \, k \, l' \\ t \, q \, t' \end{pmatrix} |\alpha l t> <\alpha' l' t'|
\end{aligned} \tag{31}$$

where Eq. (27) has been used. By substitution of Eq. (10) in the form

$$[|\alpha l\} \times \{\alpha' l'|]^k_q = \sum_{t t'} \sqrt{2k+1} \, (-1)^{l+l'+k} \begin{pmatrix} l \, l' \, k \\ t \, t' \, q \end{pmatrix} |\alpha l t> <\alpha' l' t'| \tag{32}$$

the set of operators $\mathfrak{O}^k$ becomes

$$\mathfrak{O}^k = \sum_{\alpha l \alpha' l'} <\alpha l || \mathfrak{O}^k || \alpha' l'> [k]^{-\frac{1}{2}} [|\alpha l\} \times \{\alpha' l'|]^k \tag{33}$$

This equation differs from the analogous one, Eq. (14.17) in Ref. (*4*), since with the real bases the transformation from a set to the contragredient one is effected by the unit matrix and this is left out of Eq. (33).

We consider now a system which consists of two independent sub-systems, denoted by the sub-indices 1 and 2. In the sub-systems sets of irreducible tensor operators $\mathfrak{P}^{k_1}$ and $\mathfrak{Q}^{k_2}$ are characterized by their reduced matrices $<\alpha_1 l_1 || \mathfrak{P}^{k_1} || \alpha_1' l_1'>$ and $<\alpha_2 l_2 || \mathfrak{Q}^{k_2} || \alpha_2' l_2'>$ of the eigenstates $|\alpha_1 l_1\}$ and $|\alpha_2 l_2\}$, respectively. Therefore, use of Eq. (10) generates the irreducible product of the two operator sets, $[\mathfrak{P}^{k_1} \times \mathfrak{Q}^{k_2}]^k$. By means of Eq. (33) the irreducible product of the operator sets is also represented by the reduced matrix of the eigenstates $[|\alpha_1 l_1\} \times |\alpha_2 l_2\}]^l$ defined in the combined system. Then, by equating the two expressions and using the recoupling transformation, Eq. (24), the reduced matrix desired is

$$
\begin{aligned}
&<\alpha_1 l_1 \alpha_2 l_2 l || [\mathfrak{P}^{k_1} \times \mathfrak{Q}^{k_2}]^k || \alpha_1' l_1' \alpha_2' l_2' l'> \\
&= <\alpha_1 l_1 || \mathfrak{P}^{k_1} || \alpha_1' l_1'> <\alpha_2 l_2 || \mathfrak{Q}^{k_2} || \alpha_2' l_2'> [l, l', k]^{\frac{1}{2}} \begin{bmatrix} l_1 & l_1' & k_1 \\ l_2 & l_2' & k_2 \\ l & l' & k \end{bmatrix} \qquad (34)
\end{aligned}
$$

in terms of the reduced matrices of the separate sets (*4*, Eq. (15.4)).

From this relation and by means of the known properties (*15*, Eq. (3.19)) of the 9-l symbol, important relations occur in which either one of the operators is the unit operator 1. We shall make extended use of the case where $\mathfrak{P}^{k_1}_{q_1} = 1^0_\sigma$,

$$
\begin{aligned}
&<\alpha_1 l_1 \alpha_2 l_2 l || [1^0 \times \mathfrak{Q}^{k_2}]^k || \alpha_1 l_1 \alpha_2' l_2' l'> \\
&= <\alpha_2 l_2 || \mathfrak{Q}^{k_2} || \alpha_2' l_2'> [l, l']^{\frac{1}{2}} (-1)^{l_1 + l_2' + l + k_2} \begin{pmatrix} l & l' & k_2 \\ l_2' & l_2 & l_1 \end{pmatrix} \delta(k_2 k) \qquad (35)
\end{aligned}
$$

Also the matrix element of the scalar product which is proportional to the irreducible product of the degree zero,

$$
\mathfrak{P}^k \cdot \mathfrak{Q}^k = [k]^{\frac{1}{2}} [\mathfrak{P}^k \times \mathfrak{Q}^k]^0_\sigma \qquad (36)
$$

will be used. Again from Eq. (34) one obtains

$$
\begin{aligned}
&<\alpha_1 l_1 \alpha_2 l_2 l t | \mathfrak{P}^k \cdot \mathfrak{Q}^k | \alpha_1' l_1' \alpha_2' l_2' l' t'> \\
&= (-1)^{l_1' + l_2 + l + k} <\alpha_1 l_1 || \mathfrak{P}^k || \alpha_1' l_1'> <\alpha_2 l_2 || \mathfrak{Q}^k || \alpha_2' l_2'> \begin{pmatrix} l_1 & l_1' & k \\ l_2' & l_2 & l \end{pmatrix} \\
&\times \delta(l l') \, \delta(t t') \qquad (37)
\end{aligned}
$$

When the two operator sets operate on the same system there is only one kind of eigenstate $|\alpha\, l\}$ and one must take into account the orthogonality of these eigenstates. One derives

$$
\begin{aligned}
&<\alpha\, l||[\mathfrak{P}^{k_1} \times \mathfrak{Q}^{k_2}]^k||\alpha'\, l'> \\
&= (-1)^{l+l'+k}\, [k]^{\frac{1}{2}} \sum_{\alpha'' l''} <\alpha\, l||\mathfrak{P}^{k_1}||\alpha''\, l''> <\alpha''\, l''||\mathfrak{Q}^{k_2}||\alpha'\, l'> \\
&\times \begin{pmatrix} l & l' & k \\ k_2 & k_1 & l'' \end{pmatrix}
\end{aligned}
\tag{38}
$$

and similarly the analog of Eq. (37) is

$$
\begin{aligned}
&<\alpha\, l\, t|\,\mathfrak{P}^k \cdot \mathfrak{Q}^k\,|\alpha'\, l'\, t'> \\
&= \delta(l\, l')\, \delta(t\, t') \sum_{\alpha'' l''} \frac{(-1)^{l''+l+k}}{2\, l+1} <\alpha\, l||\mathfrak{P}^k||\alpha''\, l''> <\alpha''\, l''||\mathfrak{Q}^k||\alpha'\, l'>
\end{aligned}
\tag{39}
$$

Judd (*7*, p. 73) has shown how one may work back from Eq. (38) to Eq. (34).

6. Irreducible Sets of Spherical Harmonic Operators

A few examples of the application of the 3-l symbols to calculations with real spherical harmonics will be shown. Frequently surface harmonics occur that are normalized like Legendre polynomials; such harmonics will be denoted by the letter $\mathfrak{C}$,

$$\mathfrak{C}^l_t = \sqrt{\frac{4\pi}{2\, l+1}}\, \mathfrak{Z}^l_t \tag{40a}$$

and the corresponding solid harmonic denoted by $\mathfrak{H}$ is

$$\mathfrak{H}^l_t = r^l\, \mathfrak{C}^l_t \tag{40b}$$

When harmonics act as operators the labels $lt = l\lambda\varsigma$ will be replaced by $kq = k\varkappa\varsigma$. The reduced matrix of a $\mathfrak{C}^k$ operator may be deduced from the value in the complex standard representation (*4*, Eq. (14.12)) by use of Eqs. (29) and (20),

$$<l||\mathfrak{C}^k||l'> = \sqrt{[l, l']} \begin{pmatrix} l & k & l' \\ \sigma & \sigma & \sigma \end{pmatrix} \tag{41}$$

where the 3-l symbol is given by Eq. (21).

Important examples of these expressions are the calculations of matrix elements of operators which are expanded by means of the spherical harmonic addition theorem (*6*, p. 141)

$$\mathsf{P}^k(\cos\omega) = \sum_q \mathfrak{C}_q^k(\theta_1\,\varphi_1)\,\mathfrak{C}_q^k(\theta_2\,\varphi_2) \tag{42}$$

Here the left-hand side is the Legendre polynomial of the degree k in the cosine to the angular distance $\omega = \text{Arccos}(\cos\theta_1\cos\theta_2 + \sin\theta_1 \sin\theta_2\cos(\varphi_1 - \varphi_2))$ between two points with the coordinates $(\theta_1\,\varphi_1)$ and $(\theta_2\,\varphi_2)$.

Although Eq. (41) is derived for surface harmonics the application may be extended to solid harmonics as well. Thus the product of two solid harmonics is derived in a way analogous to Eq. (31)

$$(r^k\,\mathfrak{C}_q^k)\,r^{l'}\,\mathfrak{Z}_{t'}^{l'} = r^{k+l'}\sum_{lt}\mathfrak{Z}_t^l\sqrt{[l,l']}\begin{pmatrix} l & k & l' \\ \sigma & \sigma & \sigma\end{pmatrix}\begin{pmatrix} l & k & l' \\ t & q & t'\end{pmatrix} \tag{43}$$

For $r = 1$ Eq. (43) reduces to the coupling rule for surface harmonics (e.g. *5*, Appendix 2, Table A 21). If Eq. (43) is multiplied by $r^l\,\mathfrak{Z}_t^l$ and integrated over the surface of the unit sphere, one has

$$\begin{aligned}\iiint_{r^2=1}(r^l\,\mathfrak{Z}_t^l)\,\mathfrak{H}_q^k\,(r^{l'}\,\mathfrak{Z}_{t'}^{l'})\,dx\,dy\,dz &= \langle l\|\mathfrak{H}^k\|l'\rangle\begin{pmatrix} l & k & l' \\ t & q & t'\end{pmatrix}\\ &= \langle l\|\mathfrak{C}^k\|l'\rangle\begin{pmatrix} l & k & l' \\ t & q & t'\end{pmatrix}\end{aligned} \tag{44}$$

In the following we shall always understand integrals of solid harmonics to be performed over the surface of the unit sphere.

A different kind of operator is obtained if the variables (x, y, z) of the solid harmonic $\mathfrak{H}_q^k = r^k\,\mathfrak{C}_q^k$ are replaced by $\left(\frac{\partial}{\partial x}, \frac{\partial}{\partial y}, \frac{\partial}{\partial z}\right)$. By this substitution a spherical harmonic differential operator is generated. This operator will be denoted $\mathfrak{D}_q^k$ to indicate that it is an element of an irreducible tensorial set (*25*, *20*, *21*).

When the differential operator $\mathfrak{D}_q^k$ operates on a solid harmonic $r^l\,\mathfrak{Z}_t^l$, Eq. (5a) will yield the operator in a particularly simple form. Since the solid harmonic is subject to the Laplace equation $\nabla^2(r^l\mathfrak{Z}_t^l) = 0$, all but the first term can be excluded from the expression of the operator. Having decided on the analytical form of $\mathfrak{D}_q^k$, the reduced matrix will be deduced. With the solid harmonic in the form Eq. (5b), one has

$$\frac{\partial^k}{\partial z^k}(r^l\,\mathfrak{Z}_{\lambda\varsigma}^l) = \sqrt{\frac{(l+\lambda)!\quad(l-\lambda)!\quad(2l+1)}{(l+\lambda-k)!\,(l-\lambda-k)!\,[2(l-k)+1]}}\,(r^{l-k}\,\mathfrak{Z}_{\lambda\varsigma}^{l-k}) \tag{45}$$

for $k \leqq l - \lambda$. The expression is zero for $k > l - \lambda$. It then follows that

$$\mathfrak{D}_{\sigma}^{k} (r^l \mathfrak{Z}_{\lambda\varsigma}^{l}) = \frac{(2\,k)\,!}{2^k\,(k\,!)^2}\,\frac{\partial^k}{\partial z^k}\,(r^l \mathfrak{Z}_{\lambda\varsigma}^{l})$$
$$= \frac{(2\,k)\,!}{2^k\,k\,!} \begin{pmatrix} l+\lambda \\ k \end{pmatrix}^{\frac{1}{2}} \begin{pmatrix} l-\lambda \\ k \end{pmatrix}^{\frac{1}{2}} \left(\frac{2\,l+1}{2\,(l-k)+1}\right)^{\frac{1}{2}} (r^{l-k}\,\mathfrak{Z}_{\lambda\varsigma}^{l-k}) \qquad (46)$$

If both sides of this equation are multiplied by $r^{l-k}\,\mathfrak{Z}_{\lambda\varsigma}^{l-k}$ and integrated over the surface of the unit sphere, one can use the Wigner-Eckart theorem and Eq. (21) to obtain:

$$<(l-k)\,\sigma|\,\mathfrak{D}_{\sigma}^{k}\,|l\,\sigma>$$
$$= <(l-k)\,||\,\mathfrak{D}^k\,||l> \begin{pmatrix} l \\ k \end{pmatrix} \left(\frac{(2\,k)\,!\,[2\,(l-k)]\,!}{(2\,l+1)\,!}\right)^{\frac{1}{2}} \qquad (47)$$
$$= \frac{(2k)\,!}{2^k\,k\,!} \begin{pmatrix} l \\ k \end{pmatrix} \left(\frac{2\,l+1}{2\,(l-k)+1}\right)^{\frac{1}{2}}$$

Hence, the reduced matrix of $\mathfrak{D}^k$ is

$$<(l-k)\,||\,\mathfrak{D}^k\,||l> = \frac{(2\,k)\,!}{2^k\,k\,!} \begin{pmatrix} 2\,l+1 \\ 2\,k \end{pmatrix}^{\frac{1}{2}} \sqrt{2\,l+1} \qquad (48)$$

The reduced matrix can be used to write down the result of the action of the differential operator $\mathfrak{D}_q^k$ on the solid harmonic $r^l\,\mathfrak{Z}_t^l$,

$$\mathfrak{D}_q^k (r^l\,\mathfrak{Z}_t^l) = \sum_{t'} (r^{l-k}\,\mathfrak{Z}_{t'}^{l-k}) \sqrt{[l]}\,\frac{(2\,k)\,!}{2^k\,k\,!} \begin{pmatrix} 2\,l+1 \\ 2\,k \end{pmatrix}^{\frac{1}{2}} \begin{pmatrix} l-k & k & l \\ t' & q & t \end{pmatrix}, \; k \leqq l \qquad (49)$$

The compact form of this result is obtained by the explicit use of the 3-l symbol; for comparison we refer to *Hobson* [(6), Sect. 88] where three expressions cover the particular case of Eq. (49) where all the ς are sigmas or cosines.

The properties of the differential operator $\mathfrak{D}_q^k$ have been derived by means of solid harmonics. However, in virtue of Eq. (33) one can define an operator set $\mathfrak{D}^k$ through the reduced matrix, Eq. (48),

$$\mathfrak{D}^k = \sum_{\alpha'\alpha l} <\alpha'\,(l-k)\,||\mathfrak{D}^k||\,\alpha l>[k]^{-\frac{1}{2}}\,[\,|\alpha'\,(l-k)\} \times \{\alpha l\,|\,]^k \qquad (50)$$

This set operates on the basis functions of the three-dimensional rotation group and, when the functions are represented by surface harmonics, any matrix element has the same value as would be calculated by operating on solid harmonics and reducing to surface harmonics before the integration.

The set of operators $\mathfrak{C}^k$ and $\mathfrak{D}^k$, both defined by their reduced matrices, Eqs. (41) and (48), respectively, may be combined according to the rules, Eqs. (38) and (39). This leads to an important type of operator generally denoted $[\mathfrak{C}^{k_1} \times \mathfrak{D}^{k_2}]^k$. One special case of this operator $(k_1 = k_2\ ;\ k = 0)$ has been utilized in the derivation of analytical expressions *(21)* of the operator of the Angular Overlap Model. Another special case has $k_1 = k_2 = k = 1$. By means of the tables in appendix 2 one obtains,

$$[\mathfrak{C}^1 \times \mathfrak{D}^1]^1_\sigma = (-1)\sqrt{3}\begin{pmatrix} 1 & 1 & 1 \\ 1c & 1s & \sigma \end{pmatrix} \mathfrak{C}^1_{1c}\,\mathfrak{D}^1_{1s} + (-1)\sqrt{3}\begin{pmatrix} 1 & 1 & 1 \\ 1s & 1c & \sigma \end{pmatrix} \mathfrak{C}^1_{1s}\,\mathfrak{D}^1_{1c}$$

$$= -\frac{1}{\sqrt{2}}\, x \frac{\partial}{\partial y} + \frac{1}{\sqrt{2}}\, y \frac{\partial}{\partial x} \tag{51}$$

The reduced matrix of this operator follows from a direct calculation, e.g.

$$\langle \alpha 21s | [\mathfrak{C}^1 \times \mathfrak{D}^1]^1_\sigma | \alpha 21c \rangle = \langle \alpha 2 \| [\mathfrak{C}^1 \times \mathfrak{D}^1]^1 \| \alpha 2 \rangle \frac{1}{\sqrt{30}}$$

$$= \iiint\limits_{r^2=1} \sqrt{\frac{15}{4\pi}}\, yz \left(-\frac{1}{\sqrt{2}}\, x \frac{\partial}{\partial y} + \frac{1}{\sqrt{2}}\, y \frac{\partial}{\partial x}\right) \sqrt{\frac{15}{4\pi}}\, zx\, dx\, dy\, dz = \frac{1}{\sqrt{2}}$$

$$\Rightarrow \langle \alpha 2 \| [\mathfrak{C}^1 \times \mathfrak{D}^1]^1 \| \alpha 2 \rangle = \sqrt{3 \cdot 5} \tag{52}$$

in agreement with the general expression, Eq. (54). It is noted that the sigma component of the angular momentum operator, $\mathfrak{L}^1$, is proportional to $[\mathfrak{C}^1 \times \mathfrak{D}^1]^l_\sigma$,

$$\mathfrak{L}^1_\sigma = -i\left(x \frac{\partial}{\partial y} - y \frac{\partial}{\partial x}\right) = i\sqrt{2}[\mathfrak{C}^1 \times \mathfrak{D}^1]^1_\sigma \tag{53a}$$

and in general

$$\mathfrak{L}^1 = i\sqrt{2}[\mathfrak{C}^1 \times \mathfrak{D}^1]^1 \tag{53b}$$

Eq. (53b) can be used to derive the reduced matrix of the angular momentum operator. For $k_1 = k_2 = k = 1$, Eq. (38) gives

$$\begin{aligned} &\langle l' \| [\mathfrak{C}^1 \times \mathfrak{D}^1]^1 \| l \rangle \\ &= (-1)^{2l+1}\sqrt{3}\, \langle l \| \mathfrak{C}^1 \| l-1 \rangle \langle l-1 \| \mathfrak{D}^1 \| l \rangle \\ &\times \begin{pmatrix} l & l & 1 \\ 1 & 1 & (l-1) \end{pmatrix} \\ &= \frac{1}{\sqrt{2}} \sqrt{l(l+1)(2l+1)}\, \delta(l'l) \end{aligned} \tag{54}$$

To obtain the last equality the closed expression for the 6-l symbol (*15*, Eq. (2.23))

$$\begin{pmatrix} l_1 & l_2 & l_3 \\ 1 & l_3-1 & l_2 \end{pmatrix} = (-1)^{\mathrm{L}} \left[\frac{\mathrm{L}\,(\mathrm{L}+1)\,(\mathrm{L}-2l_1)\,(\mathrm{L}-2l_2)\,(\mathrm{L}-2l_3+1)}{2l_2\,(2l_2+1)\,(2l_2+2)\,(2l_3-1)\,2l_3\,(2l_3+1)}\right]^{\frac{1}{2}},$$

$$\mathrm{L} = l_1 + l_2 + l_3$$

together with Eq. (21) have been used. From Eqs. (53b) and (54) the reduced matrix of $\mathfrak{L}^1$ follows,

$$<l||\mathfrak{L}^1||l> = i\sqrt{l(l+1)\,(2\,l+1)} \tag{55}$$

Eq. (55) has previously been derived by other means, e.g. (*4*, Eq. (14.9)).

In the discussion of Eq. (5) it was mentioned that the surface harmonics, $\mathfrak{Z}^l_{\lambda\varsigma}$, form the bases for the irreducible representations of the three-dimensional rotation group, R_3, on real standard form. In the rotation-inversion group, R_{3i}, which is R_3 augmented with the inversion operation I, surface harmonics of even degree form bases for the gerade representations, whereas surface harmonics of odd degree form bases for the ungerade representations. Thus, irreducible tensorial sets of the types $|\alpha l g\}$, l odd, and $|\alpha l\, u\}$, l even, are not found among the ordinary spherical harmonics. However, the irreducible set $[\mathfrak{C}^1 \times \mathfrak{D}^1]^1$ of Eq. (51) is an example of an operator belonging to these representations since it must be written as $[\mathfrak{C}^{1u} \times \mathfrak{D}^{1u}]^{1g}$ in R_{3i}.

To examine this example somewhat more closely, we form the direct product of the two irreducible sets $|\alpha_1\, l\}$ and $|\alpha_2\, l\}$ which transform in identically the same way in R_3,

$$|\alpha_1\, l\} \times |\alpha_2\, l\} = |\,\alpha(2\,l)\} + |\alpha\,(2\,l-1)\} + \cdots + |\alpha\,1\} + |\alpha 0\} \tag{56}$$

Here, the symmetrized direct product $[l^2]$ consists of the terms in the direct sum which are characterized by an even l, whereas the antisymmetrized direct product (l^2) consists of those characterized by an odd l. If the initial sets are surface harmonics, $\mathfrak{C}^l$ say, then $\alpha_1 = \alpha_2$ and all terms belonging to (l^2) vanish in agreement with Eq. (43). On the other hand, if the first set is of $\mathfrak{C}^k$ type and the other of $\mathfrak{D}^k$ type, the sets are distinguishable even though they operate in the same *xyz* space. Therefore one can form operators of odd degree which are gerade in the group R_{3i}.

Similarly, when the initial sets differ in l, the number of terms in the direct sum is generally determined by the triangular conditions. However, when surface harmonics are coupled, the rule, Eq. (43) with

$r^2 = 1$, applies, and every other term vanishes. This does not happen for the product $[\mathfrak{C}^{k_1} \times \mathfrak{D}^{k_2}]^k$ which therefore can account for operators of even degree which are ungerade in R_{3i}.

Further, it follows from the Wigner-Eckart theorem that two tensorial sets of operators of the same degree, $\mathfrak{P}^k$ and $\mathfrak{Q}^k$ have their matrix elements proportional to one another,

$$<\alpha l t | \mathfrak{P}^k_q | \alpha' l' t'> = \frac{<\alpha l \| \mathfrak{P}^k \| \alpha' l'>}{<\beta l \| \mathfrak{Q}^k \| \beta' l'>} <\beta l t | \mathfrak{Q}^k_q | \beta' l' t'> \tag{57}$$

By use of this relation one can express any tensorial operator desired in terms of the operators $[\mathfrak{C}^{k_1} \times \mathfrak{D}^{k_2}]^k_q$. As an example, we use appendix 2 to form two of the operators that are equivalent to $\mathfrak{B}^{2u}_{1c}$ (*25*, Eq. (84)) belonging to the tensorial set $\mathfrak{B}^{2u}$ of R_{3i}.

$$\begin{aligned} &<\alpha l t | \mathfrak{B}^{2u}_{1c} | \alpha' l' t'> \\ &= \frac{\delta(l, 2u, l')}{<\beta l \| [\mathfrak{C}^{2g} \times \mathfrak{D}^{1u}]^{2u} \| \beta' l'>} \\ &\quad \times <\beta l t | \frac{1}{\sqrt{2}} (z^2 - x^2) \frac{\partial}{\partial y} - \frac{1}{\sqrt{2}} y \left(z \frac{\partial}{\partial z} - x \frac{\partial}{\partial x} \right) | \beta' l' t'> \\ &= \frac{\delta(l, 2u, l')}{<\beta l \| [\mathfrak{C}^{1u} \times \mathfrak{D}^{2g}]^{2u} \| \beta' l'>} \\ &\quad \times <\beta l t | \frac{1}{\sqrt{2}} \left(z \frac{\partial}{\partial z} - x \frac{\partial}{\partial x} \right) \frac{\partial}{\partial y} - \frac{1}{\sqrt{2}} y \left(\frac{\partial^2}{\partial z^2} - \frac{\partial^2}{\partial x^2} \right) | \beta' l' t'> \end{aligned} \tag{58}$$

II. Application to Ligand-Field Matrices

1. The Ligand-Field Model

The ligand-field model is a semi-empirical scheme of calculation that deals with the interpretation of the properties of complexes of metal ions with a partly filled l-shell. In this model the experiments are described in terms of empirically determined parameters whose coefficients are exactly calculated.

Griffith has presented a detailed analysis of the ligand-field model in his monumental monograph, Ref. (*5*). In particular, he has discussed the fundamental problems associated with the definition of the ligand-field operator (*loc. cit.* Chapt. 7, § 2 and 3). He actually uses the expanded radial function model (*8*; *9*, p. 56) where the number of electrons is the same as in the free metal ion and where the basis set $\mathfrak{Z}^l = |\alpha l\}$ of orbitals constitutes an irreducible tensorial set in spherical symmetry. The expanded radial function model is the Slater-Condon-Shortley theory[3]) of atomic spectra to which a first-order perturbation accounting for the effect of the ligands is superimposed. Therefore all concepts of the theory of atomic spectra are relevant and Racah's methods based upon irreducible tensorial sets can be applied with great advantage.

It follows from these remarks that the main structure of the energy pattern is accounted for by three independent first-order perturbations on the l^n-configuration. They are: The interelectronic repulsion which is represented by the two-electron operator

$$\sum_{i<j}^{n} 1/r_{ij} ,$$

the spin-orbit interaction represented by the one-electron operator

$$\sum_{i=1}^{n} \xi(r_i)\, \mathfrak{s}^{[1]}(i) \cdot \mathfrak{l}^{[1]}(i) ,$$

and the effects of the ligands which are represented by the one-electron operator

$$\sum_{i=1}^{n} \mathsf{v}(i)$$

to be discussed in the next section.

In the present paper the so-called weak-field coupling scheme is adopted, *i.e.* the matrices of the above-mentioned three perturbation operators are constructed using function bases SL (Russell-Saunders case). These bases may be specified either in the M_SM_L or the JM scheme. In the second scheme the spin-orbit coupling is diagonal in the JM quantum numbers. In the first scheme real functions characterized by $\lambda\varsigma$ may, instead of the M_L functions, be taken as the components of L, and in the second scheme $\lambda\varsigma$, instead of M, as components of J. However, in the latter case this is only possible when J is an integer *i.e.* when the system has an even number of electrons.

The real scheme often gives the matrix of the ligand field operator a smaller number of non-diagonal elements than does the complex one.

[3]) The Slater-Condon-Shortley theory is a first-order perturbation treatment based upon irreducible tensorial sets of l-orbitals (*2*).

It should be noted that the four different forms of matrices involving the three perturbations will have the same eigenvalues which would also be the same as the corresponding matrices in a strong-field coupling scheme. In this case, at least part of the ligand field matrix is diagonalized first so that the diagonals correspond to functions $S\Gamma$ where Γ is an irreducible representation of some sub-group of R_{3i}. The strong-field coupling scheme will not be considered in the present paper.

2. The Ligand-Field Operator and Its One-electron Matrix Elements

The ligand-field operator is a Hamiltonian containing the kinetic energy of the electrons plus three types of potential energy terms, the nuclear-nuclear and electron-electron repulsion terms and the electron-nuclear attraction terms. It may be defined as the Hamiltonian of the central-ion plus ligand system minus that of the central-ion system itself. It may be written as $\mathsf{H}_1 - \mathsf{H}_2$,

$$\mathsf{H}_1 = \sum_{\alpha=\sigma} \frac{\mathsf{p}_\alpha^2}{2\,\mathrm{M}_\alpha} + \frac{1}{2} \sum_{(\alpha,\beta)=(\sigma,\sigma)}' \frac{Z_\alpha Z_\beta}{R_{\alpha\beta}} + \sum_{i=s} \frac{\mathsf{p}_i^2}{2\,m_i} - \sum_{(\alpha,i)=(\sigma,s)} \frac{Z_\alpha}{r_{\alpha i}} + \frac{1}{2} \sum_{(i,j)=(s,s)}' \frac{1}{r_{ij}}$$

$$\mathsf{H}_2 = \sum_{i=k,l} \frac{\mathsf{p}_i^2}{2\,m_i} - \sum_{(\alpha,i)=(\varkappa,k),(\varkappa,l)} \frac{Z_\alpha}{r_{\alpha i}} + \frac{1}{2} \sum_{(i,j)=(k,k),(k,l),(l,l)}' \frac{1}{r_{ij}}$$

$$\mathsf{H}_1 - \mathsf{H}_2 = \sum_{\alpha=\sigma} \frac{\mathsf{p}_\alpha^2}{2\,\mathrm{M}_\alpha} + \frac{1}{2} \sum_{(\alpha,\beta)=(\sigma,\sigma)}' \frac{Z_\alpha Z_\beta}{R_{\alpha\beta}} + \sum_{i=n} \frac{\mathsf{p}_i^2}{2\,m_i} - \sum_{(\alpha,i)=(\varkappa,n),(\nu,s)} \frac{Z_\alpha}{r_{\alpha i}} + \frac{1}{2} \sum_{(i,j)=(n,s)}' \frac{1}{r_{ij}} \tag{59}$$

where Greek letters refer to nuclei and Latin letters to electrons. σ is a collective symbol for $\varkappa$ and ν, referring to central ion and ligands, respectively. Similarly, s is a collective symbol for k, l, and n, referring to the filled shells of the central ion, the partly filled shells of the central ion, and the filled shells of the ligands, respectively. In general the nuclei are assumed fixed so that their kinetic energy term vanishes and the energies become vertical energies[4]) corresponding to Franck-Condon transitions.

Our problem is to evaluate matrix elements of $H_2 + (H_1 - H_2)$. In the application of the operator H_2 by the methods of *Condon* and *Shortley* (2) this operator is reduced to some constant energy terms for the l^n-configuration plus the two-electron operator

$$\frac{1}{2} \sum_{(i,j)=(l,l)}' \frac{1}{r_{ij}}$$

acting within this configuration. Similarly, as discussed by *Griffith* (5) the operator $H_1 - H_2$ may be reduced to some constant terms within the l^n configuration plus a one-electron operator acting within this configuration. This one-electron operator may be expanded onto the spherical harmonics,

$$V = \sum_{i=1}^{n} v(i)$$

$$v(i) = v(r_i, \theta_i, \varphi_i) = \sum_{kq} \mathfrak{A}_q^k \, \mathfrak{H}_q^k \tag{60}$$

The vanishing of the potential parameters $\mathfrak{A}_q^k$ is determined by the symmetry of the environments and the position of the coordinate system. In fact, only the terms of Eq. (60), which contain the totally symmetric representation of the molecular symmetry group, will be non-vanishing. The expansion coefficients $\mathfrak{A}_q^k$ are essentially the empirical ligand-field parameters to be determined by experiment. This interpretation is clarified by consideration of a matrix element[5]) of $v(i)$ within the set $|\alpha l\}$,

$$\begin{aligned} &<\alpha l u|v(r_i, \theta_i, \varphi_i)|\alpha l v> \\ &= \sum_{kq} \mathfrak{A}_q^k <\mathrm{R}(\alpha l)|r^k|\mathrm{R}(\alpha l)> <\alpha l||\mathfrak{C}^k||\alpha l> \begin{pmatrix} l & k & l \\ u & q & v \end{pmatrix} \\ &= \sum_{kq} \mathrm{E}_q^k \begin{pmatrix} l & k & l \\ u & q & v \end{pmatrix} \end{aligned} \tag{61}$$

---

4) The empirical ligand-field parameters derived from absorption and emission spectra are different. This is a consequence of the fact that the model is concerned with vertical energies so that the same molecule is represented by two different ligand-field operators in the two cases. Similarly, in absorption spectra from excited states the ligand-field operators will vary with the state from which excitation takes place.

5) Instead of the labels $|\alpha l t>$, $|\alpha l t'>$ we shall use $|\alpha l u>$, $|\alpha l v>$ here in order to conform with previous notation (*19*).

The expanded radial functions $\mathrm{R}(\alpha l)$ determine the radial integral $\langle \mathrm{R}(\alpha l)|r^k|\mathrm{R}(\alpha l)\rangle$ which is independent of q. The reduced element of $\mathfrak{C}^k$ is given by Eq. (41) and the coefficient $\mathfrak{A}_q^k$ is a parameter whose value may vary with q.

The three constants are collectively replaced by the symbol E_q^k. The E_q^k parameters are not true reduced matrix elements since q as well as k are necessary for their characterization; they will be termed "reduced ligand-field parameters". The factor $\begin{pmatrix} l & k & l \\ \sigma & \sigma & \sigma \end{pmatrix}$, which is implicit in the E_q^k parameters through Eq. (41), prevents any contribution from odd k-values. The same selection rule is also obtained by utilizing the parity of the basis functions $\mathfrak{Z}^l = |\alpha l\}$. Thus at most

$$\sum_{k \text{ even}}^{2l} (2k+1) = (2l+1)(l+1)$$

reduced ligand-field parameters occur and this happens only when the complex considered belongs to one of the symmetry point groups C_1 or S_2. Because of the well-defined parity of the basis functions, the number of parameters associated with a given point group G is smaller than or equal to that required by symmetry for the direct product group $G \times S_2$ [holohedrized symmetry, (*17*) and (*20*)].

The reduced ligand-field parameters will now be expressed in terms of the Angular Overlap Model (AOM) parameters (*17, 18, 19, 20, 21*). The latter parameters are taken to represent the weak covalent antibonding effect of the ligands on the partly filled l-shell of the metal ion and they are proportional to the square of the angular overlap integrals between ligand and metal orbitals.

A detailed description of AOM and its assumptions is given in Ref. (*19*, p. 82—83). Here we shall point out that the ideas of the AOM are essentially based upon the properties of the real[6]) orbitals $\mathfrak{Z}^l$, Eq. (5), which form the bases of the group $C_{\infty v}$. The AOM-

[6]) In Ref. (*19*) the perturbation from a single ligand was considered first; next, it was assumed that the perturbations from different ligands were additive. When dealing with a single ligand, the symmetric ligand-field perturbation matrix can be made diagonal by a orthogonal matrix which represents a pure rotation in the three-dimensional space. The elements of this diagonal matrix are called e_t, with $t = \sigma, \pi s, \pi c, \ldots, \lambda s, \lambda c, \ldots, ls, lc$. The rotation matrix is assumed (*19*) to correspond to the transformation of the basis functions into a set, whose zonal harmonics (σ-functions) have the metal-ligand axis as symmetry axis. The same interpretation is not possible in the complex standard basis when non-linearly ligating ligands are present. Here, the Hermitian matrix that describes the perturbation from a single ligand can be diagonalized by a unitary matrix. The result is the same diagonal matrix as previously mentioned, but in this case the unitary matrix is not a matrix representative of the three-dimensional rotation group.

matrix element of the ligand-field operator in the $|\alpha l\}$ basis has the form (*19*),

$$\begin{aligned}&<\alpha l u|\,V(r_i\,\theta_i\,\varphi_i)\,|\alpha l v>\\&=\sum_{j=1}^{N}\sum_{t}\boldsymbol{D}^l_{ut}\,(\varphi_j\,\theta_j\,\psi_j)\,\boldsymbol{D}^l_{vt}\,(\varphi_j\,\theta_j\,\psi_j)\,e_t(j)\end{aligned} \tag{62}$$

Here, the matrix $\boldsymbol{D}^l(\varphi_j\,\theta_j\,\psi_j)$ is the real representative of the three-dimensional rotation group, as previously mentioned[6]). $(\varphi_j\,\theta_j)$ specifies the position of the ligand no. j and ψ_j its rotational orientation about the central-ion to ligator bond axis, relative to a given standard orientation. $e_t(j)$ denotes the AOM parameters of the j'th ligand, and the sum over the ligands runs to N, the number of ligands. There are at most $(2\,l+1)\,(l+1)$ elements represented by Eq. (62), and by equating the expressions of Eqs. (61) and (62) one obtains the relations desired between the reduced ligand-field parameters and the parameters of the Angular Overlap Model. One such relation is obtained from the trace of the perturbation matrix,

$$\begin{aligned}&\sum_{u}<\alpha l u|V(r_i\,\theta_i\,\varphi_i)\,|\alpha l u>=\sum_{ukq}\mathrm{E}^k_q\begin{pmatrix}l & k & l\\ u & q & u\end{pmatrix}=\mathrm{E}^0_\sigma\sqrt{[l]}\\&=\sum_{j,t}\sum_{u}\boldsymbol{D}^l_{ut}\,(\varphi_j\,\theta_j\,\psi_j)\,\boldsymbol{D}^l_{ut}(\varphi_j\,\theta_j\,\psi_j)\;e(j)=\sum_{j=1}^{N}\sum_{t}e_t(j)\end{aligned} \tag{63}$$

Since only energy differences are considered, this relation is redundant (see footnote to p. 290) and it merely expresses the necessary adjustment of the zero point of the energy scale when the two descriptions are compared.

3. Calculation of Ligand-Field Matrices

The calculation of the ligand-field matrices within an l^n electron configuration is made by standard tensorial methods. The independent perturbations are: the interelectronic repulsion, the ligand-field, and the spin-orbit interaction, and they will be discussed in this order using, as mentioned above, the Russell-Saunders case of the theory of atomic spectra.

The matrix representing the interelectronic repulsion is taken from published tables (*2, 11*). For d^n configurations both references use Slater-Condon parameters (*2*) which may be transformed into Racah parameters (*13*). The matrices are diagonal except possibly when several terms with identical SL numbers occur, e.g. the two ^{2}D terms, $^2_1$D and $^2_3$D, of d^3 which are further characterized by their seniority numbers.

The ligand-field operator is a spatial operator and is a special case of a one-electron operator. Such operators are conveniently treated in the formalism developed by *Racah* (*14*). He has shown that the matrix elements of a one-electron irreducible tensorial operator

$$\mathfrak{F}^k_q = \sum_{i=1}^{n} \mathfrak{f}^k_q(i) \tag{64}$$

between states $|l^n \alpha SLM_SM_L>$ of an l^n configuration may be evaluated as n times the element of the operator $\mathfrak{f}^k_q(n)$ operating on the n'th electron. The states of the configuration l^n are expressible in terms of those of the configuration l^{n-1} to which one l-electron has been added

$$|l^n \alpha SL\} = \sum_{\alpha_1 S_1 L_1} |l^{n-1}(\alpha_1 S_1 L_1)\, l\, SL\} <l^{n-1}(\alpha_1 S_1 L_1)\, l\, SL|\} l^n \alpha SL> \tag{65}$$

where $<l^{n-1}(\alpha_1 S_1 L_1)\, l\, SL|\}\, l^n\, \alpha SL>$ is a coefficient of fractional parentage (*14*).

One then obtains

$$\begin{aligned} &<l^n\, \alpha SLM_SM_L|\mathfrak{F}^k_q|l^n\, \alpha' S'L'M_S'M_L'> \\ &= \sum_{\alpha_1 S_1 L_1} n <l^n\, \alpha SL\{|l^{n-1}\, (\alpha_1 S_1 L_1)\, l\, SL> \\ &\quad \times <\alpha_1 S_1 L_1\, l_n\, SLM_SM_L|\, \mathfrak{f}^k_q(n)\, |\alpha_1 S_1 L_1\, l_n\, S'L'M_S'M_L'> \\ &\quad \times <l^{n-1}\, (\alpha_1 S_1 L_1)\, l\, S'L'|\}\, l^n\, \alpha' S'L') \end{aligned} \tag{66}$$

The matrix element is given also by the Wigner-Eckart theorem, Eqs. (27) or (28b), and therefore one concentrates upon the evaluation of the reduced matrix. This is conveniently calculated in terms of the unit tensorial operator (*13*) of the degree k, which in view of Eq. (33) is defined through the reduced matrix

$$<\alpha\, l\|\mathfrak{U}^k\|\, \alpha'\, l'> = \delta(\alpha\alpha')\, \delta(ll')\, \delta(lkl') \tag{67}$$

When $\mathfrak{F}^k_q$ is a term of the ligand-field operator, the factors $\delta(SS')$ $\delta(M_SM_S')$ occur in the matrix element of Eq. (66), and the operator $\mathfrak{f}^k_q(n)$ becomes a spatial operator of the type $[1^0 \times \mathfrak{f}^k(n)]^k_q$ where 1^0 refers to the first $(n-1)$ electrons. When the reduced matrix of this operator,

Eq. (35), is introduced into the reduced form of Eq. (66), one has the succinct expression

$$<l^n \alpha SL \| \mathfrak{F}^k \| l^n \alpha' SL'> = <l \| \mathfrak{f}^k(n) \| l> <l^n \alpha SL \| \mathfrak{U}^k \| l^n \alpha' SL'> \quad (68)$$

where the first factor on the right-hand side is the reduced matrix[7]) of the n'th electron and the second factor is the reduced matrix of the one-electron tensorial operator,

$$\mathfrak{U}^k = \sum_{i=1}^{n} \mathfrak{u}^k(i):$$

$$<l^n \alpha SL \| \mathfrak{U}^k \| l^n \alpha' SL'>$$
$$= \sum_{\alpha_1 S_1 L_1} n(-1)^{L_1+L+l+k} [L, L']^{\frac{1}{2}} \begin{pmatrix} L & L' & k \\ l & l & L_1 \end{pmatrix} <l^n \alpha SL\{|l^{n-1} (\alpha_1 S_1 L_1)\, l\, SL>$$
$$\times <l^{n-1} (\alpha_1 S_1 L_1)\, l\, SL' |\} l^n \alpha' SL'> \quad (69)$$

This function is tabulated in Ref. (*11*).
Matrix elements of the one-electron operator

$$\sum_{i=1}^{n} \xi(r_i)\, \mathfrak{s}^1(i) \cdot \mathfrak{l}^1(i)$$

which represents the spin-orbit interaction, are best discussed in the SLJM-coupling scheme since the operator is diagonal in the JM quantum numbers. First, the double tensorial operator (*13*)

$$\mathfrak{V}^{11} = \sum_{i=1}^{n} \mathfrak{s}^1(i) \cdot \mathfrak{U}^1(i) \quad (70)$$

is defined and introduced in the matrix element

$$<l^n \alpha SLJM | \mathfrak{F}^0_\sigma | l^n \alpha' S'L'JM>$$

---

7) This could for example be the reduced ligand-field parameter, in which case Eq. (66) is

$$<l^n \alpha SLM_S T \mid V \mid l^n \alpha SL' M_S T'>$$
$$= \sum_{kq} E^k_q < l^n \alpha SL \| \mathfrak{U}^k \| l^n \alpha SL' > \begin{pmatrix} L & k & L' \\ T & q & T' \end{pmatrix}$$

in the real representation.

to be calculated. Next, this matrix element is factorized according to Eq. (37), and finally, Eq. (35) is applied to each of the reduced matrices so obtained. In terms of the reduced one-electron elements Eq. (55),

$$<\frac{1}{2}\|\mathfrak{s}^1\|\frac{1}{2}> = i\sqrt{\frac{1}{2}\left(\frac{1}{2}+1\right)\left(2\times\frac{1}{2}+1\right)} = i\sqrt{\frac{3}{2}}$$

and

$$<\alpha l\|\xi(r)\,\mathfrak{l}^1\|\alpha l> = \zeta_{\alpha l}\times i\sqrt{l(l+1)\,(2\,l+1)},$$

one has

$$\begin{aligned}&<l^n\,\alpha\mathrm{SLJM}|\sum_{i=1}^{n}\xi(r_i)\,\mathfrak{s}^1(i)\cdot\mathfrak{l}^1(i)\,|l^n\,\alpha'\mathrm{S'L'JM}>\\&=(-1)^{\mathrm{S'+L+J}}\sqrt{l(l+1)\,(2\,l+1)}\,\zeta_{\alpha l}\begin{pmatrix}\mathrm{S} & \mathrm{S'} & 1\\ \mathrm{L'} & \mathrm{L} & \mathrm{J}\end{pmatrix}\\&\quad\times<l^n\,\alpha\mathrm{SL}\|\mathfrak{V}^{11}\|l^n\,\alpha'\mathrm{S'L'}>\end{aligned}\tag{71}$$

where

$$\begin{aligned}&<l^n\,\alpha\mathrm{SL}\|\mathfrak{V}^{11}\|l^n\,\alpha'\mathrm{S'L'}>\\&=\sum_{\alpha_1,\mathrm{S}_1,\mathrm{L}_1}n(-1)^{\mathrm{S_1+L_1+S+L+\frac{1}{2}}+l}\sqrt{\frac{3}{2}}\,[\mathrm{S},\mathrm{S'},\mathrm{L},\mathrm{L'}]^{\frac{1}{2}}\begin{pmatrix}\mathrm{S} & \mathrm{S'} & 1\\ \frac{1}{2} & \frac{1}{2} & \mathrm{S}_1\end{pmatrix}\begin{pmatrix}\mathrm{L} & \mathrm{L'} & 1\\ l & l & \mathrm{L}_1\end{pmatrix}\\&\quad\times<l^n\,\alpha\mathrm{SL}\{|l^{n-1}\,(\alpha_1\mathrm{S}_1\mathrm{L}_1)\,l\;\mathrm{SL}>\\&\quad\times<l^{n-1}\,(\alpha_1\,\mathrm{S}_1\,\mathrm{L}_1)\,l\;\mathrm{S'L'}|\}\,l^n\,\alpha'\mathrm{S'L'}>\end{aligned}\tag{72}$$

The reduced matrices of $\mathfrak{V}^{11}$ are tabulated in Ref. (*11*). To conform with Refs. (*13*) and (*14*), the two factors of $i=\sqrt{-1}$ that occur in the reduced elements of the angular momentum operators are included in the common phase $(-1)^{\mathrm{S'+L+J+1}}\times i^2$ on the right-hand side of Eq. (71).

4. Conclusions

The irreducible tensor method was originally developed by *G. Racah* in order to make possible a systematic interpretation of the spectra of atoms. In the present paper this method has been extended to irreducible sets of real functions that have the same transformation properties as the usual real spherical harmonics. Such an extension is particularly useful in the discussion of the spectra of molecules which belong to the finite point groups or to the continuous groups with axial symmetry. There are several reasons for this.

Most important is the fact that the real spherical harmonics, more often than the complex ones, are symmetry adapted to the point groups. This means for these cases that the group $R_{3\iota}$ is reduced with respect to the point groups, except possibly for a reordering transformation and a phase shift on some (sine) functions (*25*). Thus the full advantage of the symmetry is used and by a reordering transformation the energy matrices break up into a block-diagonal form by which the eigenvalue problem is reduced before the actual diagonalization.

When it is impossible to use real functions, the complex description is easily introduced since the reduced matrices and the 6-l and 9-l symbols are invariant in the two representations. It is further important that the reduced ligand-field parameters are the same, even though they have been defined on the basis of the real orbitals. In this connection it may be mentioned that Racah's formulae for the 3-j and 6-j symbols (*4, 15*), which are convenient for computer work, make it possible to generate ligand-field matrices by a rather simple algorithm.

The real orbitals are attractive from the chemist's point of view because of their pictorial significance. They can be drawn and one can see their nodal surfaces and their whole symmetry, and its relation to the geometry of the molecules. It is their angular properties which in this respect are particularly important. In connection with the Angular Overlap Model of the ligand field, the qualitative aspects of these properties have been made quantitative. This has been done by the introduction of the concept of the angular overlap integral, whose absolute value is equal to unity for orbitals which are completely symmetry-adapted with respect to the point group $C_{\infty v}$ of the bond axis and vanishing for orbitals of different symmetry (*20*, Sect. 4d).

The fundamental quantities in the theory of real irreducible tensorial sets are the 3-l symbols. They fall into two classes, the even ones which are invariant under permutations of their columns, and the odd ones which change their sign under odd permutations. Although their values are well defined through Eq. (20), it is noted that the even 3-l symbols may be calculated as an integral (25 p. 213) over three surface harmonics using the expression, Eq. (21), for the reduced matrix, Eq. (41). The odd 3-l symbols may be calculated by recursion using Eq. (26) with $l_3 = 1$, $l = l_{12} - 1$, and $l_2 = l_{23}$. As the 6-l symbol in this equation does not depend on the form in which the irreducible representations are chosen, it follows that a phase common to all (25 p. 216) the odd 3-l symbols is undetermined in the real representation. This phase is fixed through Eqs. (2) and (9) which also secures that the reduced matrices are invariant in the two representations.

III. Appendices

1. Alternative Evaluation of 3-l Symbols

The values of the 3-l symbols may be obtained from table 1 in terms of the 3-j symbols from published tables (*15*). For computer work the 3-j symbols may also be calculated by Racah's closed formula, Eq. (10.14) of (*4*), with the phase difference between $\bar{V}(j)$ and 3-j taken into account.

Table 1 is used as follows: First, permute the columns of the 3-l symbol to be evaluated so as to fit the λ equation, Eq. (73). Next, consider to which number $\equiv$ (mod 4) the sum $l_1 + l_2 - l_3$ in the permuted symbol belongs, and finally, use the master formula (74) with the phase determined from the table by the appropriate combination of the ς's.

Table 1

$$\lambda_1 + \lambda_2 = \lambda_3\ ; \quad \text{all} \quad \lambda > 0 \tag{73}$$

$$\begin{pmatrix} l_1 & l_2 & l_3 \\ \lambda_1\varsigma_1 & \lambda_2\varsigma_2 & \lambda_3\varsigma_3 \end{pmatrix} = [\text{phase}]\frac{(-1)^{l_3-\lambda_3}}{\sqrt{2}}\begin{pmatrix} l_1 & l_2 & l_3 \\ \lambda_1 & \lambda_2 & -\lambda_3 \end{pmatrix} \tag{74}$$

			$l_1+l_2-l_3$	
ς_1	ς_2	ς_3	$\equiv 0$	$\equiv 2$
c	c	c	[+]	[−]
s	s	c	[−]	[+]
s	c	s	[+]	[−]
c	s	s	[+]	[−]

			$l_1+l_2-l_3$	
ς_1	ς_2	ς_3	$\equiv 1$	$\equiv 3$
s	s	s	[−]	[+]
c	c	s	[+]	[−]
c	s	c	[−]	[+]
s	c	c	[−]	[+]

If in Eqs. (73) and (74) $\lambda_1 = 0$, or $\lambda_2 = 0$, multiply the right-hand side of Eq. (74) by $+\sqrt{2}$ and identify a σ-function by c. $\lambda_3 = 0$ is not allowed unless $\lambda_1 = \lambda_2 = \lambda_3 = 0$.

2. A Short Table of 3-l Symbols

l_1	l_2	l_3	t_1	t_2	t_3	3-l
1	1	0	σ	σ	σ	$\sqrt{(1/3)}$
1	1	0	$1c$	$1c$	σ	$\sqrt{(1/3)}$
1	1	0	$1s$	$1s$	σ	$\sqrt{(1/3)}$
1	1	1	σ	$1c$	$1s$	$\sqrt{(1/6)}$
2	1	1	σ	σ	σ	$\sqrt{(2/15)}$
2	1	1	σ	$1c$	$1c$	$-\sqrt{(1/30)}$
2	1	1	σ	$1s$	$1s$	$-\sqrt{(1/30)}$

l_1	l_2	l_3	t_1	t_2	t_3	3-l
2	1	1	$1c$	σ	$1c$	$\sqrt{(1/10)}$
2	1	1	$1s$	σ	$1s$	$\sqrt{(1/10)}$
2	1	1	$2c$	$1c$	$1c$	$\sqrt{(1/10)}$
2	1	1	$2c$	$1s$	$1s$	$-\sqrt{(1/10)}$
2	1	1	$2s$	$1c$	$1s$	$\sqrt{(1/10)}$
2	2	0	σ	σ	σ	$\sqrt{(1/5)}$
2	2	0	$1c$	$1c$	σ	$\sqrt{(1/5)}$
2	2	0	$1s$	$1s$	σ	$\sqrt{(1/5)}$

l_1	l_2	l_3	t_1	t_2	t_3	3-l
2	2	0	$2c$	$2c$	σ	$\sqrt{(1/5)}$
2	2	0	$2s$	$2s$	σ	$\sqrt{(1/5)}$
2	2	1	σ	$1c$	$1s$	$\sqrt{(1/10)}$
2	2	1	σ	$1s$	$1c$	$-\sqrt{(1/10)}$
2	2	1	$1c$	$1s$	σ	$\sqrt{(1/30)}$
2	2	1	$2c$	$1c$	$1s$	$-\sqrt{(1/30)}$
2	2	1	$2c$	$1s$	$1c$	$-\sqrt{(1/30)}$
2	2	1	$2c$	$2s$	σ	$\sqrt{(2/15)}$
2	2	1	$2s$	$1c$	$1c$	$\sqrt{(1/30)}$
2	2	1	$2s$	$1s$	$1s$	$-\sqrt{(1/30)}$
2	2	2	σ	σ	σ	$\sqrt{(2/35)}$
2	2	2	$1c$	σ	$1c$	$\sqrt{(1/70)}$
2	2	2	$1s$	σ	$1s$	$\sqrt{(1/70)}$
2	2	2	$2c$	σ	$2c$	$-\sqrt{(2/35)}$
2	2	2	$2c$	$1c$	$1c$	$\sqrt{(3/70)}$
2	2	2	$2c$	$1s$	$1s$	$-\sqrt{(3/70)}$
2	2	2	$2s$	σ	$2s$	$-\sqrt{(2/35)}$
2	2	2	$2s$	$1c$	$1s$	$\sqrt{(3/70)}$
3	2	1	σ	σ	σ	$\sqrt{(3/35)}$
3	2	1	σ	$1c$	$1c$	$-\sqrt{(1/35)}$
3	2	1	σ	$1s$	$1s$	$-\sqrt{(1/35)}$
3	2	1	$1c$	σ	$1c$	$\sqrt{(2/35)}$
3	2	1	$1c$	$1c$	σ	$\sqrt{(8/105)}$
3	2	1	$1c$	$2c$	$1c$	$-\sqrt{(1/210)}$
3	2	1	$1c$	$2s$	$1s$	$-\sqrt{(1/210)}$
3	2	1	$1s$	σ	$1s$	$\sqrt{(2/35)}$
3	2	1	$1s$	$1s$	σ	$\sqrt{(8/105)}$
3	2	1	$1s$	$2c$	$1s$	$\sqrt{(1/210)}$
3	2	1	$1s$	$2s$	$1c$	$-\sqrt{(1/210)}$
3	2	1	$2c$	$1c$	$1c$	$\sqrt{(1/21)}$
3	2	1	$2c$	$1s$	$1s$	$-\sqrt{(1/21)}$
3	2	1	$2c$	$2c$	σ	$\sqrt{(1/21)}$
3	2	1	$2s$	$1c$	$1s$	$\sqrt{(1/21)}$
3	2	1	$2s$	$1s$	$1c$	$\sqrt{(1/21)}$
3	2	1	$2s$	$2s$	σ	$\sqrt{(1/21)}$
3	2	1	$3c$	$2c$	$1c$	$\sqrt{(1/14)}$
3	2	1	$3c$	$2s$	$1s$	$-\sqrt{(1/14)}$
3	2	1	$3s$	$2c$	$1s$	$\sqrt{(1/14)}$
3	2	1	$3s$	$2s$	$1c$	$\sqrt{(1/14)}$
3	2	2	σ	$1c$	$1s$	$\sqrt{(2/35)}$
3	2	2	σ	$2c$	$2s$	$-\sqrt{(1/70)}$
3	2	2	$1c$	σ	$1s$	$-\sqrt{(1/35)}$
3	2	2	$1c$	$2c$	$1s$	$\sqrt{(3/140)}$
3	2	2	$1c$	$2s$	$1c$	$-\sqrt{(3/140)}$
3	2	2	$1s$	σ	$1c$	$\sqrt{(1/35)}$
3	2	2	$1s$	$1c$	$2c$	$-\sqrt{(3/140)}$
3	2	2	$1s$	$1s$	$2s$	$-\sqrt{(3/140)}$
3	2	2	$2c$	σ	$2s$	$-\sqrt{(1/14)}$
3	2	2	$2c$	$1c$	$1s$	0
3	2	2	$2s$	σ	$2c$	$\sqrt{(1/14)}$
3	2	2	$2s$	$1c$	$1c$	0
3	2	2	$2s$	$1s$	$1s$	0
3	2	2	$3c$	$1c$	$2s$	$-\sqrt{(1/28)}$
3	2	2	$3c$	$1s$	$2c$	$-\sqrt{(1/28)}$
3	2	2	$3s$	$1c$	$2c$	$\sqrt{(1/28)}$
3	2	2	$3s$	$1s$	$2s$	$-\sqrt{(1/28)}$
4	2	2	σ	σ	σ	$\sqrt{(2/35)}$
4	2	2	σ	$1c$	$1c$	$-\sqrt{(8/315)}$
4	2	2	σ	$1s$	$1s$	$-\sqrt{(8/315)}$
4	2	2	σ	$2c$	$2c$	$\sqrt{(1/630)}$
4	2	2	σ	$2s$	$2s$	$\sqrt{(1/630)}$
4	2	2	$1c$	σ	$1c$	$\sqrt{(1/21)}$
4	2	2	$1c$	$1c$	$2c$	$-\sqrt{(1/252)}$
4	2	2	$1c$	$1s$	$2s$	$-\sqrt{(1/252)}$
4	2	2	$1s$	σ	$1s$	$\sqrt{(1/21)}$
4	2	2	$1s$	$1c$	$2s$	$-\sqrt{(1/252)}$
4	2	2	$1s$	$1s$	$2c$	$\sqrt{(1/252)}$
4	2	2	$2c$	σ	$2c$	$\sqrt{(1/42)}$
4	2	2	$2c$	$1c$	$1c$	$\sqrt{(2/63)}$
4	2	2	$2c$	$1s$	$1s$	$-\sqrt{(2/63)}$
4	2	2	$2s$	σ	$2s$	$\sqrt{(1/42)}$
4	2	2	$2s$	$1c$	$1s$	$\sqrt{(2/63)}$
4	2	2	$3c$	$1c$	$2c$	$\sqrt{(1/36)}$
4	2	2	$3c$	$1s$	$2s$	$-\sqrt{(1/36)}$
4	2	2	$3s$	$1c$	$2s$	$\sqrt{(1/36)}$
4	2	2	$3s$	$1s$	$2c$	$\sqrt{(1/36)}$
4	2	2	$4c$	$2c$	$2c$	$\sqrt{(1/18)}$
4	2	2	$4c$	$2s$	$2s$	$-\sqrt{(1/18)}$
4	2	2	$4s$	$2c$	$2s$	$\sqrt{(1/18)}$

3. Notes on the 6-l Symbol, Eq. (23)

The 6-l symbol is closely connected to the recoupling transformation between triple products. In order to obtain the expression, Eq. (22), one forms the two triple products that have the initial sets in the same order (compare Eq. (14a)),

$$[[|l_1\} \times |l_2\}]^{l_{12}} \times |l_3\}]^l_t$$
$$= \sum_{t_1 t_2 t_{12} t_3} \sqrt{[l_{12}, l]}\, (-)^{\,l_1+l_2+l_3+l} \begin{pmatrix} l_1 & l_2 & l_{12} \\ t_1 & t_2 & t_{12} \end{pmatrix} \begin{pmatrix} l_{12} & l_3 & l \\ t_{12} & t_3 & t \end{pmatrix} \quad (75)$$
$$\times\, |l_1\, t_1> |l_2\, t_2> |l_3\, t_3>$$

and an analogous expression for $[|l_1\} \times [|l_2\} \times |l_3\}]^{l_{23}}]^l_t$. Then the expansion coefficients in the recoupling transformation

$$[|l_1\} \times [|l_2\} \times |l_3\}]^{l_{23}}]^l$$
$$= \sum_{l_{12}} <((l_1\, l_2)\, l_{12}\, l_3)\, l\,|\,(l_1\, (l_2\, l_3)\, l_{23})\, l> [[|l_1\} \times |l_2\}]^{l_{12}} \times |l_3\}]^l \quad (76)$$

are given by

$$<((l_1\, l_2)\, l_{12}\, l_3)\, l\,|\,(l_1\, (l_2\, l_3)\, l_{23})\, l>$$
$$= \sum_{\substack{t_1 t_2 t_{12} \\ t_3\; t_{23}}} [l] \sqrt{[l_{12}, l_{23}]} \begin{pmatrix} l_1 & l_2 & l_{12} \\ t_1 & t_2 & t_{12} \end{pmatrix} \begin{pmatrix} l_1 & l_{23} & l \\ t_1 & t_{23} & t \end{pmatrix} \begin{pmatrix} l_2 & l_3 & l_{23} \\ t_2 & t_3 & t_{23} \end{pmatrix} \begin{pmatrix} l_{12} & l_3 & l \\ t_{12} & t_3 & t \end{pmatrix} \quad (77)$$

From this derivation of Eq. (22) it might seem to be more convenient not to introduce the phase on the right-hand side of Eq. (22), but to keep the columns of the two middle 3-l symbols in the definition, Eq. (23), in the directly derived order. However, this would lead to a less symmetric symbol, the W function of *Racah* (*13*).

The sequence of l's occurring in each of the four 3-l symbols on the right-hand side of Eq. (23) is easily remembered: Each 3-l contains an l from each column in the 6-l. One 3-l uses the top row of the 6-l and the others contain one l from the top row and two from the bottom row, in cyclic order.

Although extensive tables of 6-j symbols exist (*15*), it may be convenient to refer to simple algebraic expressions for symbols containing at least one $j \leqq 2$. With one l equal to zero, the expression is (*4*)

$$\begin{pmatrix} l_1 & l_2 & l_3 \\ l_2 & l_1 & 0 \end{pmatrix} = \frac{(-1)^{l_1+l_2+l_3}}{\sqrt{[l_1, l_2]}} \quad (78)$$

6-j symbols with $j = 1$ are listed in Ref. (*3*) and W functions with a $j \leqq 2$ are given in Ref. (*1*).

The most useful symmetry properties of the 6-l symbols are (*4*) that they stand invariant under even as well as odd permutations of their columns. Further, they are left invariant if any pair of columns is turned upside down.

As an example of the use of the 6-l symbols, we refer to the detailed derivation of the coefficients of fractional parentage for the ^{2}D term of the p^3 configuration given by *Racah* (*14*, p. 369).

4. Calculation of Ligand-Field Matrices. An Example

In order to show the way the calculations of ligand-field matrices run in the present scheme, we shall work through a simple example by briefly indicating the steps involved in such a calculation.

We consider a complex of a transition metal ion with two electrons in the partly filled $3d$ shell: $(3d)^2$. There are three planar identical ligands placed with the ligator atoms at the corners of an equilateral triangle with the metal ion at the midpoint. The ligands are supposed to form σ and π bonds to the metal ion. However, the two π orbitals are not equivalent; one, πs say, is used for binding inside the planar ligand molecule whereas the other, (πc), is available for interaction with the metal ion. The molecular plane of all three ligands forms an angle of $\frac{\pi}{4}$ with the plane of the triangle, as indicated in Fig. 1. The complex has the point-

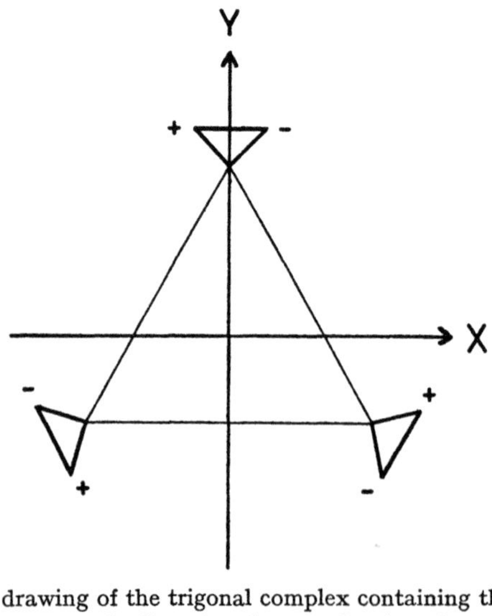

Fig. 1. Projection drawing of the trigonal complex containing three planar ligands. The heavy lines represent the ligands, whose molecular planes have been rotated by 45° relative to the XY-plane, defined by the central ion and the ligating atoms. The (+) and (−) mean above and below the XY-plane. The projection in this way has the full symmetry D_3 of the complex.

group symmetry D_3. The Z-axis coincides with the molecular three-fold axis and, for reasons which will be apparent later (p. 294), the Y-axis has been chosen as the two-fold axis. Despite the relative simplicity of this example, it actually exhausts the principal problems in setting up the ligand-field matrices and even in this case the present methods turn out to be quicker *and* safer than more elementary considerations. If desired, it is a rather simple matter to extend the example to more interesting cases, *e.g.* trigonal bipyramidal Nickel(II) complexes.

Consider first the ligand-field operator, Eq. (60). Owing to the molecular symmetry and to the position of the coordinate system, one has

$$\begin{aligned} \mathsf{v}(i) = {} & \mathfrak{A}^0_\sigma\, \mathfrak{C}^0_\sigma\,(\theta_i, \varphi_i) + \mathfrak{A}^2_\sigma\, r^2\, \mathfrak{C}^2_\sigma\,(\theta_i, \varphi_i) + \mathfrak{A}^4_\sigma\, r^4\, \mathfrak{C}^4_\sigma\,(\theta_i, \varphi_i) \\ & + \mathfrak{A}^4_{3c}\, r^4\, \mathfrak{C}^4_{3c}\,(\theta_i, \varphi_i) \end{aligned} \tag{79}$$

Matrix elements of this operator within the set of d-orbitals, $|\alpha 2\}$, are calculated as indicated in Eq. (61) by use of the tables in Appendix 2. These matrix elements are also expressible in the angular overlap model (*19*) as shown in Eq. (62). The elements of the $\boldsymbol{D}^2$ matrix are calculated from Table 2b of Ref. (*19*) with the angular coordinates $(\varphi_j\, \theta_j\, \psi_j)$ for the three ligands equal to $\left(\frac{\pi}{2}, \frac{\pi}{2}, \frac{\pi}{4}\right)$, $\left(\frac{7\,\pi}{6}, \frac{\pi}{2}, \frac{\pi}{4}\right)$ and $\left(\frac{11\,\pi}{6}, \frac{\pi}{2}, \frac{\pi}{4}\right)$ as seen from Fig. 1.

The non-vanishing matrix elements are

$$\begin{aligned}
<\alpha\, 2\sigma|\mathsf{v}|\alpha\, 2\sigma> &= \frac{1}{\sqrt{5}}\, \mathrm{E}^0_\sigma + \sqrt{\frac{2}{35}}\, \mathrm{E}^2_\sigma + \sqrt{\frac{2}{35}}\, \mathrm{E}^4_\sigma = \frac{3}{4}\, e_\sigma \\
<\alpha\, 2\pi s|\mathsf{v}|\alpha\, 2\pi s> &= <\alpha\, 2\pi c|\mathsf{v}|\alpha\, 2\pi c> \\
&= \frac{1}{\sqrt{5}}\, \mathrm{E}^0_\sigma + \frac{1}{\sqrt{70}}\, \mathrm{E}^2_\sigma - \sqrt{\frac{8}{315}}\, \mathrm{E}^4_\sigma = \frac{3}{4}\, e_{\pi s} + \frac{3}{4}\, e_{\pi c} \\
<\alpha\, 2\delta s|\mathsf{v}|\alpha\, 2\delta s> &= <\alpha\, 2\delta c|\mathsf{v}|\alpha\, 2\delta c> \\
&= \frac{1}{\sqrt{5}}\, \mathrm{E}^0_\sigma - \sqrt{\frac{2}{35}}\, \mathrm{E}^2_\sigma + \frac{1}{\sqrt{630}}\, \mathrm{E}^4_\sigma \\
&= \frac{9}{8}\, e_\sigma + \frac{3}{4}\, e_{\pi s} + \frac{3}{4}\, e_{\pi c} \\
-<\alpha\, 2\pi s|\mathsf{v}|\alpha\, 2\delta s> &= + <\alpha\, 2\pi c|\mathsf{v}|\alpha\, 2\delta c> \\
&= \frac{1}{6}\, \mathrm{E}^4_{3c} = \frac{3}{4}\, e_{\pi s} - \frac{3}{4}\, e_{\pi c}
\end{aligned} \tag{80}$$

where it may be noted that the baricenter rule applies to the E_σ^2 and E_σ^4 elements. From Eq. (80) the relations between the reduced ligand field parameters and the AOM parameters follow[8]:

$$E_\sigma^0 = \frac{1}{\sqrt{5}} \times 3\,(e_\sigma + e_{\pi s} + e_{\pi c})$$

$$E_\sigma^2 = \sqrt{\frac{10}{7}}\left(-\frac{3}{2}e_\sigma - \frac{3}{4}e_{\pi s} - \frac{3}{4}e_{\pi c}\right)$$

$$E_\sigma^4 = \sqrt{\frac{18}{35}}\left(\frac{27}{8}e_\sigma - \frac{9}{4}e_{\pi s} - \frac{9}{4}e_{\pi c}\right)$$

$$E_{3c}^4 = \frac{9}{2}e_{\pi s} - \frac{9}{2}e_{\pi c} \tag{81}$$

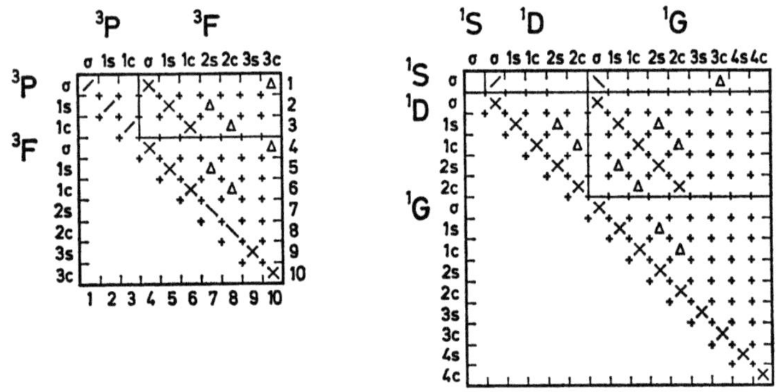

Fig. 2. The upper half of the symmetric ligand field matrices for the complex depicted in Fig. 1. The SLMsT coupling scheme for the d^2 configuration is used and the non-vanishing elements are indicated with the symbols: $E_\sigma^2 \diagup$; $E_\sigma^4 \diagdown$; $E_{3c}^4 \triangle$. The ligand-field* parameters E_σ^0 and the Racah parameter A, both of which appear in the diagonals only, may be discarded and the Racah parameters B and C, also appearing only in the diagonals, are not indicated. There are three identical ten by ten triplet matrices belonging to the M_S quantum numbers 1, 0, and −1.

* We note the relation

$$\langle l^n;\alpha SL \| \mathfrak{U}^0 \| l^n;\alpha' S' L'\rangle = n\sqrt{\frac{2L+1}{2l+1}}\,\delta(\alpha\alpha')\,\delta(SS')\,\delta(LL')$$

which is not stated in Ref. (*11*)

8) The e_δ parameters have been considered small here and neglected. As mentioned on p. 280, the parameter E_σ^0 is without a physical meaning. However, this is unimportant for the model because it occurs only in the diagonal elements and always with the same coefficient. So E_σ^0 will disappear when only energy differences which are the observable quantities in a ligand-field context, are considered. The top equality of eq. (81) is therefore without much interest. The physical problem associated with such an equality has been discussed previously (*20*) where AOM was considered to represent essentially the non-spherical part of the ligand field. This problem will be considered in more detail in a subsequent paper (*26*).

The first of the equalities (81) is a restatement of Eq. (63). Since only energy differences are of interest, the four symmetry-determined E_q^k parameters reduce to three: E_σ^2, E_σ^4, and E_{3c}^4. The presentation of the example suggests that only the two AOM parameters e_σ and $e_{\pi c}$ are significant, a statement which must be justified by a comparison of the results of the calculations with the experiments.

We shall first work out the ligand-field matrices in the case where the spin-orbit coupling interaction is not taken into account. In this case the Russell-Saunders coupling scheme in the real representation may be used. The $(\alpha d)^2$ configuration gives rise to two triplet terms, ^3P and ^3F and to three singlet terms, ^1S, ^1D, and ^1G. Since the spin-orbit coupling is neglected, there is no interaction between triplet and singlet states and one obtains two independent matrices, as outlined in Fig. 2.

In the matrices in Fig. 2 the diagonal elements generally contain the Racah parameters A, B, and C and the reduced ligand-field parameters E_σ^0, E_σ^2, and E_σ^4. For example, the diagonal element with the function $|d^2;\, ^1D\, 2s\rangle$, which is calculated as indicated in section II.3., is

$$\langle d^2;\, ^1D\, 2s| \left(\sum_{i<j}^{2} \frac{1}{r_{ij}} + \sum_{i=1}^{2} v(i) \right) |d^2;\, ^1D\, 2s\rangle$$

$$= F^0 - \frac{3}{49} F^2 + \frac{4}{49} F^4 + \sum_{k=0,\,2,\,4} E_q^k \langle d^2;\, ^1D \| \mathfrak{U}^k \| d^2;\, ^1D\rangle \begin{pmatrix} 2 & k & 2 \\ 2s & \sigma & 2s \end{pmatrix}$$

$$= A - 3\,B + 2\,C + E_\sigma^0(2) \left(\frac{1}{\sqrt{5}}\right) + E_\sigma^2 \left(-\frac{3}{7}\right) \left(-\sqrt{\frac{2}{35}}\right)$$

$$+ E_\sigma^4 \left(\frac{4}{7}\right) \left(\frac{1}{\sqrt{630}}\right) \qquad (82)$$

Here, the elements of the first operator are (in Slater-Condon parameters) taken from Ref. (*11*) p. 53 and translated into Racah parameters (*13*). The matrix element of the ligand-field operator is calculated using first the Wigner-Eckart theorem, Eq. (27), and then Eq. (68), taking the reduced matrix elements of the $\mathfrak{U}^k$ operator from Ref. (*11*), p. 64.

Off-diagonal elements which are diagonal in the $\lambda\varsigma$ quantum numbers cannot contain the reduced ligand-field parameters E_{3c}^4 and the remaining type of off-diagonal elements cannot contain the parameters E_σ^2 and E_σ^4. An example is

$$\langle d^2;\, ^1D\, 2s| \left(\sum_{i<j}^{2} \frac{1}{r_{ij}} + \sum_{i=1}^{2} v(i) \right) |d^2;\, ^1G\, 1s\rangle$$

$$= 0 + E_{3c}^4 \langle d^2;\, ^1D \| \mathfrak{U}^4 \| d^2;\, ^1G\rangle \begin{pmatrix} 2 & 4 & 4 \\ 2\,s & 3\,c & 1\,s \end{pmatrix}$$

$$= E_{3c}^4 \left(\frac{\sqrt{110}}{7}\right) \left(\sqrt{\frac{3}{220}}\right) \qquad (83)$$

It follows from the properties of the molecular symmetry group that the matrices in Fig. 2 break up into minor independent parts which can be obtained by a simple reordering transformation. Thus the triplet matrix gives one block of $^3A_1$ symmetry, namely the element (9,9), and one block of $^3A_2$ symmetry consisting of the rows and columns with numbers 1, 4, and 10. There are two ^3E blocks, one of *s*-symmetry (rows and columns nos. 2, 5, and 7) and one of *c*-symmetry (rows and columns nos. 3, 6, and 8), which differ in the signs of some of the non-diagonal elements. They have, of course, identical eigenvalues. Using the standardization proposed in Ref. (*25*) for the group D_3, a sign change on the πs-functions yields identical ^3E matrices. Then the *s*-functions transform as the standard e_1 component of ^3E and the *c*-functions transform as $^3E\,e_2$ under the symmetry operations of the group.

In the case where the spin-orbit coupling is included in the Hamiltonian, the SLJM coupling scheme in the complex representation is better suited for the calculations since the spin-orbit coupling operator is diagonal in the JM quantum numbers. We shall denote the states $^{2S+1}L_J M$ and have now an interaction between the triplet and the singlet states furnished by the matrix elements of the spin-orbit coupling operator. In order to derive the matrix elements of the ligand-field operator by means of the Wigner-Eckart theorem, Eq. (28b), the expression Eq. (79) is transformed through Eq. (9) into[9].

$$\mathsf{v}(i) = \mathfrak{A}^0_\sigma\, \mathfrak{C}^0_0\,(\theta_i\,\varphi_i) + \mathfrak{A}^2_\sigma\, r^2\,(-\mathfrak{C}^2_0\,(\theta_i\,\varphi_i)) + \mathfrak{C}^4_\sigma\, r^4\, \mathfrak{C}^4_0\,(\theta_i\,\varphi_i)$$

$$+ \mathfrak{A}^4_{3c} r^4 \left(\frac{-1}{\sqrt{2}}\, \mathfrak{C}^4_3\,(\theta_i\,\varphi_i) + \frac{1}{\sqrt{2}}\, \mathfrak{C}^4_{-3}\,(\theta_i\,\varphi_i)\right) \qquad (84)$$

One can now set up the forty-five by forty-five ligand-field matrix as shown in Fig. 3.

As an example, the diagonal element with the function $^{2S+1}L_J\,M = {}^1D_2\,2$ is

---

[9]) The ligand field may also be expanded onto the complex harmonics, Eq. (4), [see also Ref. (*4*), Eqs. 5.19 and 6.12], in which case Eq. (84) has the form

$$\mathsf{v}(i) = \mathfrak{A}^{(0)}_0\, \mathfrak{C}^{[0]}_0 + \mathfrak{A}^{(2)}_0\, r^2\, \mathfrak{C}^{[2]}_0 + \mathfrak{A}^{(4)}_0\, r^4\, \mathfrak{C}^{[4]}_0 + \mathfrak{A}^{(4)}_3\, r^4\, \mathfrak{C}^{[4]}_3 + \mathfrak{A}^{(4)}_{-3}\, r^4\, \mathfrak{C}^{[4]}_{-3}$$

It is seen that

$$\mathfrak{A}^{(0)}_0 = \mathfrak{A}^0_\sigma,\ \mathfrak{A}^{(2)}_0 = -\,\mathfrak{A}^2_\sigma,\ \mathfrak{A}^{(4)}_0 = \mathfrak{A}^4_\sigma,$$

and

$$\mathfrak{A}^{(4)}_3 = -\,\mathfrak{A}^{(4)}_{-3} = -\,\mathfrak{A}^4_{3c}/\sqrt{2}$$

so that the number of independent parameters is the same.

$$<d^2;{}^1D_2 2|\left(\sum_{i<j}^{2}\frac{1}{r_{ij}}+\sum_{i=1}^{2}\mathsf{v}(i)+\sum_{i=1}^{2}\xi(r_i)\,\mathfrak{s}^1(i)\cdot\mathfrak{l}^1(i)\right)|d^2;{}^1D_2 2>$$

$$= F^0-\frac{3}{49}F^2+\frac{4}{49}F^4+E_\sigma^0<d^2;{}^1D_2\|\mathfrak{U}^0\|d^2;{}^1D_2>(-1)^{2-2}\begin{pmatrix}2&0&2\\-2&0&2\end{pmatrix}$$

$$+(-1)\,E_\sigma^2<d^2;\;{}^1D_2\|\mathfrak{U}^2\|\,d^2;\;{}^1D_2>(-1)^{2-2}\begin{pmatrix}2&2&2\\-2&0&2\end{pmatrix}$$

$$+\qquad E_\sigma^4<d^2;\;{}^1D_2\,\|\mathfrak{U}^4\|\,d^2;\;{}^1D_2>(-1)^{2-2}\begin{pmatrix}2&4&2\\-2&0&2\end{pmatrix}$$

$$+(-1)^{0+2+2}\sqrt{30}\,\zeta_{\alpha 2}\begin{Bmatrix}0&0&1\\2&2&2\end{Bmatrix}<d^2;\;{}^1D\,\|\mathfrak{V}^{11}\|\,d^2;\;{}^1D>$$

$$= A-3B+2C+E_\sigma^0(2)\,(+1)\left(\frac{1}{\sqrt{5}}\right)+(-1)\,E_\sigma^2\left(-\frac{3}{7}\right)(+1)\left(\sqrt{\frac{2}{35}}\right)$$

$$+E_\sigma^4\left(\frac{4}{7}\right)(+1)\left(\frac{1}{\sqrt{630}}\right)+(+1)\sqrt{30}\,\zeta_{\alpha 2}\,(0)\,(0) \qquad (85)$$

where the 3-j symbols are taken from Ref. (*15*) and where the 6-j symbol vanishes for symmetry reasons.

In Eq. (85) the reduced matrix elements of the $\mathfrak{U}^k$ operator are evaluated from the general expression

$$<l^n\,S\,L\,J\,\|\mathfrak{U}^k\|\,l^n\,S\,L'\,J'>$$

$$=\sqrt{[J,J']}\,(-1)^{S+L'+J+k}\begin{Bmatrix}J&J'&k\\L'&L&S\end{Bmatrix}<l^n\,S\,L\,\|\mathfrak{U}^k\|\,l^n\,S\,L'> \qquad (86)$$

which is obtained by use of Eq. (35). On the right hand side of Eq. (86) the 6-j symbol may be taken from Ref. (*15*) and the reduced matrix from Ref. (*11*).

An element connecting terms of different spin multiplicity is e.g.

$$<d^2;\;{}^3F_2\,1|\left(\sum_{i<j}^{2}\frac{1}{r_{ij}}+\sum_{i=1}^{2}\mathsf{v}(i)+\sum_{i=1}^{2}\xi(r_i)\,\mathfrak{s}^1(i)\cdot\mathfrak{l}^1(i)\right)|d^2;\;{}^1D_2\,1>$$

$$=0+0+(-1)^{0+3+2}\sqrt{30}\,\zeta_{\alpha 2}\begin{Bmatrix}1&0&1\\2&3&2\end{Bmatrix}<d^2;\;{}^3F\|\mathfrak{V}^{11}\|\,d^2;\;{}^1D> \qquad (87)$$

$$=(-1)\sqrt{30}\,\zeta_{\alpha 2}\left(\frac{1}{\sqrt{15}}\right)\left(\sqrt{\frac{6}{5}}\right)$$

where the 6-j symbol may be calculated from Eq. (78) and where the reduced matrix of $\mathfrak{V}^{11}$ is found in Ref. (*11*), p. 64.

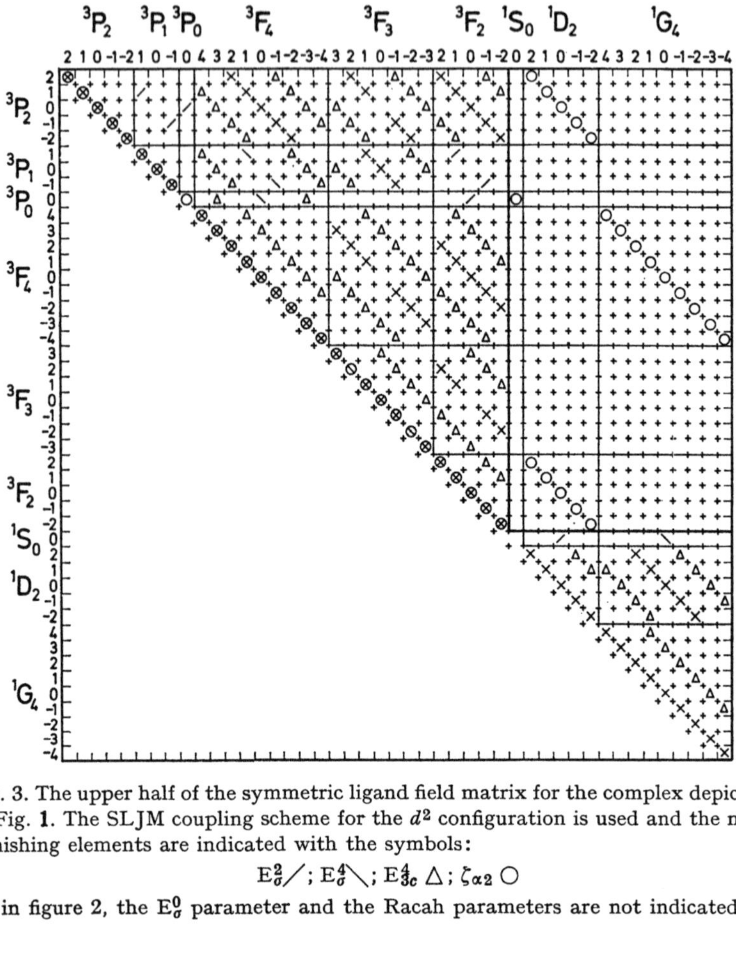

Fig. 3. The upper half of the symmetric ligand field matrix for the complex depicted in Fig. 1. The SLJM coupling scheme for the d^2 configuration is used and the non-vanishing elements are indicated with the symbols:

$E^2_\sigma \diagup$; $E^4_\sigma \diagdown$; E^4_{3c} $\triangle$; $\zeta_{\alpha 2}$ $\bigcirc$

As in figure 2, the E^0_σ parameter and the Racah parameters are not indicated

The choice of the Y-axis as the two-fold axis in a complex is actually a fortunate one, which in general prevents the imaginary unit $i = \sqrt{-1}$ from showing up in the potential expression analogous to Eq. (84). Thus the Hermitian ligand-field matrix becomes a real symmetric one.

References

1. *Biedenharn, L. C., Blatt, J. M., Rose, M. E.:* Rev. Mod. Phys. *24*, 249 (1952).
2. *Condon, E. U., Shortley, G. H.:* The Theory of Atomic Spectra. Reprint. Cambridge: Univ. Press 1963.
3. *Edmonds, A. R.:* Angular Momentum in Quantum Mechanics. Princeton: Univ. Press 1957.

4. *Fano, U., Racah, G.:* Irreducible Tensorial Sets. 2. Print. New York: Academic Press 1967.
5. *Griffith, J. S.:* The Theory of Transition Metal Ions. Cambridge: Univ. Press 1961.
6. *Hobson, E. W.:* The Theory of Spherical and Ellipsoidal Harmonics. Cambridge: Univ. Press 1931.
7. *Judd, B. R.:* Operator Techniques in Atomic Spectroscopy. New York: McGraw-Hill 1963.
8. *Jørgensen, C. K.:* Discussions Faraday Soc. *26,* 110 (1958).
9. — Absorption Spectra and Chemical Bonding in Complexes. Second Impression. Oxford: Pergamon Press 1964.
10. *Löwdin, P. O.:* Rev. Mod. Phys. *39,* 259 (1967).
11. *Nielson, C. W., Koster, G. F.:* Spectroscopic Coefficients for the p^n, d^n, and f^n Configurations. Cambridge, Mass.: The M. I. T. Press 1963.
12. *Racah, G.:* Phys. Rev. *61,* 186 (1942).
13. — Phys. Rev. *62,* 438 (1942).
14. — Phys. Rev. *63,* 367 (1943).
15. *Rotenberg, M., Bivins, R., Metropolis, N., Wooten, J. K., Jr.:* The 3-*j* and 6-*j* Symbols. Cambridge, Mass.: The M. I. T. Press 1959.
16. *Schäffer, C. E., Jørgensen, C. K.:* J. Inorg. Nucl. Chem. *8,* 143 (1958).
17. — Mat. Fys. Medd. Kgl. Dan. Vid. Selsk. *34,* No. 13 (1965).
18. *Schäffer, C. E.:* Proc. Roy. Soc. (London) *A 297,* 96 (1967).
19. — Struct. Bonding *5,* 68 (1968).
20. — Pure Appl. Chem. *24,* 361 (1970).
21. — Intern. J. Quantum Chem. *5,* 379 (1971).
22a. *Wigner, E. P.:* Gruppentheorie und Ihre Anwendung auf die Quantenmechanik der Atomspektren. Braunschweig: Friedr. Vieweg 1931.
22b. — Group Theory. New York–London: Academic Press 1959.
23. — Am. J. Math. *63,* 57 (1941).
24. *Wybourne, B. G.:* Spectroscopic Properties of Rare Earths. New York: Interscience 1965.
25. *Harnung, S. E., Schäffer, C. E.:* Struct. Bonding *12,* 201 (1972).
26. *Schäffer, C. E.:* Struct. Bonding *14,* (1972).

Received April 19, 1972

Structure and Bonding: Index Volume 1-12

Ahrland, S.: Factors Contributing to (b)-behaviour in Acceptors. Vol. 1, pp. 207—220.

— Thermodynamics of Complex Formation between Hard and Soft Acceptors and Donors. Vol. 5, pp. 118—149.

Allen, G. C., and *Warren, K. D.*: The Electronic Spectra of the Hexafluoro Complexes of the First Transition Series. Vol. 9, pp. 49—138.

Babel, D.: Structural Chemistry of Octahedral Fluorocomplexes of the Transition Elements. Vol. 3, pp. 1—87.

Bayer, E., und *Schretzmann, P.*: Reversible Oxygenierung von Metallkomplexen. Vol. 2, pp. 181—250.

Bearden, A. J., and *Dunham, W. R.*: Iron Electronic Configurations in Proteins: Studies by Mössbauer Spectroscopy. Vol. 8, pp. 1—52.

Braterman, P. S.: Spectra and Bonding in Metal Carbanyls, Part A: Bonding. Vol. 10, pp. 57—86.

Bray, R. C., and *Swann, J. C.*: Molybdenum-Containing Enzymes. Vol. 11, pp. 107—144.

van Bronswyk, W.: The Application of Nuclear Quadrupole Resonance Spectroscopy to the Study of Transition Metal Compounds. Vol. 7, pp. 87—113.

Buchanan, B. B.: The Chemistry and Function of Ferredoxin. Vol. 1, pp. 109—148.

Ciampolini, M.: Spectra of 3d Five-Coordinate Complexes. Vol. 6, pp. 52—93.

Fajans, K.: Degrees of Polarity and Mutual Polarization of Ions in the Molecules of Alkali Fluorides, SrO, and BaO. Vol. 3, pp. 88—105.

Feeney, R. E., and *Komatsu, S. K.*: The Transferrins. Vol. 1, pp. 149—206.

Fraga, S., and *Valdemoro, C.*: Quantum Chemical Studies on the Submolecular Structure of the Nucleic Acids. Vol. 4, pp. 1—62.

Gillard, R. D., and *Mitchell, P. R.*: The Absolute Configuration of Transition Metal Complexes. Vol. 7, pp. 46—86.

Griffith, J. S.: On the General Theory of Magnetic Susceptibilities of Polynuclear Transition-metal Compounds. Vol. 10, pp. 87—126.

Gutmann, V., and *Mayer, U.*: Thermochemistry of the Chemical Bond. Vol. 10, pp. 127—151.

Harnung, S. E., and *Schäffer, C. E.*: Phase-fixed 3-Γ Symbols and Coupling Coefficients for the Point Groups. Vol. 12, pp. 201—255.

— — Real Irreducible Tensorial Sets and their Application to the Ligand-Field Theory. Vol. 12, pp. 257—295.

von Herigonte, P.: Electron Correlation in the Seventies. Vol 12, pp. 1—47.

Hill, H. A. O., *Röder, A.*, and *Williams, R. J. P.*: The Chemical Nature and Reactivity of Cytochrome P—450. Vol. 8, pp. 123—151.

Hudson, R. F.: Displacement Reactions and the Concept of Soft and Hard Acids and Bases. Vol. 1, pp. 221—233.

Hulliger, F.: Crystal Chemistry of Chalcogenides and Pnictides of the Transition Elements. Vol. 4, pp. 83—229.

Iqbal, Z.: Intra- und Inter-Molecular Bonding and Structure of Inorganic Pseudohalides with Triatomic Groupings. Vol. 10, pp. 25—55.

Jerome-Lerutte, S.: Vibrational Spectra and Structural Properties of Complex Tetracyanides of Platinum, Palladium and Nickel. Vol. 10, pp. 153—166.

Jørgensen, C. K.: Electric Polarizability, Innocent Ligands and Spectroscopic Oxidation States. Vol. 1, pp. 234—248.

— Recent Progress in Ligand Field Theory. Vol. 1, pp. 3—31.

— Relations between Softness, Covalent Bonding, Ionicity and Electric Polarizability. Vol. 3, pp. 106—115.

— Valence-Shell Expansion Studied by Ultra-violet Spectroscopy. Vol. 6, pp. 94—115.

Kimura, T.: Biochemical Aspects of Iron Sulfur Linkage in None-Heme Iron Protein, with Special Reference to "Adrenodoxin". Vol. 5, pp. 1—40.

König, E.: The Nephelauxetic Effect. Calculation and Accuracy of the Interelectronic Repulsion Parameters I. Cubic High-Spin d^2, d^3, d^7, and d^8 Systems. Vol. 9, pp. 175—212.

Krumholz, P.: Iron(II) Diimine and Related Complexes. Vol. 9, pp. 139—174.

Lindskog, S.: Cobalt(II) in Metalloenzymes. A Reporter of Structure-Function Relations. Vol. 8, pp. 153—196.

Lucken, E. A. C.: Valence-Shell Expansion Studied by Radio-Frequency Spectroscopy. Vol. 6, pp. 1—29.

Maggiora, G. M., and *Ingraham, L. L.:* Chlorophyll Triplet States. Vol. 2, pp. 126—159.

Mayer, U., and *Gutmann, V.:* Phenomenological Approach to Cation-Solvent Interactions. Vol. 12, pp. 113—140.

Moreau-Colin, M. L.: Electronic Spectra and Structural Properties of Complex Tetracyanides of Platinum, Palladium and Nickel. Vol. 10, pp. 167—190.

Morris, D. F. C.: Ionic Radii and Enthalpies of Hydration of Ions. Vol. 4, pp. 63—82.

— An Appendix to Structure and Bonding Vol. 4 (1968). Vol. 6, pp. 157—159.

Neilands, J. B.: Naturally Occurring Non-porphyrin Iron Compounds. Vol. 1, pp. 59—108.

— Evolution of Biological Iron Binding Centers. Vol. 11, pp. 145—170.

Oelkrug, D.: Absorption Spectra and Ligand Field Parameters of Tetragonal 3*d*-Transition Metal Fluorides. Vol. 9, pp. 1—26.

Reinen, D.: Ligand-Field Spectroscopy and Chemical Bonding in Cr^{3+}-Containing Oxidic Solids. Vol. 6, pp. 30—51.

— Kationenverteilung zweiwertiger $3d^n$-Ionen in oxidischen Spinell-, Granat- und anderen Strukturen. Vol. 7, pp. 114—154.

Schäffer, C. E.: A Perturbation Representation of Weak Covalent Bonding. Vol. 5, pp. 68—95.

Schutte, C. J. H.: The Ab-Initio Calculation of Molecular Vibrational Frequencies and Force Constants. Vol. 9, pp. 213—263.

Shriver, D. F.: The Ambident Nature of Cyanide. Vol. 1, pp. 32—58.

Smith, D. W.: Ligand Field Splittings in Copper(II) Compounds. Vol. 12. pp. 49—112.

—, and *Williams, R. J. P.:* The Spectra of Ferric Haems and Haemoproteins. Vol. 7, pp. 1—45.

Speakman, J. Clare: Acid Salts of Carboxylic Acids, Crystals with some "Very Short" Hydrogen Bonds. Vol. 12, pp. 141—199.

Spiro, G., and *Saltman, P.:* Polynuclear Complexes of Iron and their Biological Implications. Vol. 6, pp. 116—156.

Strohmeier, W.: Problem und Modell der homogenen Katalyse. Vol. 5, pp. 96—117.

Thompson, D. W.: Structure and Bonding in Inorganic Derivatives of β-Diketones. Vol. 9, pp. 27—47.

Thomson, A. J., Williams, R. J. P., and *Reslova, S.:* The Chemistry of Complexes Related to *cis*-$Pt(NH_3)_2Cl_2$. An Anti-Tumour Drug. Vol. 11, pp. 1—46.

Weissbluth, M.: The Physics of Hemoglobin. Vol. 2, pp. 1—125.

Weser, U.: Chemistry and Structure of some Borate Polyol Compounds. Vol. 2, pp. 160—180.

— Reaction of some Transition Metals with Nucleic Acids and their Constituents. Vol. 5, pp. 41—67.

Williams, R. J. P., and *Hale, J. D.:* The Classification of Acceptors and Donors in Inorganic Reactions. Vol. 1, pp. 249—281.

Winkler, R.: Kinetics and Mechanism of Alkali Ion Complex Formation in Solution. Vol. 10, pp. 1—24.

Wood, J. M., and *Brown, D. G.:* The Chemistry of Vitamin B_{12}-Enzymes. Vol. 11, pp. 47—105.

Wüthrich, K.: Structural Studies of Hemes and Hemoproteins by Nuclear Magnetic Resonance Spectroscopy. Vol. 8, pp. 53—121.

GPSR Compliance

The European Union's (EU) General Product Safety Regulation (GPSR) is a set of rules that requires consumer products to be safe and our obligations to ensure this.

If you have any concerns about our products, you can contact us on ProductSafety@springernature.com

In case Publisher is established outside the EU, the EU authorized representative is:

Springer Nature Customer Service Center GmbH
Europaplatz 3
69115 Heidelberg, Germany

Batch number: 10704707

Printed by Printforce, the Netherlands